BIOLOGICAL AND MEDICAL PHYSICS,
BIOMEDICAL ENGINEERING

BIOLOGICAL AND MEDICAL PHYSICS, BIOMEDICAL ENGINEERING

The fields of biological and medical physics and biomedical engineering are broad, multidisciplinary and dynamic. They lie at the crossroads of frontier research in physics, biology, chemistry, and medicine. The Biological & Medical Physics, Biomedical Engineering Series is intended to be comprehensive, covering a broad range of topics important to the study of the physical, chemical and biological sciences. Its goal is to provide scientists and engineers with textbooks, monographs, and reference works to address the growing need for information.

Books in the series emphasize established and emergent areas of science including molecular, membrane, and mathematical biophysics; photosynthetic energy harvesting and conversion; information processing; physical principles of genetics; sensory communications; automata networks, neural networks, and cellular automata. Equally important will be coverage of applied aspects of biological and medical physics and biomedical engineering such as molecular electronic components and devices, biosensors, medicine, imaging, physical principles of renewable energy production, advanced prostheses, and environmental control and engineering.

For other volumes in this series, go to:
www.springer.com/series/3740

David D. Zhou
Elias Greenbaum
Editors

Implantable Neural Prostheses 2

Techniques and Engineering Approaches

Springer

Editors
David D. Zhou
Second Sight Medical
Products, Inc.
12744 San Fernando Road
Sylmar, CA 91342
USA
dzhou@2-sight.com

Elias Greenbaum
Oak Ridge National Laboratory
P.O. Box 2008
Oak Ridge, TN 37831-6194
USA
greenbaum@ornl.gov

ISSN 1618-7210
ISBN 978-0-387-98119-2 e-ISBN 978-0-387-98120-8
DOI 10.1007/978-0-387-98120-8
Springer New York Dordrecht Heidelberg London

Library of Congress Control Number: 2009935455

Printed on acid-free paper

Springer is part of Springer Science+Business Media (www.springer.com)

Preface

Significant progress has been made in the development of neural prostheses for restoration of human functions and improvement of the quality of life. Biomedical engineers and neuroscientists around the world are working to improve the design and performance of existing devices and to develop novel devices for artificial vision, artificial limbs, and brain-machine interfaces.

This book, *Implantable Neural Prostheses 2: Techniques and Engineering Approaches*, is part two of a two-volume sequence that describes state-of-the-art advances in techniques associated with implantable neural prosthetic devices. The techniques covered include biocompatibility and biostability, hermetic packaging, electrochemical techniques for neural stimulation applications, novel electrode materials and testing, thin-film flexible microelectrode arrays, in situ characterization of microelectrode arrays, chip-size thin-film device encapsulation, microchip-embedded capacitors and microelectronics for recording, stimulation, and wireless telemetry. The design process in the development of medical devices is also discussed.

Advances in biomedical engineering, microfabrication technology, and neuroscience have led to improved medical-device designs and novel functions. However, many challenges remain. This book focuses on the engineering approaches, R&D advances, and technical challenges of medical implants from an engineering perspective. We are grateful to leading researchers from academic institutes, national laboratories, as well as design engineers and professionals from the medical device industry who have contributed to the book. Part one of this series covers designs of implantable neural prosthetic devices and their clinical applications. Devices covered include visual implants, cochlear implants, auditory midbrain implants, spinal cord stimulators, deep brain stimulators, Bion microstimulators, the brain control and sensing interface, cardiac electrostimulation devices, and magnetic stimulation devices. Regulatory approval of implantable medical devices in the United States and Europe is also discussed. We hope a better understanding of design issues, techniques, and challenges may encourage innovation and interdisciplinary efforts to expand the frontiers of R&D of implantable neural prostheses.

Los Angeles, California
Oak Ridge, Tennessee
March 2009

David D. Zhou
Elias Greenbaum

Contents

Contributors

Orlando Auciello Materials Science Division, Argonne National Laboratory, Argonne, IL 60439, USA; Argonne National Laboratory, Center for Nanoscale Materials, Argonne, IL 60439, USA, auciello@anl.gov

Richard Castro Second Sight Medical Products, Inc., Sylmar Biomedical Park, Sylmar, CA 91342, USA, richardc@2-sight.com

Carlos J. Cela Department of Electrical and Computer Engineering, North Carolina State University, Raleigh, NC, USA, cjcela@ncsu.edu

Moo Sung Chae Department of Electrical Engineering, University of California, Santa Cruz, CA 95064, USA, gomdori07@gmail.com

Karen C. Cheung Department of Electrical and Computer Engineering, University of British Columbia, Vancouver, BC, Canada, kcheung@ece.ubc.ca

Mike Colvin Boston Scientific Neuromodulation Corporation, Valencia, CA 91355, USA, michael.colvin@bsci.com

X. Tracy Cui Department of Bioengineering, University of Pittsburgh, Pittsburgh, PA 15260, USA, xic11+@pitt.edu

Sanjay Gaikwad Second Sight Medical Products, Inc., Sylmar Biomedical Park, Sylmar, CA 91342, USA, sgaikwad@2-sight.com

Ira B. Goldberg Department of Electrical Engineering, University of California, Los Angeles, Los Angeles, CA 90095, ibgoldbe@ee.ucla.edu

Elias Greenbaum Molecular Bioscience and Biotechnology, Chemical Sciences Division, Oak Ridge National Laboratory, Oak Ridge, TN, USA, greenbaum@ornl.gov

Robert J. Greenberg Second Sight Medical Products, Inc., Sylmar Biomedical Park, Sylmar, CA 91342, USA, robert.greenberg@2-sight.com

Amy Hines Second Sight Medical Products, Inc., Sylmar Biomedical Park, Sylmar, CA 91342, USA, amyc@2-sight.com

Andy Hung Boston Scientific Neuromodulation Corporation, Valencia, CA 91355, USA, andy.hung@bsci.com

Guangqiang Jiang Alfred E. Mann Foundation for Scientific Research, Santa Clarita, CA 91355, USA, jiang@aemf.org

Jack W. Judy Department of Electrical Engineering, University of California, Los Angeles, CA 90095, USA, jack.judy@ucla.edu

Vincent Kandagor Molecular Bioscience and Biotechnology, Chemical Sciences Division, Oak Ridge National Laboratory, Oak Ridge, TN, USA, kandagorvc@ornl.gov

Jim Little Second Sight Medical Products, Inc., Sylmar Biomedical Park, Sylmar, CA 91342, USA, jiml@2-sight.com

Wentai Liu Department of Electrical Engineering, University of California, Santa Cruz, CA 95064, USA, wentai@soe.ucsc.edu

Gianluca Lazzi Department of Electrical and Computer Engineering, North Carolina State University, Raleigh, NC, USA, lazzi@ncsu.edu

Daniel R. Merrill Alfred E. Mann Foundation for Scientific Research, Santa Clarita, CA 91355, USA, danm@aemf.org

Charlene A. Sanders Molecular Bioscience and Biotechnology, Chemical Sciences Division, Oak Ridge National Laboratory, Oak Ridge, TN, USA, sandersca@ornl.gov

Bing Shi Argonne National Laboratory, Materials Science Division, Argonne, IL 60439, USA, shi@anl.gov

Ken Stokes Brady Leads Research, Cardiac Rhythm Disease Management Division, Medtronic, Inc., Brainerd, MN 56401, USA, kbstokes@mlecwb.net

Zhi Yang Department of Electrical Engineering, University of California, Santa Cruz, CA 95064, USA, yangzhi@soe.ucsc.edu

David D. Zhou Second Sight Medical Products, Inc., Sylmar Biomedical Park, Sylmar, CA 91342, USA, dzhou@2-sight.com

Acronyms

AAMI	Association for the Advancement of Medical Instrumentation
ADC	Analog-to-digital converter
AES	Auger electron spectroscopy
AFM	Atomic force microscopy
Ag/AgCl	Silver/silver chloride
AIROF	Activated iridium oxide film
ASK	Amplitude shift keying
ALD	Atomic layer deposition
ALT	Accelerated life testing
ANSI	American National Standards Institute
ASIC	Application specific integrated circuit
ASTM	American Society for Testing and Materials
BMI	Brain machine interface
BPSK	Binary phase shift keying
C_{dl}	Double layer capacitor
CE	Counter electrode
CHLD	Cumulative helium leak detection
CLS	Chronic lead study statistics
CMFB	Common mode feedback
CMOS	Complementary metal oxide semiconductor technology
CMRR	Common mode rejection ratio
CNS	Central nervous system
CP	Commercially pure
CSC	Charge storage capacity
CSPGs	Chondroitin sulfate proteoglycans
CTE	Coefficient of thermal expansion
CV	Cyclic voltammetry
CVD	Chemical vapor deposition
CUP	Circuit-under-pad
DAC	Digital-to-analog converter
DBS	Deep brain stimulation
DIR	Design input requirements
DMR	Device master record

DNR	Dynamic range to noise ratio
DPSK	Differential phase shift keying
DWL	Direct write laser
ECF	Extracellular fluid
ECG	Electrocardiogram
ECM	Extracellular matrix
ECoG	Electrocorticogram
EDX	Energy-dispersive x-ray analysis
EEG	Electroencephalographic
EIS	Electrochemical impedance spectroscopy
ESC	Environmental stress cracking
FDA	Food and Drug Administration
FEM	Finite element modeling
FE-SEM	Field emission scanning electron microscope
FET	Field effect transistor
FMEA	Failure modes and effects analysis
FSK	Frequency shift keying
GDC	Global digital controller
GFAP	Glial fibrillary acidic protein
GLP	Good laboratory practices
HTCC	High-temperature co-fired ceramic
HOMO	Highest occupied molecular orbital
HPF	High pass filter
IC	Integrated circuit
ICD	Implantable cardioverter defibrillator
ICP	Inductively-coupled plasma
ICP	Inherently conducting polymer
IPLAS	Innovative plasma systems
IrOx	Iridium oxide
ISO	International Organization for Standardization
LCP	Liquid crystal polymer
LDC	Local digital controller
LFP	Local field potential
LNA	Low-noise amplifier
LTCC	Low-temperature co-fired ceramic
LTD	Low temperature degradation
LUMO	Lowest unoccupied molecular orbital
MCD	Microcrystalline diamond
MEA	Micro electrode array
MEMS	Micro-electromechanical systems
MIO	Metal ion oxidation
MPCVD	Microwave plasma-enhanced chemical vapor deposition
NaDBS	Sodium dodecylbenzenesulfonate
NCD	Nanocrystalline diamond
NEO	Nonlinear energy operator

NGF	Neural growth factor
OEC	Olfactory ensheathing cell
OOK	On-off keying
OTA	Operational transconductance amplifier
PANi	Polyaniline
PBS	Phosphate-buffered saline
PCA	Principal component analysis
PDP	Product development process
PECVD	Plasma-enhanced chemical vapor deposition
PEDOT	Poly(3,4-ethylenedioxythiophene)
PLGA	Poly(lactic-co-glycolic acid)
PLLA	Poly(L-lactic acid)
PNS	Peripheral nervous system
POT	Poly-3-octylthiophene
PPM	Pulse-position modulation
PSS	Polystyrenesulfonate
PSRR	Power supply rejection ratio
PPy	Polypyrrole
PVA	Polyvinyl alcohol
PVD	Physical vapor deposition
Q	Charge
Q/A	Charge density
Qinj	Injected charge
RE	Reference electrode
Redox	Reduction/oxidation
RF	Radio frequency
RGA	Residual gas analysis
RHE	Reversible hydrogen electrode
ROC	Receiver operating characteristic
RP	Retinitis pigmentosa
RPA	Returned products analysis
SAM	Scanning acoustic microscopy
SAM	Self-assembled monolayer
SAR	Successive approximation register
SCE	Saturated calomel electrode
SEM	Scanning-electron microscopy
SIMS	Secondary ion mass spectroscopy
SIROF	Sputtered iridium oxide film
SNR	Signal-to-noise ratio
SOI	Silicon-on-insulator
SUMMiT	Sandia's ultraplanar multi-level MEMS technology
TEM	Transmission electron microscopy
TGF	Transforming growth factor
TIG	Tungsten inert gas welding
TIROF	Thermally prepared iridium oxide film

TZP	Tetragonal zirconia polycrystal
UNCD	Ultrananocrystalline diamond
USP	United States Pharmacopeia standards
VGCM	Variable gain current mirror
WDS	Wavelength-dispersive x-ray spectrometry
WE	Working electrode
WPAN	Wireless personal area network
XPS	X-ray photoelectron spectroscopy

The Biocompatibility and Biostability of New Cardiovascular Materials and Devices

Ken Stokes

Abstract Evaluating a new material for use in an implantable device is a complicated business. ISO 14971 is designed to assist in determining device-risk assessment. ISO 10993 is intended to help steer one through the evaluation of materials for implantable devices. An FDA Guidance is available specifically for pacemaker leads, but may be helpful for other devices as well. However, completing a battery of in vitro and in vivo tests does not necessarily qualify a material for implant, because the in vivo environment cannot be duplicated in vitro. In vivo materials testing helps, but is still insufficient because the device may have its own issues. Device implants in animals may get one to human clinical studies and market release. Even after this stage, appropriate postmarket surveillance is necessary to know for sure how the device is really performing.

Contents

K. Stokes (✉)
Brady Leads Research, Cardiac Rhythm Disease Management Division, Medtronic, Inc., MN 11052; Minneapolis, Robinwood Lane, Brainerd, MN 56401-5947, USA
e-mail: kbstokes@mlecwb.net

D.D. Zhou, E. Greenbaum (eds.), *Implantable Neural Prostheses 2*,
Biological and Medical Physics, Biomedical Engineering, DOI 10.1007/978-0-387-98120-8_1,

1 Introduction

The term "biocompatibility" has been defined as the ability of a material to perform with an appropriate host response in a specific application [1]. The term "biostability" can be defined as the ability of a material to adequately resist the degrading effects of the host response. These two terms are interrelated since a biocompatible material that degrades may become nonbiocompatible. Consider polyester polyurethane-coated breast implants used from the 1950s into the 1980s. The material was initially quite biocompatible, but it hydrolyzed over time, apparently releasing a known carcinogen in small amounts [2]. Another example is the Dalkon shield. This intrauterine birth control device had a tail string made of nylon 6 as a coating over nylon 6 multifilaments. This coating degraded (via hydrolysis), cracked, and became a pathway for microbes to migrate into the uterus resulting in pelvic inflammatory disease [3]. Even biocompatible materials that do not degrade in vivo can be used in devices that are not biocompatible. The same is true for materials that are biostable per se, but, due to interactions or manufacturing processes, can make a nonbiostable device [4]. How do we evaluate materials to avoid or preclude untoward surprises? Once clinical trials produce a favorable outcome, can we be assured that the device is biocompatible, biostable, and functions as intended? How do we know that the marketed device is actually performing as expected?

In this chapter we will first briefly summarize an experience where, in spite of initial state-of-the-art materials testing, three previously unknown failure mechanisms were discovered clinically in cardiac pacemaker leads [5]. The details will be included by reference only, due to space limitations. Then some protocols for material and device testing will be briefly reviewed, with the ultimate goal of discovering the unexpected and precluding a repeat of our experience. Of course the testing protocols for various materials and devices will vary accordingly. However, as an example, we will cover the testing used today to establish biocompatibility and biostability of polymeric materials used in cardiovascular pacemaker leads.

2 Background

2.1 Learning from the Past

Our present-day requirements are affected by our understanding of modern science as well as our history. For example, dual-chamber pacemakers were developed in the early 1970s to stimulate both the right atrium and right ventricle (Fig. 1). While these had significant physiological benefits, they were sparingly used. The reason was simply that the silicone rubber-insulated leads available at that time were relatively large in diameter and were somewhat sticky in blood. The relatively large diameter was the result of poor physical properties, especially low elastic modulus and low tear resistance. This made the preferred insertion of two leads in one vein difficult. Using two veins, such as the internal and external jugulars, often produced morbidities, such as congestion due to restricted venous drainage from the head. Thus, an unmet medical need was for leads that were smaller, more maneuverable, and slippery in blood to facilitate the implant of dual-chamber devices.

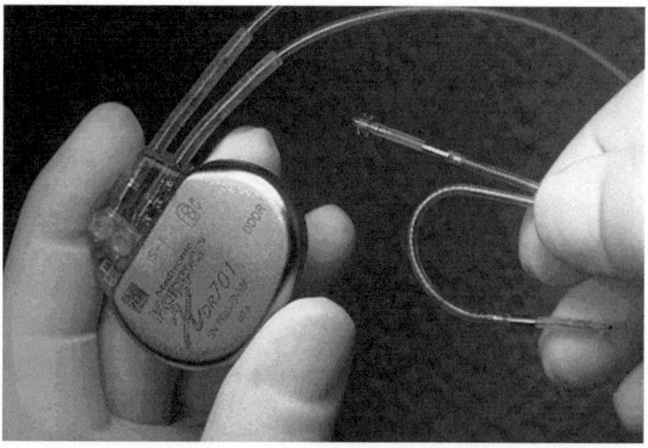

Fig. 1 A dual-chamber pulse generator and two leads. The pulse generator is a hermetically sealed titanium "can" containing an LiI power source, circuitry (including microprocessor), niobium or tantalum/sapphire feed throughs, and a connector to receive two leads. The lead with the straight distal tip is inserted in the right ventricle while that with the "J" shape is placed in the right atrial appendage. Both leads are bipolar (two electrodes within the heart). The tiny projections (tines) at the tip help provide fixation for the distal stimulating electrode

Certain polyether polyurethanes possess ideal properties to make relatively small, maneuverable, and slippery (in blood) leads. Of those available at the time, only the Pellethane 2363 series met our needs.[1] In 1975, the first polyurethane-insulated lead was implanted in a human as part of a spinal stimulator. The softer Shore 80A and stiffer Shore 55D polymers were tested extensively in vivo over the ensuing 5 years for cardiac use. This may leave the impression that we implanted neuro leads in humans first, then tested the material. In fact, the neuro implants were based on the

[1] Initially Upjohn, later Dow Chemical, Midland Michigan.

state of the art testing at the time, but we wanted more long term in vivo data for cardiovascular. In vitro and ex vivo testing demonstrated that the materials were biocompatible and stable in the intended environment (as it was understood at the time) [6]. Just to be sure, however, we implanted the materials in the subcutis of rabbits for 2 years, conducting extensive characterization tests as a function of time. No untoward biocompatibility or biostability issues were revealed. Device tests were conducted in canines for 12 weeks, which had been shown to be sufficient time to characterize acute and chronic performance. Human clinical evaluations over 1–2 years (depending on the models) demonstrated that the devices had superior performance compared to their predecessors [7, 8]. Four polyurethane-insulated lead models were market released in the United States in April 1980 with FDA approval.

2.2 Environmental Stress Cracking (ESC)

On May 15, 1981, an atrial lead was returned from Europe with a large gap in its insulation (Fig. 2). This turned out to be the first of many polyurethane-insulated leads experiencing what we later determined was a form of stress cracking. Analysis demonstrated that the tissue-contacting surfaces were oxidized, but the underlying bulk was not affected within our instrumental limits. With the exception of ozone cracking of natural rubber, stress-cracking mechanisms were unknown in elastomers. Indeed it was "well known" at the time that there was no ozone in the tissues adjacent to the leads. In fact, it was "well known" that the oxygen tension in the venous and subcutaneous tissues adjacent to the lead was extremely low. Given a lack of oxygen, light, or excessive heat as found in the worldly environment, the polymer could not be subjected to photo or thermal oxidative processes. The medical and (journal) polymer literature shed no light on the issue. Indeed, literature on degradation of polymers was uncommon. We eventually discovered that because degradation is a very slow process, it was not a good subject for academia (one cannot keep grad students forever to do real-time studies). Industry, however, did have the time and motivation to do the appropriate studies and publish, but the work was

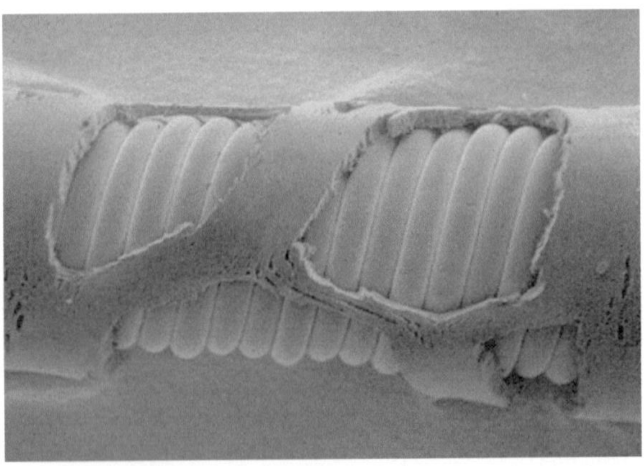

Fig. 2 A portion of an atrial lead explanted from a human after only about 7 months implant was received in May 1981. The large gaps in the insulation have exactly matching opposite edges. The surface is oxidized per FTIR, but the bulk is unaffected. This was the first known case of environmental stress cracking in an implanted elastomer

to be found in relatively obscure (to the medical-device industry) books, not journals [9, 10]. These books provided insight into polymer-degradation mechanisms in the worldly environment, but shed no light on the in vivo environment. It was only by studying what appeared to be unrelated literature (autoimmune disease, space shuttle environment, pathology, etc.) that we learned that it was just being discovered that oxygen free radicals were produced as a result of the foreign body process that encapsulated implanted devices [11]. This involves the adhesion and activation of monocyte/macrophages on the device surface, followed by fibroblasts and collagenous encapsulation. This was most revealing, because the prior pacemaker literature uniformly indicated that the encapsulation was just collagen, with no significant cellular components [12].

Of course, to understand a previously unknown mechanism and evaluate corrective measures, we had to develop unique tests that did not exist before. These are accelerated in vivo tests, since ESC has not yet been exactly duplicated in vitro (Fig. 3). We now know that ESC is a four-factor interaction, all four of which

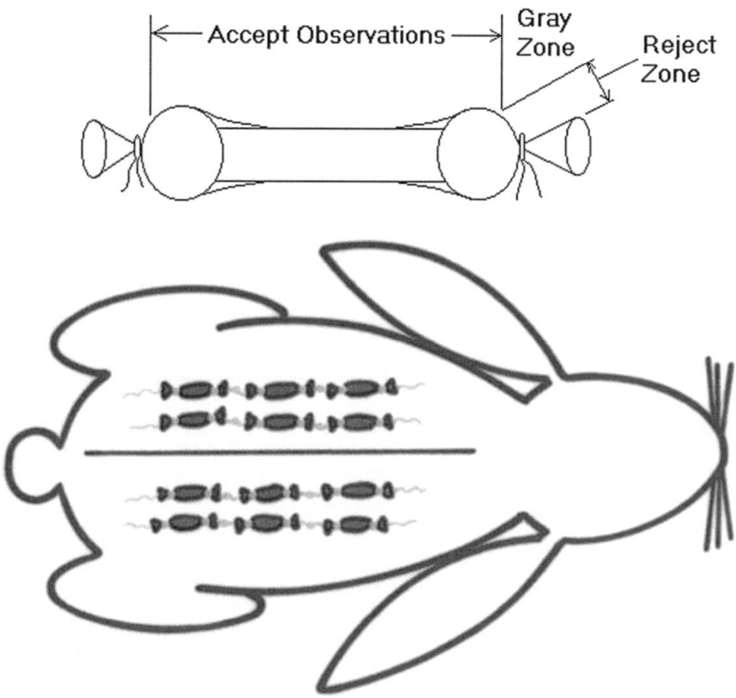

Fig. 3 Accelerated in vivo ESC test. At the *top* is a tubing specimen stretched over a 0.5-inch-long dumbbell mandrel, with the strain fixed at each end by polyester ligatures. Samples are tied together in strings, identified with a colored glass bead, and implanted in the subcutis of rabbits as shown *above*. After explant, observations made in the "accept zone" are accepted. Observations in the "reject zone," where the strain is uncontrolled, are rejected. A coating of medical adhesive in heptane can be applied to the ends of the samples to cover the knots, etc., isolating the tissue exposure to the center portion of the specimens where the strain is ideally controlled. Samples may then be analyzed by FTIR, HPLC molecular weight, SEM, and other suitable techniques

must be active: residual stress (strain) in the polymer, autoxidation of the *surface* (there is no bulk degradation) due to lysosomal release of superoxide anion and hydroxyl radical from macrophages on the surface, a degradable linkage (in this case, oxidation initiated at a hydrogen atom on the ether's α carbon), and a ubiquitous compound that serves as the crack driver, α_2-macroglobulin [13]. It is clear that with the exception of unusual dumb luck it would have been highly unlikely for one to discover unknown mechanisms without test methods to find them. It should be noted that ESC is not a significant failure mechanism today due to the selection of more oxidation-resistant versions of the polymer (Shore 55D hardness), manufacturing process designed to eliminate residual stress, and numerous other factors.

2.3 Metal Ion Oxidation (MIO)

In late 1982 or early 1983, we discovered that there was *bulk* autoxidation affecting some leads, from the inside out (Fig. 4). We were beginning to appreciate the fact that biologically produced oxygen radicals could be present, but they react at the tissue-contacting surfaces, and thus cannot penetrate into the lead. Hydrogen peroxide is also produced by macrophages and can permeate the material, but it has been shown that by itself it does not cause any measurable degradation. Again, the ordinary accelerators present in the worldly environment, light and excessive heat, were absent. Certainly an isothermal environment of $37 \pm 3^{\circ}$C would not serve to produce relatively rapid autoxidation in vivo when it did not in vitro. Who would have guessed that there could be a degradative interaction between the polymer insulation and the lead's metallic conductors? The conductor coils are made from MP35N

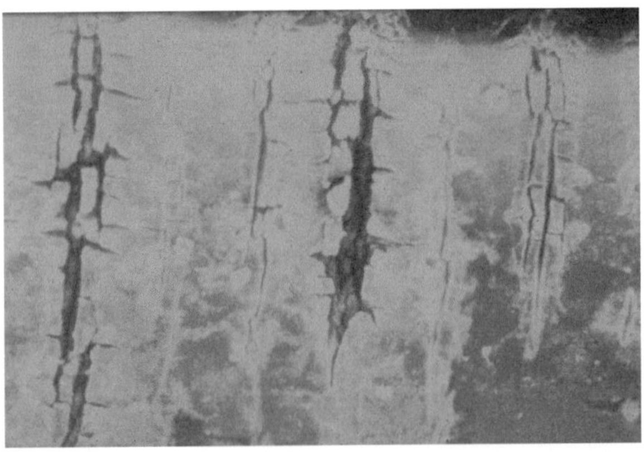

Fig. 4 Scanning electron micrograph of the conductor-contacting inner surface of a lead's insulation showing MIO degradation where the tubing contacted the metal coils

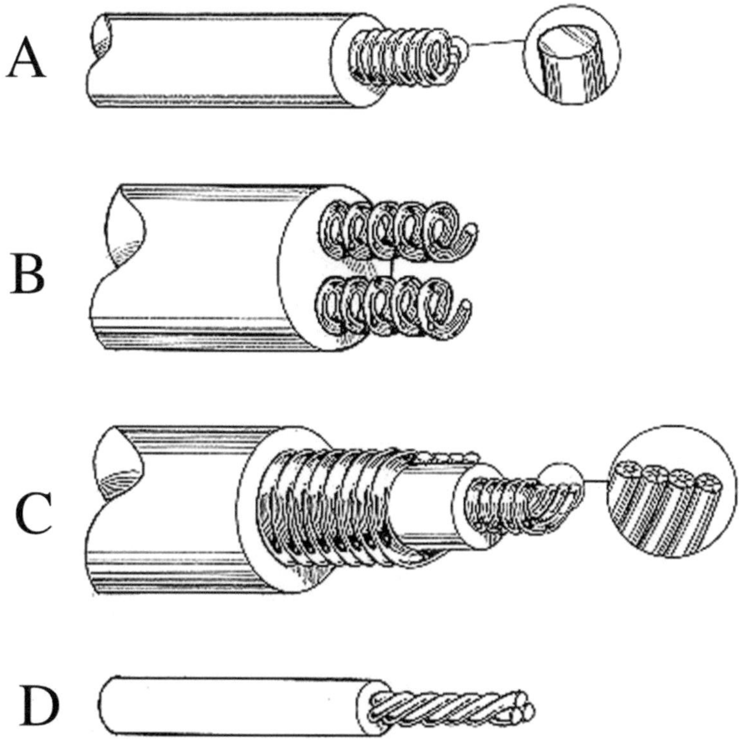

Fig. 5 Examples of various types of lead conductors and their insulation: **A**. a unipolar single-filar conductor coil, **B**. parallel single-filar conductor coils, **C**. coaxial multifilar conductor coils, and **D**. a multifilament microcable

(Ni, Co, Cr, Mo, Fig. 5).[2] We now know that transition metal ions can react with polymers. In this case the culprit is cobalt ion (complexes) released from the conductor via trace amounts of corrosion. We also now know that H_2O_2 permeating the insulation can interact with the cobalt in the MP35N to aid its corrosion and to form complexes that can react directly with the polymer via redox [14]. It can also catalyze decomposition of H_2O_2 to reactive oxidants and oxygen free radicals within the device itself [15, 16].

Again, there were no test methods for this phenomenon so we had to develop them as is described in detail elsewhere [14, 15]. Numerous in vitro tests immersing tubing samples containing cobalt rods in H_2O_2 solution or immersing polymer samples in Co^{2+}/H_2O_2 have been reported, but the correlation with in vivo reality is poor to unclear (Fig. 6) [15, 16]. An in vivo test using tubing samples containing pure cobalt rod appears to greatly accelerate degradation and is believed to be

[2]Dupont

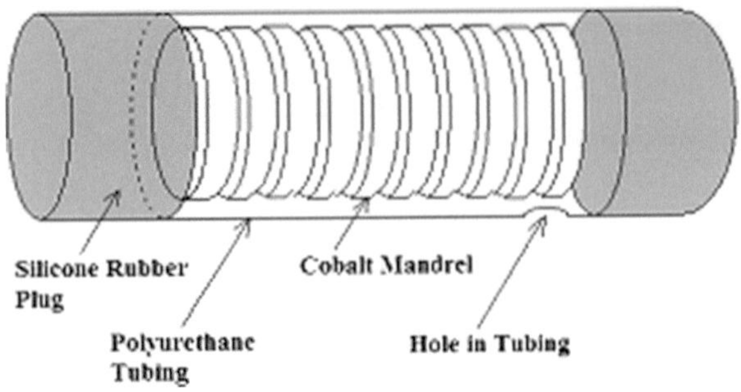

Fig. 6 A 0.5-inch-long grooved cobalt mandrel is inserted into tubing, which is sealed on both ends with medical adhesive. The grooves are intended to simulate the spaces between pacemaker lead conductor coils. A 1-mm-diameter hole is made in one end to assure ingress of exudate/transudate. Samples are tied together in strings, identified with a colored glass bead, and implanted in the subcutis of rabbits as shown in Fig. 3. After optical microscopic examination, the samples may be analyzed by FTIR, HPLC molecular weight, SEM, and other suitable techniques

more accurate, but requires substantial implant time [15]. Unfortunately, no cobalt-free conductor material has been developed yet that has the necessary properties required for implantable leads. The mechanism has been controlled through the use of barrier coatings, and other factors.

2.4 Subclavian Crush

Still a third failure mechanism was not seen in animal studies, but was discovered clinically. Explanted and returned leads with crushed, flattened, and fractured conductor coils began to show up in the mid 1980s (Fig. 7) [17]. When polyurethane leads were introduced for human use, a new implant technique was developed. Instead of inserting the leads through a cephalic or jugular vein cut down, they were

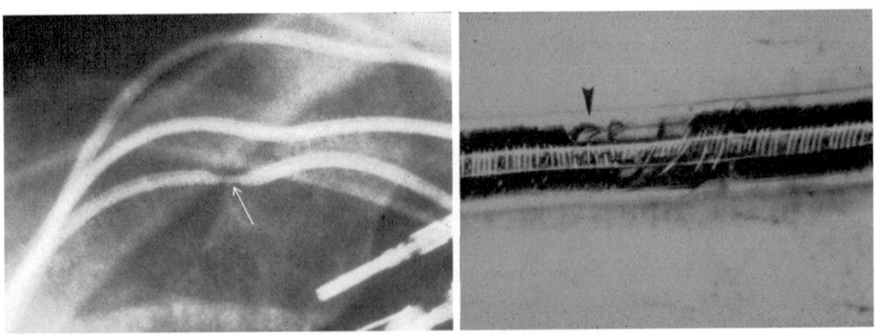

Fig. 7 An x-ray showing several leads implanted via the subclavian vein. One lead has fractured conductors due to crush (at the *arrow*, *left*). A photograph of an explanted lead rendered inoperable by subclavian crush (*right*)

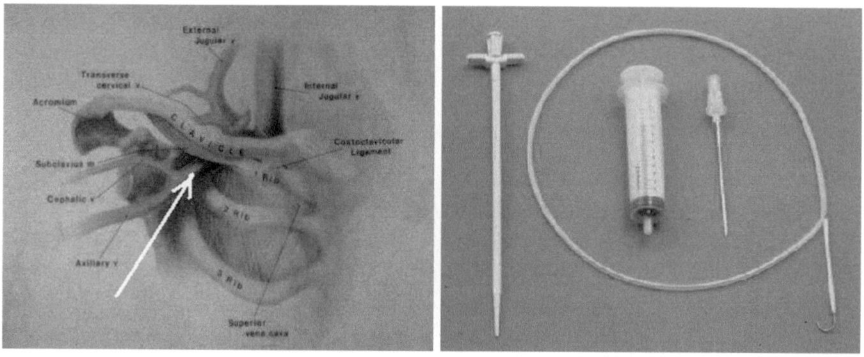

Fig. 8 Subclavian stick implant kit (*right*) includes a syringe, which is inserted into the subclavian vein (*left*), preferably before the vein crosses the first rib and clavicle. Next the syringe is removed and a guidewire inserted through the needle, into the target cardiac chamber. The needle is removed over the guidewire. The introducer (containing a stiffening obturator) is threaded over the guidewire, into the venous system. The obturator and guidewire are removed, the lead inserted through the introducer, and finally the sheath is removed by splitting it while pulling it back.

implanted with an introducer through the subclavian vein (Fig. 8) [18]. We now know that if the lead is inserted into the subclavian vein before it crosses between the first rib and clavicle, it will not be crushed. If the lead body is inserted between the first rib and clavicle, which then enters the vein within the thorax (most commonly done initially), then it is at risk of being crushed by the normal movement of the two bones [19, 20]. Subclavian crush is no longer a clinically significant issue because of improved design in some cases and the development of improved surgical insertion techniques [21].

2.5 *Why Were These Mechanisms Not Discovered Before Market Release?*

We were criticized for not discovering subclavian crush in animal experiments. A plaintiff's expert claimed that if we had implanted 500 dogs for 3 years prior to clinical use, it would have been obvious. The "expert" making this claim, however, did not apparently realize that dogs, cats, sheep, etc., do not have clavicles, so detection of subclavian crush would be impossible in animal tests, no matter how many were used.

We were also criticized for implanting materials, but not devices for 2 years prior to clinical use. In fact, it seemed quite reasonable at that time that a biocompatible and biostable material would make a biocompatible and biostable lead. In fact, we now know that even 2–5-year implants of devices in canines may or may not necessarily produce measurable ESC or MIO (depending on the lead model). Nonetheless, lead failure due to ESC, MIO, and crush in humans was reported in those lead models in as little as 1–2 years (see postmarket surveillance below) [22]. We believe that this species difference is due to differences in degradation rate, not mechanism.

2.6 Chronic Removability of Transvenous Cardiac Leads

Prior to the discovery of the above failure mechanisms, it was unusual to explant chronic leads. Patients were quite elderly and ill with relatively short longevity. If a lead had to be explanted (for example, because of infection or fracture), open-heart surgery was typically indicated. Improved pacemakers, easier to implant leads (both polyurethane and silicone rubber), and expanded indications have resulted in ever-younger patients and greater longevities. Because of product recalls and advisories across the pacemaker industry, chronic lead explant became more frequent, but was not easily done. The culprit was encapsulation in the cardiovascular system that took several years to develop, making "short term" (<2 years) animal tests misleading. In the 1980s we began to conduct long-term (up to 5 year) canine studies of chronic leads to evaluate new materials and verify our accelerated biostability results. In conjunction with this, we were able to study and analyze the encapsulation process. Details of the mechanism of transvenous lead encapsulation are provided elsewhere [23]. In summary, as the leads are inserted through the vasculature, they scrape the endothelium producing a few intermittent thrombi attached to the heart wall (Fig. 9A). These lyse and disappear within a week or two. For the ensuing 2+ years, the lead bodies are typically clean and free of thrombus or encapsulation (Fig. 9B). There is a thin, translucent capsule extending proximally a few millimeters from the electrode sites and a thick capsule extending about a centimeter distally from the venous insertion site where the lead is ligated. By 3 years, 75% of the leads have substantial, intermittent encapsulation of varying thickness (Fig. 9C). By 5 years almost all leads have intermittent bands of encapsulation of varying thicknesses. By 8 years the leads are fully encapsulated (Fig. 9C). The mechanism is a creeping thrombosis/organization process [23].

Clearly the lead materials are not thrombogenic in their venous environment or they would not remain thrombus/encapsulation free for several years. The mechanism is mechanical (microstasis of blood causing thrombosis), not a reaction to the material per se. Thorough studies over many years have demonstrated that new materials, coatings, etc., have no effect on encapsulation. Is this a biocompatibility issue? It is not for functioning leads that continue to work well even if encapsulated. It is when chronic lead extraction is considered. The mechanical design of the device can minimize encapsulation or accelerate it, but not stop it. If a lead becomes infected, it is mandatory that it be removed. If the lead remains functional and does not become infected, then there is usually no need to remove it. In many cases even nonfunctional leads have been left in place, with a new one inserted alongside. Fortunately, new surgical techniques have been developed to

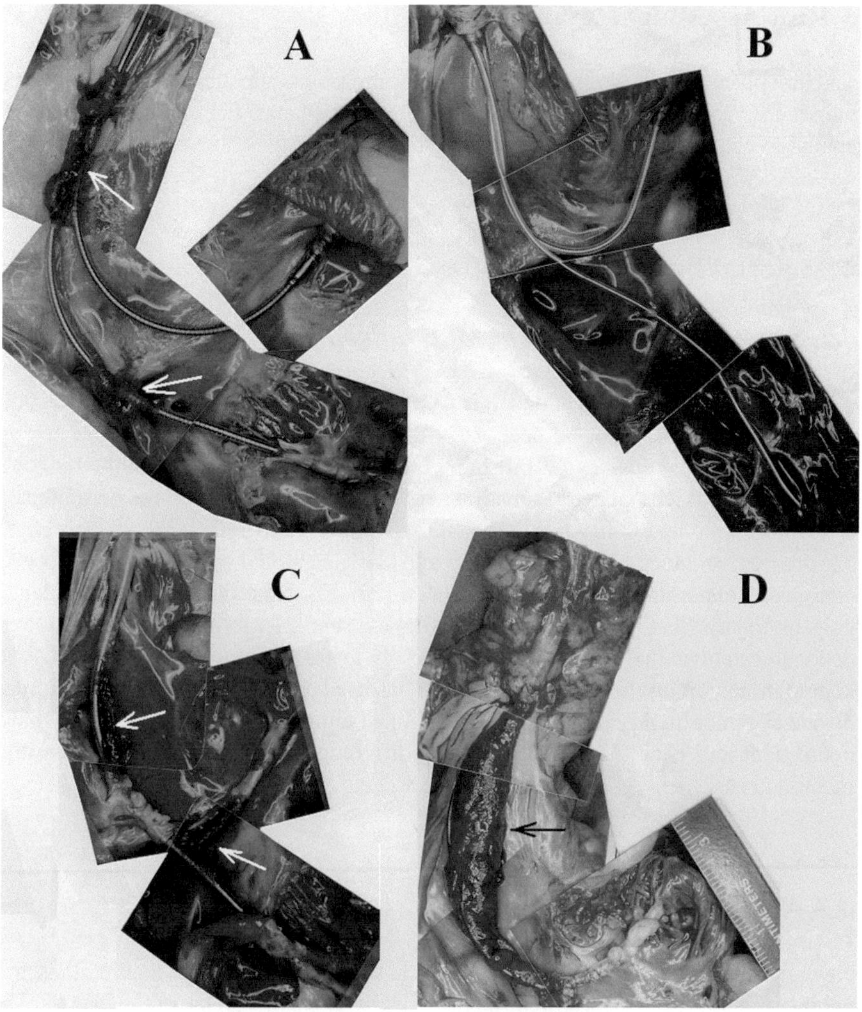

Fig. 9 Photographs of lead implant sites in canines. **A**. After 3–7 days implant, several thrombi (arrows) caused by endothelial injury during implant adhere to the heart wall, not the lead. **B**. Six weeks to more than 2 years after implant, the lead bodies are free and clear of thrombus or fibrotic encapsulation; short thin capsules are found at the endocardium/electrode sites and the venous entry site. **C**. Three to five years post implant, the leads are encapsulated in a discontinuous and sometimes corregated collagenous capsule, again with fresh (not postmortem) thrombi (*arrows*). **D**. After 8 years, both leads are fully encapsulated. A large partially organized thrombus is shown at the *arrow*

extract chronic leads safely without open-heart surgery, using catheters, snares, and even devices using ablating lasers or cautery [24]. Thus, until some means of preventing encapsulation in the cardiovascular system (short of anticoagulating the patient) is developed, we must live with it, and make sure we do not make it worse by design.

3 Risk Assessment

The first place one should look before developing a biocompatibility/biostability test plan is ISO/DIS 14971 [25]. Annex C, Annex I, and tables E.1 and E.2 in Annex E appear to be especially useful for biomaterials [26–28].

4 Material Biocompatibility Testing

4.1 Substantially Equivalent Materials

To paraphrase the FDA's Guidance for implantable cardiac pacing leads, biocompatibility testing may not be necessary if a material has a long history of use in currently marketed devices [29]. If there is sufficient knowledge about the biocompatibility and toxicity of every constituent of the device, then it need not be subjected to further biocompatibility tests. The FDA Guidance and ISO 10993-1, paragraph 3.7 provide guidance as to what factors must be considered to determine if a previously used material needs to be reevaluated [29, 30]. Note that the FDA Guidance goes on to say "It is incumbent upon the device submitter to provide sufficient evidence to establish that further biocompatibility testing is not necessary." It must be kept in mind that just because a competitor has used a material clinically in a similar device does not qualify it for your use as "Substantially Equivalent." You may have to consider it a "New Material" because of differences in specifications, processing, manufacturing, etc.

4.2 New Materials

New materials or previously implanted materials used in new applications are another matter, requiring a structured program of assessment. One of the first steps is to determine device category as a function of body contact (ISO 10993-1, 4.0). The categories include noncontact (4.1.1), surface contacting (including skin, mucosal membranes, and breach or compromised body surfaces (4.1.2), external communicating (4.1.3), and implantable devices (4.1.4). Categorization by duration of contact (limited, prolonged, or permanent, 4.2) is also required. One must conduct a thorough physical and chemical characterization of the material as it would be used on/in the device [29, 30]. This means that all samples tested must have been through the same processes, cleaning, sterilization, and packaging as the final device. For biocompatibility purposes, chemical characterization must pay special attention to additives, process contaminants and residues, leachable substances, degradation products, and interactions with other components of the device. All testing must be done in compliance with Good Laboratory Practices (GLP). ISO 9000 specifies the requirements for quality-management systems. A list of test methods used to characterize implantable elastomers follows as an example. Of course, different materials or uses may require substantially different test methods.

- Master file or analytically derived data (required):
  - Complete information including precursor materials, solvents, catalysts, curing agents, reinforcing agents, crosslinking agents, etc.
  - Composition reaction ratios and catalyst ratio
  - Any relevant literature and patents describing the formulation and characterization of the material

- Examples of measurements on polymeric materials processed for use in the device:
  - Molecular weight (ASTM D3593 Mw, Mn, Mw/Mn)
  - FTIR/ATR
  - T_g & T_{melt} (DMA or DSC, ASTM D3418)
  - Metal content (ASTM F1372, Pb, Cu, Sn, Sb, Hg, As, Cd, Mg, Se, Si, and Ba)
  - Thermal stability (TGA, ASTM D5023 or 5026)
  - Extractables (see ISO 10993-12)

The biocompatibility testing protocol will depend on the category of device and duration of contact as described above. For chronically implantable materials, the entire range of ISO 10993-1 is recommended, as is summarized below. Please note that ISO 10993-1 is in the process of revision. It is recommended that the reader consult the revised version when it is available. Because of cost constraints, a stepwise evaluation is recommended. The decision to conduct the more costly phase 2 tests would depend on the results of the phase 1 tests. Again, the materials must be processed, cleaned, sterilized, and packaged in the same manner as the final product. Note that the following tests are listed and briefly described in ISO 10993-1, but detailed protocols are provided in other parts of the standard for most of these (ISO 10993-2 through 12).

4.3 Phase 1 Tests (ISO 10993-1)

- Cytotoxicity (ISO 10993-5): Cell culture techniques, for example, MEM Elution Test L929 (ASTM F1027).
- Sensitization (ISO 10993-10): Estimates the potential for contact sensitization. Example, Kligmann Sensitization.
- Irritation: Estimates for irritation at appropriate sites such as skin, eye, and mucosal membranes.
- Intracutaneous reactivity (ISO 10993-10): Estimates the potential for localized reaction of tissue where exposure to skin, eye, and mucosal membranes is inappropriate. Example, USP acute intracutaneous injection.

- Systemic (acute) toxicity (ISO 10993-11): Estimates the potential harmful effects of implant $\leq$ 24 hours where absorption of leachable or degradation products is possible. Examples, USP acute systemic injection test and USP Pyrogen Test (LAL).
- Acute and subchronic toxicity (ISO 10993-11): Estimates the potential harmful effects of implant $\geq$ 24 hours to not greater than 10% of the animal's total life span (such as 13 weeks in rats). Example, intramuscular implantation test. Not required if chronic toxicity data exist (see ISO 10993-1, paragraph 5.3.1, Chronic Toxicity).
  - Genotoxicity (ISO 10993-3): Evaluates gene mutations, changes in chromosome structure, and other DNA or gene toxicities. For example: Salmonella Typhimurium, reverse mutation (Ames mutagenicity)
  - Mammalian cell transformation assay: C3H/10T1/2 or BALB/c3T3
  - Chromosomal aberrations: Test 479, in vitro sister chromatid exchange assay in mammalian cells.
- Implantation (ISO 10993-6): Assesses the local pathological effects on living tissue.
- Hemocompatibility (IS 10993-4): Evaluates the effects on blood and blood components by blood-contacting materials/devices. Example, static hemolysis test RRBC.

Products that are to be sold outside of the United States must also take into consideration any requirements unique to other countries.

4.3.1 Phase 2 Tests

If it is determined that the new material is acceptable for use in chronically implanted devices, then the following must be addressed:

- Chronic Toxicity: Estimates the potential harmful effects of implant during a major portion of the animal's total life span (such as 6 months in rats). Example, intramuscular implantation test. Note, in some cases, chronic biostability testing may serve to meet this requirement. See biostability testing below.
- Carcinogenicity (ISO 10993-3, Para 5.0): Lifetime in vivo carcinogenicity test (mice, 1.5 years). "These tests should be conducted only if there are suggestive data from other sources." For example, carcinogenicity testing must be done if genotoxicity results are positive. Alternatively, perhaps one should reconsider if the test is worth the time (about 3 years) and expense (>$1 M) if another, more biocompatible material is suitable and available.
- Reproductive and developmental toxicity (ISO 10993-3): "Reproductive/developmental toxicity tests or bioassays should only be conducted when the device has potential impact on the reproductive potential of the subject. The application site should be considered."

5 Potential for Biodegradation

ISO 10993 provides several standards for the identification and quantification of potential degradation products. Part-9 (biodegradation) states "Where the potential for resorption and/or degradation exists, corresponding tests may determine the processes of absorption, distribution, biotransformation and elimination of leachables and degradation products of medical devices, materials and/or their extracts." This standard provides guidance for protocol development, but does not provide any test protocols. Parts 13, 14, and 15 address polymers, ceramics, and metals/alloys respectively.

6 New Material Stability Testing In Vitro

Passing the above series tells us that the material as received, processed, cleaned, packaged, and sterilized is sufficiently biocompatible to use in preclinical (animal) device test. It is, however, not sufficient to justify human use for long-term implant. One must consider the biostability of the material in light of the device's intended duration of implant. There are no biostability test standards such as ISO 10993 for biocompatibility. Thus, the manufacturer must develop a test protocol that makes good scientific sense for the particular device and material in question. Of course, a literature search is an absolute "must." There may also be relevant documents available, such as the FDA's Guidance for pacemaker leads and ISO or ASTM standards that may need to be consulted. Keep in mind that the rationale for the test regimen will probably have to be explained and justified to the appropriate regulatory agency. Some typical biodegradation mechanisms are shown in Table 1 for reference.

Table 1 Some degradation mechanisms that can affect biostability of cardiac pacemaker leads

Conductor metal	Polymer
• Corrosion (pitting, fretting, crevice, intragranular, stress corrosion, galvanic) • Wear • Fatigue and fracture	• Oxidation • Hydrolysis • ESC • Mechanical (swelling, extraction, wear, creep) • Mineralization

It is absolutely necessary to consider all the materials used in the device so that interactions can be evaluated. For example, modern cardiac and neurological leads may be insulated with silicone rubber or polyether polyurethane. They also contain metallic conductors (typically MP35N, and sometimes also platinum or silver). The individual conductor wires may be covered with a redundant insulation such as eTFE. The conductors and insulation are connected to stainless steel connector pins and platinum electrodes.

6.1 Metals

For metals, FDA regulations specify ASTM F2129 (standard test method for conducting cyclic potentiodynamic polarization measurements to determine the corrosion susceptibility of small implant devices). Some devices may also require testing for stress corrosion (failure due to corrosion accelerated by tensile stress). There are many test methods for fatigue fracture, so it is incumbent on the manufacturer to select the most appropriate one(s) for device in question. Pacemaker leads, for example, are often tested in a fixture that flexes them $\pm$ 90° over a specified radius at twice the typical cardiac cycle rate (Fig. 10) [31]. Presently, FDA requires satisfactory performance over 4 hundred million cycles (equivalent to about 10 years in human use).

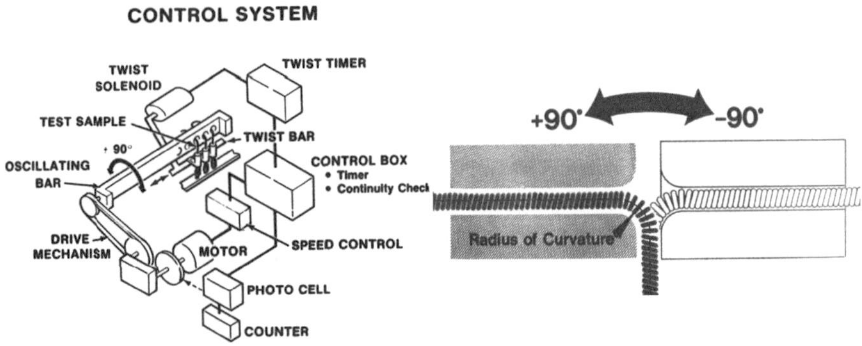

Fig. 10 A schematic view of a lead flex tester (*left*). The lead placed in the fixture is seen in the right panel, showing how it is flexed $\pm$ 90°

6.2 Polymers

The susceptibility of polymers to autoxidation can, of course, be tested in vitro at elevated temperatures, but that does not seem to be too relevant for implanted devices. Given our knowledge of the foreign body response today, it is quite reasonable to conduct immersion tests of materials or devices in H_2O_2 [11]. It remains unknown, however, just what the concentration of H_2O_2 is under the belly of a macrophage. We do know that the production of H_2O_2 rises sharply within about a week after implant, then decreases markedly, apparently continuing at a very slow trickle for several years. Acceleration would be achieved by using as high a concentration as is safe and practical. Because the H_2O_2 will decompose relatively quickly over time, it should be replaced at least several times per week to maintain its activity. Periodically removed samples should be characterized physically and chemically to facilitate determining the rate of degradation, if any. It should be noted that even high concentrations of H_2O_2 do not produce significant degradation of the silicone rubbers or polyurethanes used in pacemaker leads. One variant of this test is the immersion of the tubing containing cobalt rod shown in Fig. 6 in 37°C, 3% H_2O_2 in n-saline (non-buffered 0.9% saline). The peroxide permeates the polymer, reacts on the cobalt to both accelerate the metal's corrosion and decompose the peroxide into oxygen free radicals. This then mimics the autoxidation and redox reactions known to attack the insulation in some devices [14]. Immersion of materials in oxygen free radicals is a more complicated matter since they are typically extremely reactive and short lived. This can be achieved by adding 0.05–0.1 M $CoCl_2$ to the H_2O_2 [16]. Again, the solution needs to be changed frequently.

Immersions of polymer microtensile specimens in solutions of metal ions or lipid emulsions at elevated temperatures for 16 weeks have been reported [32]. Temperatures of 37, 70, and 90°C are recommended with sampling monthly to establish trends. For pacemaker leads, the solutions should include all the metals found within the device, a base, and an acid. For example, aqueous solutions of 1 M $AgNO_3$ or 0.1 M $CoCl_2$ (acetylacetanoate) can assess oxidation. Immersion in 1 N acetic acid, Ringer's solution, and 1.0 N HCl can assess hydrolytic resistance. Immersion in 20% intralipid (soybean) emulsion can assess the propensity to absorb lipids. However, in our experience, none of the above in vitro tests appear to be reliably predictive of performance in pacemaker lead insulation. Why? The in vivo environment cannot be duplicated in vitro. For example, the oxidation state of an ion varies as a function of what it is dissolved in. Distilled water containing a metal ion does not represent the environment within a lead. This accelerated test predicts that Ag^+ will oxidize and degrade polyether polyurethanes while Co^{2+} will not. Multiple in vivo studies clearly demonstrate exactly the opposite [14, 33]. Traces of cobalt will degrade the polymer in vivo whereas silver will not.

Hydrolytic degradation can be assessed as noted above. Keep in mind, however, that in vivo hydrolytic attack is often accelerated or catalyzed by enzymes, local pH, and/or physiologic ions. Immersion in solutions containing an esterase, cholesterol esterase, and a protease has been shown to accelerate hydrolysis in vitro [34, 35].

Wear and creep can be a little more complicated. Wear on material samples is typically evaluated by weight loss on a rotating disk exposed to abrasive wheels (Taber abrasion). This may not be too useful for in vivo implants, however. One clue as to how well a polymer will hold up to wear is the area under the stress–strain curve (the "modulus of toughness"). To evaluate the possibility of wear between pacemaker lead insulation and other leads, the pulse generator, bones, etc., one can place a weighted rod on the lead in 37°C saline. The rod can then be moved back and forth at a fixed rate. Creep is the property of viscoelastic materials to flow under a force below the ultimate tensile strength. Tensile creep is measured by hanging weights on strips of polymer, periodically measuring their length. However, in pacemaker leads, analysis shows that creep in the insulation is due to compressive, not tensile forces. Good correlation with cyclic compressive creep forces (as in subclavian crush) has been found with a fixture that exposes the lead to a rotating disk having several weighted metallic wheels set at a specified gap between them and the base plate [36]. This test can also evaluate the resistance of a finished lead body's conductor to subclavian crush.

7 In Vivo Materials Testing

Clearly one can implant microtensile specimens in animals such as rats or rabbits for several years, characterizing the materials as a function of periodic explant time. However, thin samples tend to curl or fold in vivo, affecting the reliability of the test. This can be dealt with by using thicker samples, but the rate of measurable degradation decreases as the thickness increases. Thus, we have abandoned this form of testing since it rarely gives the necessary answers in a reasonable implant time. It is more appropriate to implant the devices or subassemblies and remove characterization samples from them.

ESC has not yet been exactly duplicated in vitro. We have achieved good predictability in vivo using tubing samples with strain fixed over mandrels as shown in Fig. 3. Just how much strain is required to accelerate the mechanism must be determined for each type of polymer. For polyether polyurethane elastomers, strain must be $\geq$ 300%. This is true because in the unstrained state, the polymer is phase separated into polycrystalline "hard segment" islands in an amorphous polyether matrix [37]. Between 200 and 300% strain, this morphology is destroyed to be replaced with a linear (oriented) order, which is much more susceptible to relatively rapid degradation [38]. Up to 4 such strings (up to 20 variables) can be implanted in each rabbit's subcutis. Because of biologic variance (again a matter of rate rather than mechanism), it is necessary to implant at least 5, preferably 10 rabbits per sampling period to assure statistical reliability. Note that this test is used to evaluate both materials and their manufacturing processes. Typically, $\geq$300% strained Pellethane 2363-80A will develop significant ESC within 6–12 weeks implant. With new materials, however, the test has been run for up to 2 years to determine if there are any late-developing phenomena [39–42]. Periodic explants are evaluated by ATR/FTIR, SEM, GPC molecular weight, and other relevant techniques (see ISO 10992 and the FDA Guidance). Of course, this test is not limited to polyurethanes but has been used with other elastomer chemistries. Note that if carefully done, this 2-year test can also serve to fulfill subchronic and chronic biocompatibility testing as noted in ISO 10993-5.2.6 and 5.3.1. At termination, the animal is euthanized, the skin reflected, and photographs taken of the implant sites. Untoward biologic reactions, if any, are noted and samples taken to be stored in buffered formalin. If it is decided to submit data to regulatory agencies, the samples are retrieved and evaluated histopathologically.

In vitro MIO tests do not always appear to correlate well with in vivo results either. Short (about 0.75 inches long) tubing samples containing cobalt rod as shown in Fig. 6 are implanted in the subcutis of rabbits in a manner similar to that in the in vivo ESC test (Fig. 3). Again, because of biologic variance, it is necessary to implant at least 5, preferably 10 rabbits per sampling period to assure statistical reliability. Depending on how the polymer is processed, Pellethane 2363-80A controls will show substantial degradation in 1–2 years. Periodic explants are evaluated by ATR/FTIR, SEM, GPC molecular weight, and other relevant techniques. Again, this test may fulfill subchronic and chronic biocompatibility testing as noted in ISO 10993-5.2.6 and 5.3.1, per the methods mentioned above.

Mineralization is a phenomenon where the body deposits calcium salts, such as hydroxy apatite on/within the device (intrinsic) or in the fibrotic encapsulation (extrinsic). Note that all implanted devices have been reported to be subject to mineralization, regardless of the materials used. At present there is really no good, predictive accelerated test method, although implant in adolescent rats appears to be about as good as is available. Explants are evaluated by SEM and EDS for calcium deposition.

8 Device Implants in Animal Models

We have conducted all the appropriate testing on new materials and the device according to a scientifically based qualification plan. Are we now ready to use the material in a human use product? Not necessarily! Remember that no in vitro test can possibly exactly mimic the in vivo environment, and animals do not exactly mimic the human in vivo environment. The FDA's Guidance for the Submission of Research and Marketing Applications for Permanent Pacemaker Leads and for Pacemaker Lead Adaptor 510(k) Submissions states that the following submission strategy is required, unless "If acceptable accelerated testing protocols are available, abbreviated testing may be conducted per section II.B. Alternative Submission strategy. Acceptable accelerated testing should be supported by documentation that demonstrates that the in vitro testing can reliably predict in vivo performance."

A. Submission Strategy

- Implant leads in animal hearts to obtain data on 20 leads at the end of two years
- Historical or other suitable controls

B. Alternative Submission Strategy

- Implant leads in animal hearts with intent of obtaining data on 20 leads after 6 months
- Historical or other suitable controls
- Accelerated testing – ESC test and MIO test."

The standard specifies all the necessary parameters, including analytical techniques. The problem has been that no adequate correlation between in vitro data on new materials and in vivo device performance has been reported. Thus, the pacemaker lead industry has usually been forced to conduct 2-year implants in animals before submission. It seems quite likely, however, that the in vivo accelerated ESC and MIO tests discussed above might satisfy the "Alternative Submission Strategy," at least for polyether polyurethanes. It is strongly recommended that any implantable device using a new material follow a similar in vivo animal qualification protocol.

9 Human Clinical Implants

It is very important to note that it is considered unethical to intentionally determine biocompatibility or biostability in human clinical trials. These parameters are to have been established before regulatory submission and the first human implant. Of course, should any untoward result be found clinically, or in prior preclinical tests, it must be reported according to FDA regulations. The clinical study is to determine the human safety and reliability of the new product, not its biocompatibility or biostability. The requirements for the development of clinical study plans can be found in the FDA Guidance [29].

10 Market Release and Postmarket Surveillance

Finally, the new device has completed human clinical trials. The appropriate regulatory agency has given its approval to market the product. We're home free now, right? Unfortunately, that may not be the end of it! There are several more factors to consider. Clinical studies may not have required enough time to reveal any long-term device complications. Clinical studies also may not require large enough numbers to detect low-incidence complications (such as subclavian crush). Some new lead models revealed no significant ESC or MIO after 2–5 years implant in canines, yet began to develop decreases in actuarial survival in as little as 1–2 years in humans [22]. Again, this species difference appears to be due to differences in degradation rate, not mechanism. It is also true that animal implants are scrutinized after explant with detailed mechanochemical analytical techniques, whereas human implants are tested noninvasively via telemetry. Different kinds of data are used for device assessment in humans and animals. For another example, remember the discussion above about subclavian crush? None of the animals used for pacemaker lead performance have clavicles so subclavian crush cannot be addressed in animals. How do we discover such phenomena? How do we determine the percentage and rate of device survival? The answer is postmarket surveillance. Apparently, postmarket surveillance is still discretionary over the implantable device industry. It is now mandatory for pacemaker manufacturers. The FDA's Guidance to Sponsors on the Development of a Discretionary Postmarket Surveillance Study for Permanent Implantable Pacemakers Electrodes (Leads) is available through DSMA. It provides guidance to sponsors on the design of a postmarket study protocol, which must be submitted to FDA for approval.

10.1 Returned Products Analysis

Some devices, such as pacemaker pulse generators, are easy and safe to explant. These are returned to the manufacturer for analysis. Any adverse findings are reported to the FDA per regulations. Using the largest manufacturer (Medtronic) as an example, the statistical data on pulse generator survival are reported twice annually in a Product Performance Report [22]. Since the vast majority of pacemakers are explanted for normal battery depletion, longevity curves are useful in informing physicians about device performance and helping engineers design better power sources. Recalls and advisories, if any, are also reported along with patient management suggestions. The latter are prepared with the assistance of a physician advisory board. For an example of a product performance report, visit www.medtronic.com, then select "Information for Physicians," then "Cardiology," and click on the "CRM Product Performance Report" tab. An example of a pulse generator longevity actuarial survival curve is shown in Fig. 11. Note that the term "survival" refers to survival of the device, not the patient. Not all explanted devices are routinely returned to the manufacturer. Some, such as many heart valves, are sent to independent third party "retrieval centers" for analysis.

Elite 7074, 7075, 7076, 7077 — Product Characteristics

US Market Release	Apr-91	Malfunctions	85	NBG Code	DDD/RO
Registered US Implants	48,000			Serial Number Prefix	YE, YF, 2E, 1X
Estimated Active US Implants	400			Estimated Longevity	See page 78
Normal Battery Depletions	3,572				
Advisories	None				

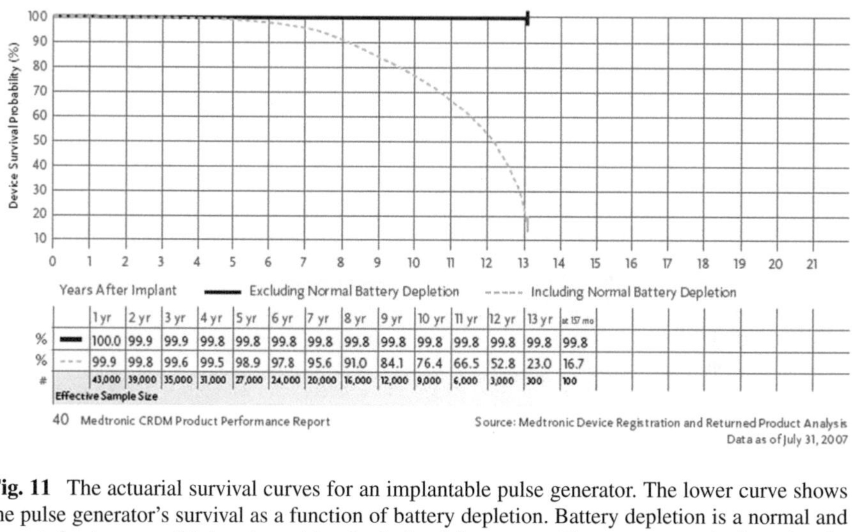

		1 yr	2 yr	3 yr	4 yr	5 yr	6 yr	7 yr	8 yr	9 yr	10 yr	11 yr	12 yr	13 yr	at 157 mo
%	▬	100.0	99.9	99.9	99.8	99.8	99.8	99.8	99.8	99.8	99.8	99.8	99.8	99.8	99.8
%	- - -	99.9	99.8	99.6	99.5	98.9	97.8	95.6	91.0	84.1	76.4	66.5	52.8	23.0	16.7
#	Effective Sample Size	43,000	39,000	35,000	31,000	27,000	24,000	20,000	16,000	12,000	9,000	6,000	3,000	300	100

40 Medtronic CRDM Product Performance Report

Source: Medtronic Device Registration and Returned Product Analysis
Data as of July 31, 2007

Fig. 11 The actuarial survival curves for an implantable pulse generator. The lower curve shows the pulse generator's survival as a function of battery depletion. Battery depletion is a normal and expected outcome so this is not considered a device failure per se. The upper curve depicts what the pulse generator's survival statistics would be if the battery's longevity were not taken into account. Thus the upper curve measures the reliability of the circuitry, hermeticity, feed throughs, etc.

10.2 Postmarket Clinical Studies

Some devices, such as transvenous pacemaker leads, are not easily removed after a few years implant. Through the 1980s, most nonfunctioning leads were either left in place with a new one implanted next to it or removed via open-heart surgery.

Many of those that were explanted, in part or in whole, were simply discarded by the hospital. To encourage return of explanted leads, companies such as Medtronic offered reimbursement. This often resulted in return of a proximal section, sometimes only the terminal assembly. Thus, the data obtained by analysis of returned products were highly inaccurate and misleading. The available returned products data suggested that all but a fraction of 1 percent of a certain lead model were performing as intended. In late 1983, Medtronic began organizing a postmarket study looking at the performance of implanted leads, aimed at determining their true clinical survivorship as well as failure modes. The initial study involved three large implanting centers. Today, the study has expanded to more than 35 centers worldwide [22]. The requirements for reporting complications are shown in Table 2. An example of a lead survival curve is shown in Fig. 12. The criterion for Medtronic to take some field action is a decrease in lead survival to <95% in 5 years. Returned leads continue to be analyzed, which provides important information about failure mechanisms, if not accurate clinical statistics.

Table 2 Chronic lead study reporting criteria. "A lead related complication is considered to have occurred if at least one of the following clinical observations is reported and at least one of the following clinical actions is made 30 days or more after implant [22]."

Clinical observations	Clinical actions
• Failure to capture (stimulate) • Failure to sense/undersensing • Oversensing • Abnormal pacing impedance (based on lead model, but normal range is typically 400–800+ ohms • Abnormal defibrillation impedance (based on lead model, but normal range is typically 20–200 ohms • Insulation breach observed visually that has degraded system performance • Clinically observed fracture, observed visually or radiographically • Extracardiac stimulation • Cardiac perforation • Lead dislodgment	• Lead surgically abandoned/capped • Lead electrically abandoned/capped • Lead explanted • Lead replaced • Polarity reprogrammed (i.e., bipolar to unipolar; unipolar to bipolar) • Lead use continued based on medical judgment despite a known clinical performance issue • Other lead-related surgery performed (i.e., lead mechanical alteration or unsuccessful repositioning) Note, successful lead repositioning is not a qualifying action.

Most pacemaker lead manufacturers continue to report (to FDA) on the basis of returned products only, and this is rarely available publicly. This produces highly misleading information, which is often used by sales people to tout the superiority of their products. Table 3 shows a comparison between some clinically and returned product-determined survival statistics from Medtronic's product performance report.

According to the RPA data, the earliest lead in Table 3, Model 4012, would never have met the 95% survival criteria for the manufacturer to take field action. But using the "Chronic Lead Study (CLS)" statistics, this lead was placed under advisory in 1991. According to the RPA data, the next generation lead, Model

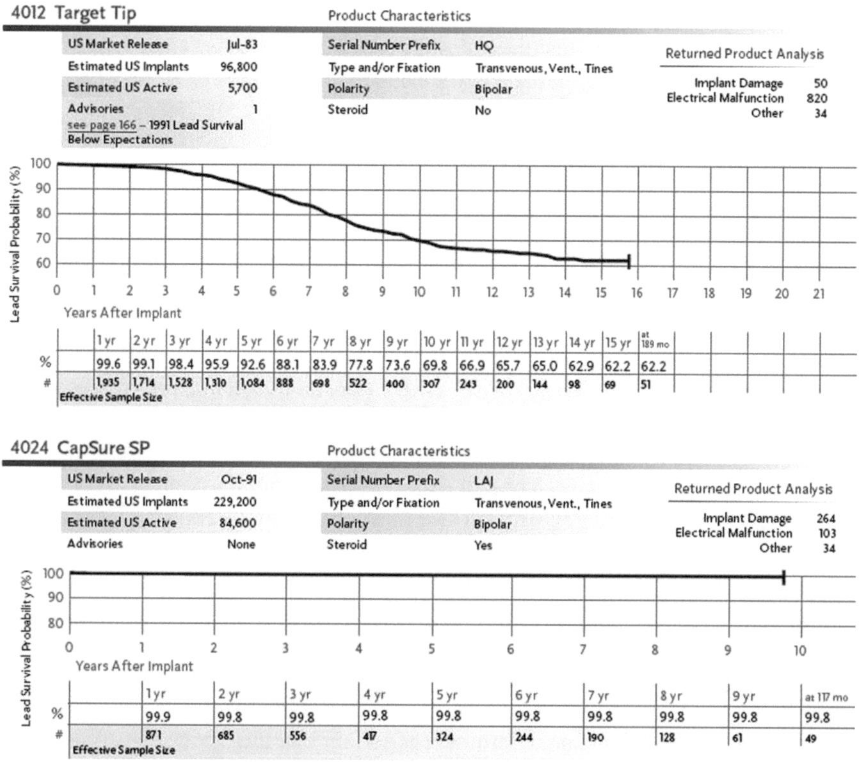

Fig. 12 Actuarial survival curves for two lead models. The upper curve for Model 4012 shows that its survival probability dropped below the trigger point of 95% after about 4.5 years. Thus this lead was put on Advisory September 26, 1991. The lower curve shows that one of its successors, Model 4024, has retained excellent reliability for nearly 10 years

Table 3 Medtronic survival statistics on three ventricular lead models, comparing analysis of returned products (ARP) with the clinical lead study (CLS) [22]

Model number	Years service	Number sold (US)	RPA % survival	Number in CLS	CLS % survival*
4004	12	74500	99.0	1640	50.6
4012	14	96800	99.8	2543	62.9
4024	10	229200	99.8	1215	99.9

* 95% Confidence interval

4004, would never have met the 95% criterion, but it was placed under advisory in 1993 based on CLS data. According to both databases, the latest model reported in Table 3, Model 4024, continues to display excellent performance after nearly 10 years in service, which demonstrates the dramatic improvement in reliability of more modern leads. Thus, to report only returned products analysis statistics is highly misleading.

The data from postmarket surveillance not only help clinicians manage their patients, but also help manufacturers design improved products. Thus, we come full circle, using postmarket surveillance to feed back into the design of new products.

11 Summary and Conclusions

To this day, we still hear people claim that in vitro testing of materials alone shows that they are suitable for use in chronically implanted devices. Others continue to say that "I proved the materials are biocompatible and biostable, so I don't have to do any device testing." This statement can be very far from the truth. In vitro testing has its place, primarily to screen materials and processes for further testing. In some cases where no suitable in vitro test exists, one may be forced to develop accelerated in vivo materials tests. Once the preliminary testing is accomplished, however, one *must* test the device per se in animals. A biocompatible material does not necessarily make a biocompatible device. The same may be said about biostability. These statements are true because shape, size, surface finish, interactions between the materials in the device, etc., all can affect its biocompatibility and biostability. But even well-performed animal studies may not unveil previously unknown mechanisms, because animals do not perfectly mimic the human in vivo environment. An excellent example of this is the subclavian crush in humans (clamping a lead between the clavicle and first rib), which is impossible to discover in animals with no clavicles. With the right protocol for the device in question, only postmarket surveillance appropriate for the device in question can determine for certain that the device does or does not meet expectations.

References

1. Williams D. (1990) Concise encyclopedia of medical and dental materials, 1st Ed. Pergamon Press, Oxford, UK.
2. Szycher M and Siciliano A. (1991) An assessment of 2,4 TDA formation from Surgitek polyurethane foam under simulated physiological conditions. J Biomater Appl 5:323–336.
3. Hudson J and Crugnola A. (1987) The in vivo biodegradation of nylon 6 utilized in a particular IUD. J Biomater Appl 1:487–501.
4. Stokes KB. (1988) Polyether polyurethanes: Biostable or not? J Biomater Appl 3(2):228–260.
5. Stokes KB (1997) Polyurethane pacemaker leads: The contribution of clinical experience to the elucidation of failure modes and biodegradation mechanisms. In KB Witkin (Ed) Clinical studies of medical devices and diagnostics: Principles and approaches. Humana Press, Totowa NJ.
6. Stokes KB and Cobian K (1982) Polyether polyurethanes for implantable pacemaker leads. Biomaterials 3:225–231.
7. Stephenson NL. (1980) Synopsis of clinical report on the Spectraflex models 6971/72 transvenous leads. Medtronic News. X(3):16.
8. Stephenson NL. (1980) Synopsis of clinical report on the Models 6990U/91U atrial J leads transvenous leads. Medtronic News X(2):16.
9. Hawkins WI. (1972) Polymer stabilization. Wiley InterScience, New York.
10. Schnabel W. (1981) Polymer Degradation. Hanser, Vienna, (Distributed in the US by Macmillan Publishing Co, New York, NY).

11. Anderson JM. (1996) Host reactions to biomaterials and their evaluation. In Ratner BD, Hoffman AS, Schoen FJ and Lemons JE (Eds) Biomaterials Science. Academic Press, San Diego, Chapter 3, pp. 165–173.
12. Parsonnet V, Zucker IR, Kannerstein ML, et al. (Jul 1966) The fate of permanent intracardiac electrodes. J Surg Res – Clin Lab Invest 6(7):285–292.
13. Zhao QH, McNally AK, Rubin KR, et al. (1993) Human plasma α_2-macroglobulin promotes in vitro oxidative stress cracking of Pellethane 2363–80A: In vivo and in vitro correlations. J Biomed Mater Res 27:379–389.
14. Stokes K, Urbanski P and Upton J. (1990) The in vivo auto-oxidation of polyether polyurethanes by metal ions. J Biomater Sci, Poly Ed 1(3):207–230.
15. Ward R, Anderson J, McVenes R, et al. (2007) In vivo biostability of polyether polyurethanes with fluoropolymer and polyethylene oxide surface modifying endgroups; resistance to metal ion oxidation. J Biomed Mater Res 80 (1):34–44.
16. Zhao Q, Casas-Bejar J, Urbanski P, et al. (1995) The glass wool $H_2O_2/CoCl_2$ test system for in vitro evaluation of biodegradative stress cracking in polyurethane elastomers. J Biomed Mater Res 29:467–475.
17. Stokes K, Staffanson D, Lessar J, et al. (1987) A possible new complication of subclavian stick: Conductor fracture. VIII World Symposium on Cardiac Pacing and Electrophysiology. Jerusalem, and PACE 10(3), Pt. II: 748 (Abst. 476).
18. Belott P and Reynolds DW (1995) Permanent pacemaker implantation. In Ellenbogen KA, Kay GN, and Wilkoff BL. (Eds) Clinical cardiac pacing. WB Saunders, Philadelphia, Chapter 27, pp. 447–490.
19. Jacobs DM, Fink AS, Miller RP, et al. (1993) Anatomical and morphological evaluation of pacemaker lead compression. PACE. 16:434–443.
20. Magney JE, Flynn DM, Parsons JA, et al. (March 1993) Anatomical mechanisms explaining damage to pacemaker leads, defibrillator leads, and failure of central venous catheters adjacent to the sternoclavicular joint. PACE 16(Pt. I):445–457.
21. Byrd CL. (1992) Safe introducer technique for pacemaker lead implantation. PACE 15:262.
22. Medtronic, Inc. (2007) Cardiac rhythm disease management, product performance report, second edition, issue No. 57. http://www.medtronic.com.
23. Stokes K, Anderson J, McVenes R, et al. (1995) The encapsulation of transvenous polyurethane insulated cardiac pacemaker leads. Cardiovasc Path 4(3):163–172.
24. Byrd CL. (2000) Management of implant complications. In Ellenbogen KA, Kay GN and Wilkoff BL (Eds) Clinical Cardiac Pacing and Defibrillation. 2nd Edition. WB Saunders, Philadelphia, pp. 669–694.
25. ISO/DIS 14971
26. ISO/DIS 14971, Annex C, pp. 24–29
27. ISO/DIS 14971, Annex I, pp. 72–73
28. ISO/DIS 14971, Annex E, pp. 45–49
29. (November 1, 2000). Guidance for the Submission of Research and Marketing Applications for Permanent Pacemaker Leads and for Pacemaker Lead Adaptor 510(k) Submissions. US Department of Health and Human Services, Food and Drug Administration, Center for Devices and Radiological Health and Interventional Cardiology Devices Branch, Division of Cardiovascular and Respiratory Devices, Office of Device Evaluation.
30. AAMI/ISO 10993-1: 1997, Biologic evaluation of medical devices-Part 1: Evaluation and testing.
31. Stokes K and Stephenson NL (1982) The implantable cardiac pacing lead – just a simple wire? In Barold SS and Mugica J (Eds) The third decade of cardiac pacing. Futura, Mount Kisko NY, pp. 365–416.
32. Coury A, Hobot C, Cahalan P, et al. (May 15, 1990) In vitro chemical stability of implanted polyurethanes. Fourth University of Minnesota Research Poster Session, Basic and Applied Research in Academia and Industry. Minneapolis, Minnesota.

33. Stokes K. (July–September, 1993). Biodegradation. Cardiovasc Path 2(3 Suppl):111S–119S.
34. Ratner BD, Gladhill KW, and Horbett TA. (1988) Analysis of in vitro enzymatic and oxidative degradation of polyurethanes. J Biomed Mater Res 22:509–527.
35. Santeere JP, Labow RS, Duguay DG, et al. (1994) Biodegradation evaluation of polyether and polyester urethanes with oxidative and hydrolytic enzymes. J Biomed Mater Res 28: 1187–1199.
36. Cobian KE, Miller J, and Ebert M (1998) New improved silicone rubber for lead insulation. Medtronic technical concept paper number UC9704728IE.
37. Bonart R. (1968) X ray investigations concerning the physical structure of crosslinking in segmented elastomers. J Macromol Sci Phys B2:115.
38. Bonart R, Morbitzer L, and Hentz G. (1969) X ray investigations concerning the physical structure of crosslinking in segmented elastomers. II. Butanediol as chain extender. J Macromol Sci Phys 3:337.
39. Ward R, Anderson J, Ebert M, et al. (2006) In vivo biostability of polysiloxane polyether polyurethanes; Resistance to metal ion oxidation. J Biomed Mater Res 77(2):380–389.
40. Ward R, Anderson J, McVenes R, et al. (2006) In vivo biostability of polysiloxane polyether polyurethanes; resistance to biologic oxidation and stress cracking. J Biomed Mater Res 77(3):580–589.
41. Ward R, Anderson J, McVenes R, et al. (2006) In vivo biostability of polyether polyurethanes with fluoropolymer surface modifying endgroups; Resistance to biologic oxidation and stress cracking. J Biomed Mater Res 79(4):827–835.
42. Ward R, Anderson J, McVenes R, et al. (2006) In vivo biostability of Shore 55D polyether polyurethanes with and without fluoropolymer surface modifying endgroups. J Biomed Mater Res 79(4):837–845.

Technology Advances and Challenges in Hermetic Packaging for Implantable Medical Devices

Guangqiang Jiang and David D. Zhou

Abstract Many implantable medical devices contain sophisticated electronic circuits. Hermetic packaging is required to provide the implant's electronic circuitry with protection from the harsh environment of the human body. This chapter provides a review of available hermetic sealing methods and their applications. General considerations of implantable medical device packaging are discussed. Various testing methods applicable to the packaging of implantable medical devices are also presented. Many issues associated with hermetic packaging are not yet completely understood, nor are any corresponding difficulties completely overcome. The continued miniaturization of future implantable medical devices provides both opportunities and challenges for packaging/materials engineers to improve the existing packaging methods, and to develop new methods. Reliable hermetic micropackaging technologies are the key to a wide utilization of microelectromechanical systems (MEMS) in miniaturized implantable medical devices.

Contents

G. Jiang (✉)
Alfred E. Mann Foundation for Scientific Research, Santa Clarita, CA 91355, USA
e-mail: jiang@aemf.org

D.D. Zhou, E. Greenbaum (eds.), *Implantable Neural Prostheses 2*,
Biological and Medical Physics, Biomedical Engineering, DOI 10.1007/978-0-387-98120-8_2,

1 Introduction

Implantable medical devices have been widely used to restore body functions, improve the quality of life, or save lives. Experts estimate that 8 to 10 percent of all Americans (some 20 million to 25 million people) [1], or about 1 in 17 people in industrialized countries [2], carry some form of implanted device. Many medical devices, such as the implantable cardiac defibrillator, cochlear implant, artificial vision prosthesis, neuromuscular microstimulator, and the like contain sophisticated electronic circuits. Such long-term implantable medical devices are susceptible to damage by body fluids over time. Hermetic packaging is required to protect the electronic circuitry of the implant from the harsh environment of the human body.

1.1 Hermetic Packaging Technology Advances

There are a variety of ways to define hermeticity. Webster's New Collegiate Dictionary defines hermeticity as "the state or condition of being airtight," or in the Microelectronics Packaging Handbook [3], it is defined as "sealed so that the object is gas tight." In the real physical world, there is no such thing as absolute or complete hermeticity because all materials are gas permeable to some degree [4].

The packaging of implantable medical devices uses various materials, including polymers, glasses, metals, and ceramics. The encapsulation method used is greatly dependent upon the technology of the electronic circuit that is to be encapsulated. Polymer encapsulation has been successfully used with relatively simple circuits assembled from discrete, low-voltage components [5]. With polymer encapsulation, the discrete components are often compactly arranged and "potted" in a mold with leads or conductive feedthrough pins penetrating through the polymeric encapsulation wall [6]. This has historically been the preferable approach to encapsulation because of its simplicity and relatively low processing temperature; however, polymers do not provide an impermeable barrier. Moisture ingress will ultimately reach the electronic components, and surface ions can allow electric shorting and degradation of the leakage-sensitive circuitry and subsequent failure of the device [7]. Therefore, in general, polymer encapsulation is unsuitable for high-density, high-voltage electronics circuits. Recent research on liquid crystal polymers (LCPs) would suggest that using LCPs for near-hermetic packages in implantable medical devices looks promising [8, 9].

Hermeticity of early experimental cochlear implants used the principles of mechanical pressure and gasket sealing [10]. Though the human implanted system permitted changes from percutaneous cable to transcutaneous telemetry for the first time, no report on the long-term performance of such a sealing mechanism used in this implant has been issued. Materials that may provide a hermetic barrier are metals, ceramics, and glasses. Metallic packaging generally uses a biocompatible metal capsule that is either machined from a solid piece of metal or deep drawn from a piece of sheet metal. Electrical signals enter and exit the package through hermetic feedthroughs. The feedthrough assembly often utilizes a ceramic or glass insulator to allow one or more conducting wires to exit the package without coming in contact with the package itself. This method has been successfully used for implantable pacemakers [11], cardioverter defibrillators [12], implantable multichannel neuromuscular stimulators [6], and cochlear implants [13].

Metal-based packaging generally requires that a power-receiving coil or communication antenna be placed outside the package to avoid significant loss of power or radio frequency signals through the walls of the implanted device, thus requiring additional space within the body to accommodate the volume of the entire implant. Bioceramics and biograde glasses possess a radio frequency transparency advantage over metallic materials. They have been used as the main packaging materials for the implantable neuromuscular microstimulators [14, 15], cochlear implants [16], and artificial retina implants [17]. Bioceramics used for structural applications include alumina (both single crystal sapphire and ruby or polycrystalline α–Al_2O_3) [18–22], zirconia (magnesium oxide partial stabilized zirconia (Mg-PSZ)) [23], yttria-stabilized tetragonal zirconia polycrystals (Y-TZP) [23], and ceria stabilized zirconia poly-crystal (Ce-TZP) [24–29]). Many different types of biocompatible glass have been successfully used for implantable medical devices, including borosilicate glass (Kimbel N51A) for the glass capsule neuromuscular microstimulator [14].

A conductive component is often the required interfacial material for an active implantable medical device for the purposes of sensing and delivering electrical signals from/to living tissue. Though the main body of the package is made of bioceramic or biocompatible glass, hermetic bonding between the ceramics or glasses and metallic/conductive components is essential. To enable a hermetic seal between similar or dissimilar metallic components, fusion welding methods, such as laser welding, tungsten inert gas welding (TIG), and electron-beam welding, are often the choices. Figure 1 shows three neuromuscular microstimulators (BIONs) in different packages. The AMI glass-packaged neuromuscular microstimulator contains three hermetic seals: two glass-to-tantalum seals produced by melting glass with an infrared laser beam; and one final hermetic seal that is obtained by melting the tantalum tube closed in a plasma needle arc welder [14]. Metal-to-ceramic brazed cases provided housing for both the AMF and ABC neuromuscular microstimulators [30–32] and the final hermetic seals were achieved by a laser-welding method.

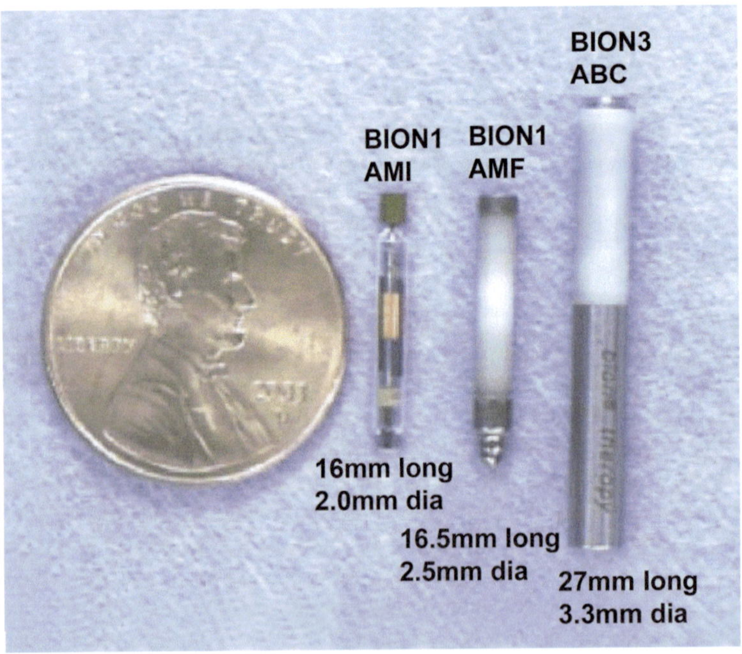

Fig. 1 Photograph depicting three neuromuscular microstimulators (BIONs) in different packages. The diameters of three BIONs are 2.0 mm from AMI-Alfred Mann Institute at University of Southern California, 2.5 mm from AMF – Alfred Mann Foundation, and 3.3 mm from ABC – Advanced Bionic Corp. (Reprinted from [33] with permission of the American Association of Neurological Surgeons)

1.2 Significance of Hermetic Packaging for Implantable Medical Devices

Despite the multitude of designs for implantable medical devices, there are usually four integral parts that must be accounted for: (1) the input or sensor, (2) the lead and lead connections, (3) the main package body, and (4) the output [34]. The principle failure points for these implant devices occur at the interfacial boundaries of adjoining surfaces, where water and ion migration proceed along the lead connections [7]. For example, an auditory reliability report, issued in 2006 by a medical device company, disclosed that roughly three out of four device failures were attributed to moisture ingression into the titanium receiver-stimulator packaging of an implant through the feedthrough (available online at http://www.bionicear.com/printables/reliabilityreport2006.pdf, accessed 10 January 2008). Water penetrating the intact polymeric encapsulants and permeating to the underlying substrates [35] is also a common cause of failure. Moisture ingress can result in failures such as open circuits [36], damage to metallics [36], surface electrical leakage [37], and electrical shorts due to moisture-promoted dendritic growth of silver and gold [37–39]. Ingress of other active gases, such as oxygen, could also cause attachment failure of solder-attached components due to solder oxidization [40].

This chapter provides a review of the available hermetic packaging methods and their applications. This has been a difficult task, as manufacturers of medical devices do not usually disclose the details of their packaging methods to protect their competitive edge. Many implantable medical devices have utilized one or more of the hermetic packaging methods. General considerations of packaging and testing methods for implantable medical devices are discussed in this chapter. Challenges associated with further advancement of implantable medical devices and future directions in the field are also examined.

2 General Packaging Considerations for Implantable Medical Devices

2.1 Biocompatibility

Biocompatibility is the first thing that the packaging engineer should consider when designing a hermetic package for an implantable medical device, as it is the package that makes direct contact with body tissue. It is critical that implantable medical devices do not elicit any undesirable local or systemic effects in the human body. In addition, the package materials should be stable and must be able to withstand attack from a harsh ionic body environment.

Biocompatible materials that have been successfully used for implantable medical device packaging include titanium and its alloys, noble metals and their alloys, biograde stainless steels, some cobalt-based alloys, tantalum, niobium, titanium-niobium alloys, Nitinol, MP35N (a nickel-cobalt-chromium-molybdenum alloy), alumina, zirconia, some biocompatible glass and polymers. A series of biocompatibility testings often suggested by the Food and Drug Administration (FDA) will be discussed in Section 4.3

2.2 Hermeticity Requirement

The degree and measure of hermeticity are a function of material choice, final seal design, fabrication processes and practices, and the use environment; so, before deciding what kind of packaging method to select for the implantable medical device, one needs to consider the level of hermeticity that is needed and the life span expectations for the product. Theoretically, all materials and all welded or joined assemblies will leak to some degree [4], whether by permeation through the bulk material or along a discontinuity path. The property of the solid that characterizes the amount of gas that can pass though the solid is called permeability [41]. The function of permeability is a combination of mass (g), distance (cm), time (s), and pressure (torr). Standard engineering practice would graph the permeability function as g/cm-s-torr. Figure 2 shows the typical helium permeability of many common classes of packaging materials and their predicted lifetime at different thicknesses. As mentioned previously, polymeric materials do not provide an impermeable barrier. Thus glasses, ceramics, metals, and single crystals can be considered for long-term implants.

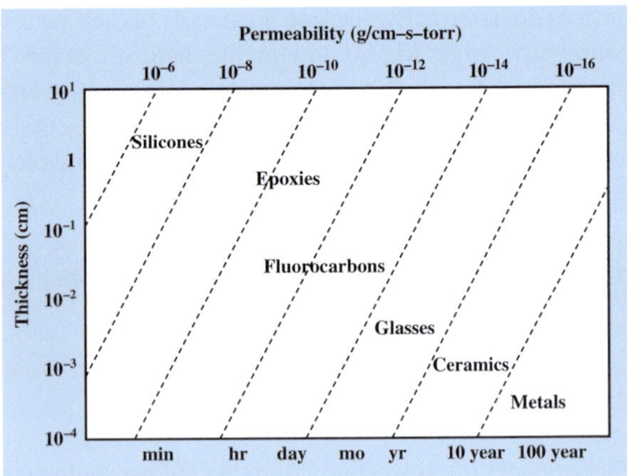

Fig. 2 Permeability chart for common classes of packaging materials (Reprinted from [4] with permission of Medical Device and Diagnostic Industry. Copyright © 2000 Canon Communications LLC)

2.3 Outgassing of Internal Materials

Even if the package materials and the sealing process(es) effectively prevent leakage through the package wall, implants still fail in some cases. The internal materials in a package, such as silicones, epoxies, and polymer coatings or insulators for chips and discrete electrical elements, often outgas which increases the vapor pressure and moisture level inside the package. The internal outgassing may induce the formation of water droplet condensation, thus compromising the device performance and eventually leading to device failure. Control of these internal sources of contaminants is essential to guarantee long-term reliability.

2.4 Wireless Communication

Wireless communication between the implantable medical device and external control unit eliminates the need for percutaneous wires, so that the risk of infection can be minimized. In some cases, the receiving coils are placed outside the metallic hermetic package and then jointly embedded with the metallic packaging using polymer encapsulation. The received signal can then be transmitted to the electronic circuitry inside the metallic package by utilizing feedthrough pins. In a polymer-encapsulated coil design, there is an issue that must be addressed: altering the inductance of the receiving coil, due to moisture ingress at the coil interface, causes the quality factor Q to decrease, thereby lowering the gain of the stage. Bioceramics and biocompatible glass provide a transparent radio frequency window for wireless communication. By employing bioceramic or biocompatible glass as the packaging material, the receiving coil can be placed inside the package and wireless communication can be established through the package.

2.5 Package Heating

When a metallic material is used as the main package body, problems such as metal heating have to be considered. As noted above, metallic materials attenuate electromagnetic and radio frequency signals. To provide reliable communication and effective charging to the implanted receiving coils encapsulated in the metallic package, strong electromagnetic signals from external transmitters are required. Eddy currents generated inside the metallic bulk generate heat when an electromagnetic field is present. International standard ISO 14708-1:2000 E requires that no outer surface of an implantable part of the active implantable medical device shall be greater than 2°C above the normal surrounding body temperature of 37°C when implanted, and when the active implantable medical device is in normal operation or in any single-fault condition. Special consideration must be given to the amount and types of metallic materials and to the package design for the implanted device to avoid excessive heating.

2.6 Coefficient of Thermal Expansion Compatibility

When bonding dissimilar materials, especially those involving high-temperature processes such as brazing or welding, a coefficient of thermal expansion (CTE) compatibility between the parts to be joined has to be considered. The larger the CTE mismatch and the bigger the parts, the higher the possible residual stress in the assembly. Such residual stress can result in failure, such as cracking, or can compromise the hermeticity of the assembly. Sometimes the failures may not appear immediately, and the consequences can be worse when discovered later. Early versions of the AMI neuromuscular stimulator package used a tubular feedthrough of 90% Pt–10% Ir. The seals produced were hermetic initially, but tended to fail catastrophically during prolonged soaking and temperature cycling in saline because of the difference in the CTE between the glass capsule wall (5.5×10^{-6}/°C) and the 90% Pt–10% Ir feedthrough (8.7×10^{-6}/°C). The excess residual stress in the walls of the sealed glass capsules can be measured using the photoelastic effect on the rotation of polarized light (Model 33 Polarimeter, Polarmetrics, Inc., Hillsborough, NH) [14]. By using tantalum (CTE = 6.5×10^{-6}/°C) as the feedthrough material, the residual stress is reduced.

3 Types of Hermetic Sealing and Their Applications

3.1 Polymer Encapsulation

There are numerous organic polymeric materials that are used as encapsulants for electronics. These materials are divided into (1) thermosetting polymers, (2) thermoplastics, and (3) elastomers. For implantable medical device applications, only a few materials in the above three groups can be made ultrapure to serve as acceptable encapsulations for implants [42]. Candidate materials include epoxies, silicones, polyurethanes, polyimides, silicone-polyimides, parylenes, polycyclicolefins, silicon-carbons, and benzocyclobutenes, as well as recently developed high-performance liquid crystal polymers (LCPs). Silicone rubber has been used as cable insulation material [43], epoxies were used in part for electronic component encapsulation [44], and Parylene C is utilized as an insulation coating on electronics in implants [45, 46]. The challenge of polymer encapsulations when applied to a long-term biomedical device primarily lies in their bio-stability within the body. Degradation of polymers includes hydrolytic, oxidative, and enzymatic mechanisms that deteriorate the chemical structure [47]. Polymer encapsulation has been successfully used with relatively simple circuits using discrete, low-voltage components. However, polymer encapsulation does not provide an impermeable barrier [48] and therefore cannot be used for packaging high-density, high-voltage electronic circuitry for long-term applications.

3.2 Glass-to-Metal Seal

Glass-to-metal seal technology is used extensively to provide a hermetic seal between a metal conductor and a metal body. Hermeticity of a glass-to-metal seal can typically be 1×10^{-8} standard cubic centimeters of gas at a pressure of one atmosphere per second (atm-cc/sec) or less, as measured by the helium-leakage rate. A typical glass-to-metal seal consists of the following elements:

- A metal bulkhead (or body) with a hole or holes in it.
- A pin(s) serving as a conductor(s) in the center of this hole(s).
- A piece of glass preformed to fit between the pin(s) and the bulkhead.

During processing, these three components are placed on a fixture which holds them in position. The entire assembly is then heated in a controlled atmosphere to the appropriate temperature for the particular set of materials. At the sealing temperature, the glass melts and fills the space between the pin and the bulkhead, and a hermetic seal is formed upon cooling.

The type of seal generated by this process is dependent upon the type of glass used and the materials used for the bulkhead and pins; there are two basic types of seals, the compression seal and the reactive seal. Early cochlear implants had used a compression glass-to-metal seal to form feedthroughs to provide the connection for 16 electrodes. Sixteen pure platinum feedthrough pins are precisely placed in position in a polycrystalline glass-ceramic substrate with a commercially pure (CP) Ti ring on the outside. The polycrystalline glass-ceramic has slightly smaller CTE than CP Ti. Upon cooling, the CP Ti ring shrinks and squeezes the glass onto the pure platinum pins to form a strong compression seal.

This reactive seal was the choice of the first-generation neuromuscular microstimulator [14]. The hermeticity of the glass-to-metal seals depends on chemical bonding between the borosilicate glass (Kimble N-51A®) and the native oxide on the tantalum electrode stem and the tantalum tubular feedthrough. Enclosures with this type of glass-to-metal seal may experience a transient loss of hermeticity and ingress of ambient gases when subjected to mechanical pressure. Integrity of glass-to-metal seals depends on a strong bond of glass-to-metal oxide at the metal/glass interface. If this bond is weakened, or otherwise compromised by inadequate oxide thickness or contamination, the seals may temporarily give way during slight distortions of the package by mechanical pressures of fixturing the unit for testing, e.g., centrifuge or fine/gross leak. Although glass-to-metal seals can be initially hermetic, their performance after temperature cycling is of suspect [14, 49]. Graeme reported that fluids and enzymes can permeate along minute pathways, or open up cracks in the glass seals through surface tension, and this was one failure mode of the early University of Melbourne's cochlear implant prototypes seen in two of the three initial patients [50].

3.3 Ceramic-to-Metal Feedthrough

A properly produced ceramic-to-metal feedthrough seal is often more robust, more durable, and has tighter hermeticity and better electrical insulation than a glass-to-metal feedthrough seal. Ninety-two percent or higher purity alumina as well as

100% pure sapphire and ruby are commonly used for hermetic seal of implantable medical devices. Other ceramics such as aluminum nitride (AlN), zirconia (ZrO_2), silicon carbide (SiC), and silicon nitride (Si_3N_4) have the potential to be used for ceramic-to-metal assemblies for implantable medical devices.

Properly designed ceramic-to-metal feedthrough seals are able to maintain hermeticity in a variety of harsh conditions, such as temperature cycling, corrosive, thermal shock, and varying pressure environments. Ceramic functions as an excellent electrical insulator at elevated temperatures, unlike glass, which conducts more electricity at high temperatures. Moreover, ceramics are less likely to fracture when subjected to high-vibration and high-acceleration conditions than glass. An additional problem for glass is the tendency of meniscus formation at the surface leading to small pieces breaking off – obviously an undesirable situation for medical implants.

Applications of ceramic-to-metal feedthroughs for implantable medical devices include implantable pacemakers [11], cardioverter defibrillators [12], implantable multichannel neuromuscular stimulator [6], and cochlear implants [16]. One commonly used ceramic-to-metal feedthrough in the implantable medical device applications is a pure platinum pin(s) in an alumina substrate [51], as illustrated in Fig. 3. This feedthrough assembly can be produced either with sputtered 99.99% gold brazing or by co-fire pure platinum pins with green alumina ceramic. In the case of brazing, a thin film of metal such as gold, platinum, niobium, or titanium can be applied on the ceramic via physical vapor deposition (PVD) to promote adhesion [52]. Insulation materials (typically silicone and polyurethane) are usually applied on the ceramic-to-metal interface (to prevent shorting due to moisture-promoted dendritic growth of gold) and between pins (to isolate the conductor).

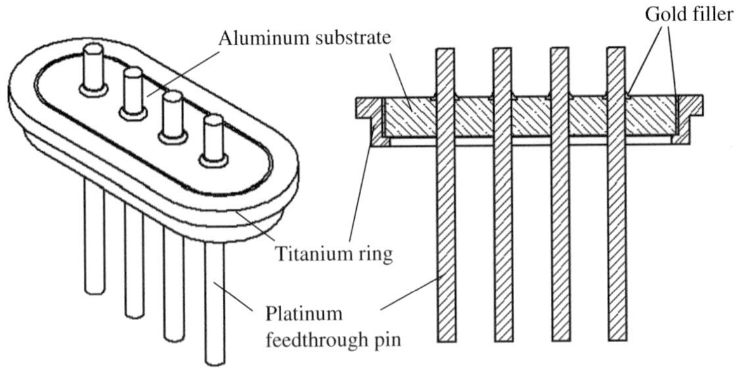

Fig. 3 A ceramic-to-metal implantable feedthrough

Section 3.4 will discuss in detail the investigation of different types of bonding methods (including active brazing, nonactive brazing, and diffusion bonding) aimed at producing hermetic ceramic-to-metal seals for implantable medical device applications. The fundamentals are applicable to the ceramic-to-metal feedthrough design as well.

3.4 Ceramic-to-Metal Seal

Sometimes a ceramic-to-metal seal is the preferable design, as in the AMF neuromuscular microstimulator, where no feedthrough pin is needed. Pure iridium and platinum-iridium components attached to the titanium metal parts act as electrodes interfacing with living tissue.

Brazing is by far the most widely used joining process for a ceramic-to-metal seal when mechanically reliable hermetic joints are required. Brazing is a process for joining two or more solid materials in close proximity to each other by introducing a filler material that melts at a high temperature (typically above 450°C) and which is below the melting points of the materials being joined. Filler materials are classified into two categories: active and nonactive. Active fillers are composed of chemically active materials or compounds that convert readily and permanently from one composition to another when subjected to sufficient energy to initiate the reaction. For the purposes of this discussion, the active fillers to be considered are often composed of active elements, such as titanium, aluminum, hafnium, zirconium, vanadium, and niobium, and the energy applied to initiate the conversion is heat. Nonactive filler materials often require prior metallization of the ceramic substrate to provide for enough wetting, so an interface (usually reactive) is formed. Physical vapor deposition (PVD), chemical vapor deposition (CVD), or mechanical metallization can be used to deposit metallic films, such as molybdenum, manganese, tungsten, or their combination onto ceramic surfaces prior to brazing. This additional metallization step can complicate the brazing process and makes quality control of the joint more difficult. Brazing with active filler materials is a relatively simple method and is generally preferred over brazing with nonactive fillers.

3.4.1 Active Brazing

Both zirconia and titanium have been widely used for biomedical applications due to their excellent mechanical properties and favorable biocompatibility [23, 30]. It is desirable to bond these two materials together for some applications [30]. Some earlier studies have selected active filler alloy brazing as the joining technique for zirconia to Ti-6Al-4 V and have successfully brazed the zirconia to Ti-6Al-4 V with Ag-Cu series filler materials [53–55]. Although this method seems very promising in terms of interfacial strength, it might meet with some objections regarding biomedical applications and the possible toxicity of Cu. Lasater disclosed a method to produce hermetically sealed zirconia-to-titanium joints using a titanium-nickel alloy filler material [56]. Fey and Jiang discovered that zirconia could be joined to titanium alloys using pure nickel brazing filler material [57]. In such a case, the titanium element from the base metal diffuses with nickel filler material and reacts to form a eutectic alloy at the interface. The Ni_2Ti_4O phase that formed upon cooling at the ceramic-to-metal interface is responsible for the bond development. Jiang et al. reported that the zirconia-to-Ti-6Al-4 V brazed joints have good biocompatibility and have been successfully used for the second-generation neuromuscular microstimulators developed at the Alfred Mann Foundation, Santa Clarita, California, USA

[30]. Other successful examples of ceramic-to-metal packages include alumina to Ti-45%Nb alloy package brazed with TiNi-50® active filler metal, for cochlear implant application; alumina to pure niobium case bonded with TiNi-50® filler metal, also for cochlear implant; and alumina to a metal assembly brazed with a modified active filler metal, for artificial retina packaging.

3.4.2 Nonactive Brazing

3.4.2.1 Metalized Ceramic Brazing

By metalizing the surface of the ceramic to be joined by depositing or embedding metal by electroplating, sputtering, ion-implanting, or some other means, brazing with a metal filler can be accomplished as is normally done with metal substrates; that is, by simply selecting a filler that is compatible with the metalized surface material.

3.4.2.2 Noble Metal Brazing

Noble metal brazing is most commonly based with silver or platinum and their alloys, and somewhat less often based with copper or nickel, and occasionally based with other noble metals (e.g., palladium and gold) [58]. Such brazing is normally done in air, or even an oxygen-rich atmosphere, with evidence that noble metal oxides form and bond with the ceramic substrate, particularly with oxide ceramics. Correia et al. have thoroughly investigated the potential use of platinum as an interlayer (25 μm) between tetragonal zirconia polycrystal (TZP) and Ti-containing blocks within a wide temperature range [59]. Though the chemical reaction is strong, the interfaces are rather weak, actually failing at the interface between the TZP and the platinum-rich zone. In TZP/(Au-25 μm)/Ti joints brazed under vacuum, the infinite supply of Ti to the interface through the liquid Au results in a continuous interface without gaps. However, the Ti-Au intermetallics formed at the interface do not hinder Ti-diffusion toward the ceramic to form unfavorable Ti-oxides and Ti_xAu_y intermetallics [60]. Silver was also tested in the form of thin foil (35 μm) for the production of TZP/Ti joints at 980°C under vacuum. The interfacial reaction seemed stronger than in the Au case. A thick zone of a Ti-oxide (assigned to Ti_3O_2) formed at the interface featuring large holes. When zirconia was brazed to titanium and its alloys with palladium in an induction furnace, bonding formed at the interface. However, the brazed joint contains pores, thus the hermeticity of such joints is a problem.

3.4.3 Diffusion Bonding of Ceramic-to-Metal

Diffusion bonding eliminates any foreign material as needed in brazing so it would be preferred for implantable medical device applications. Alumina can be diffusion bonded to a few biocompatible metals including tungsten, platinum, molybdenum, stainless steel, and niobium [58, 61]. Zirconia has been successfully diffusion bonded to niobium too. However, the poor machinability of niobium and molybdenum, and the poor mechanical properties of platinum have probably limited their applications. Zirconia-to-titanium alloy joints were also attempted by diffusion bonding [55, 59, 62] and low-strength joints were obtained, probably due to the Ti embrittlement caused by the enlargement of Ti-grains by two orders of magnitude. Diffusion bonding of TZP and Ti with a zirconium interlayer (30 μm) inserted between has been attempted by Agathopoulos et al. [62]; however, no successful joint has been reported.

3.5 Hermetic Seal with Fusion Welding

Fusion welding is often the final step in creating a hermetic seal for the implantable device. A variety of fusion welding methods used for hermetic metal-to-metal seals include laser-beam welding, electron-beam welding, resistance welding, and tungsten inert gas (TIG) welding, to name a few. The designers choose a particular welding method for their hermetic packages often based on the following considerations: the materials of the parts to be sealed, the specimen size, equipment availability, the joint configuration, and cost. Both laser-beam welding and electron-beam welding are high-energy welding processes. Laser-beam welding has become more and more popular over other methods in the recent years, most likely due to the following reasons:

1. Electron-beam welding requires a vacuum, while laser welding can be done in air or in an inert environment. Argon and helium mixtures are often the preferable inert laser-welding gas for protective purposes. Moreover, helium is the ideal gas for helium-leak testing.
2. Maintenance and operational costs for a laser-welding system are moderate.
3. A small heat-affected zone can be achieved with laser welding, which is particularly critical for the miniaturized implantable medical device, where too much heating or the close proximity of the heating zone might cause damage to the components inside the package.
4. No filler material is needed for laser welding with a properly designed weld joint.
5. Besides the challenge of designing a joint suitable for resistance welding, the process of hermetic sealing by resistance welding is often difficult, if it is even possible at all.

A successful laser weld in the application of a hermetic seal requires precision aiming stability, vibration isolation between the work surface and the environment, accurate location of the weld position, and real-time optical power feedback. Good coordination among the laser power supply, the motion control system, the vision system, the control computer, and operator is critical. Advanced laser-welding systems often have features, such as real-time power feedback, power ramping, and pulse shaping, to achieve the best weld quality possible.

Strong hermetic welds can be achieved by optimized laser parameters, proper joint design, and materials selection. Similar metallic materials are preferable for laser welding. Good welds can be obtained when welding titanium and its alloys, noble metals and their alloys, tantalum, and niobium. Welding Nitinol (a Nickel Titanium (NiTi) shape memory alloy composed of approximately 55% Nickel by weight) to itself has been successfully performed using laser welding [63], TIG welding, and resistance welding [64, 65]. 316 and 316 L stainless steels are considered among the most biocompatible of the stainless series, followed by 304 and 304 L stainless steels. A calculated Cr_{eq} to Ni_{eq} ratio of 1.52 to 1.9 is recommended to control the primary mode of solidification and prevent solidification cracks in type 304 L while the Cr_{eq}/Ni_{eq} ratio of 1.42 to 1.9 is recommended for type 316 L stainless steel [66]. Fusion welding of Tungsten tends to yield welds that are very brittle. Joining Nitinol to stainless steel is often difficult due to the formation of brittle intermetallic compounds such as FeTi and Fe_2Ti. Nitinol can be welded to other metals such as tantalum and niobium to yield acceptable joints [65]. Though joints with dissimilar metals can be achieved, issues such as galvanic corrosion have to be addressed when used in implantable medical device packages.

Frequently, more than one processing method can be used for assembling an implantable medical device. For example, ten critical joints of the first-generation neuromuscular microstimulator were accomplished by using four different technologies including infrared laser beam-assisted welding, TIG welding, resistance welding, and microsoldering [14]. Laser welding, active brazing, resistance welding, and microsoldering have been implemented for the second generation of neuromuscular microstimulators produced at The Alfred Mann Foundation, Santa Clarita, CA.

3.6 Conductive Vias on Ceramic Substrate

An alternative to the ceramic-to-metal feedthrough is the use of conductive vias on a ceramic substrate that can be produced by either high-temperature cofired ceramic (HTCC) or low-temperature cofired ceramic (LTCC). Because HTCC parts are fired at 1,400° to 1,500°C, refractory metals are often used for circuit traces, which results in high electrical resistance compared to noble metals. This resulting poor conductivity often has a detrimental effect on circuit performance. LTCC parts are fired at a lower temperature of about 950°C, so that silver and gold can be used as the conductor materials. Also, a wide variety of resistive and dielectric materials can be applied before firing to form passive components. Moreover, multiple layers with buried components can be formed, and active components with large I/O counts can be connected with wire bonding, surface mount, or flip chip techniques. These techniques allow unpackaged semiconductor device mounting, which further reduces board real estate for a given circuit configuration, so LTCC is often the preferred process. One application utilizing conductive vias on ceramic substrate is for the artificial retina where platinum-containing glass frit is used for the conductive vias

[67]. One challenge with utilizing the conductive vias on ceramic substrate parts for hermetic packaging is the limitation of high-temperature postprocessing of the ceramic substrate. In many cases, it is desirable to braze titanium and its alloys (for zirconia substrate) or niobium and its alloys or titanium-niobium (for alumina substrate) package walls to a via containing ceramic substrate to provide a means for achieving a final hermetic seal. It is possible that the hermeticity of the conductive vias could be compromised after the subsequent high-temperature processing.

4 Testing Methods for Hermetic Sealing of Implantable Medical Devices

4.1 Mechanical and Environmental Tests

An implantable active medical device should be constructed to not only withstand the mechanical forces that may occur during normal conditions of use, but also other possible environments – induced stresses, such as vibration, free fall; atmospheric pressure changes; and temperature changes during packaging, storage, transportation, and handling in an operating room. A list of mechanical tests often applicable to medical device packaging includes tensile, fatigue, vibration, shock, compression, and flexural testing. Environmental tests include temperature cycling, humidity, and corrosion tests. Refer to appropriate standards from the International Organization for Standardization (ISO), Association for the Advancement of Medical Instrumentation (AAMI), American Society for Testing and Materials (ASTM), and other agencies for detailed testing procedures.

The first-generation neuromuscular microstimulators consisted of a cylindrical glass capsule with a rigidly mounted electrode on each end, as shown in Fig. 4 [68]. It has been utilized for many applications from shoulder subluxation, to sleep apnea, to urinary incontinence, to foot drop, to knee osteoarthritis, to wrist and finger contractures, and to pressure ulcers [69, 70]. The mechanical integrity of the package has been tested by three-point bending over its long axis, tensile tests along its axial direction, free drop to steel instrument tray, five temperature cycles between autoclaving and freezing [14].

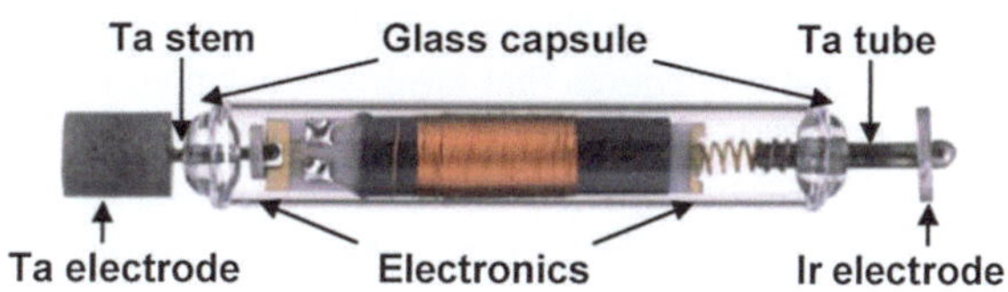

Fig. 4 The package of the first-generation neuromuscular microstimulator. (Reprinted from [71] with permission of Springer)

In 2006, Loeb et al. reported that among a total of 80 neuromuscular microstimulators that were implanted in 35 participants in five different clinical trials, four unresponsive implants were visibly broken as determined by X-ray analysis [71]. Two of the four failures occurred in adjacent neuromuscular microstimulators in one patient after several months of tetanic stimulation to treat flexion contractures of the hand. Both had been inserted in the same orientation, with the Ta stimulation electrode located deep in the interosseous membrane to target nerves to the extensor muscles of the forearm. It is hypothesized that the failures of the clinical implants occurred as a result of repeated bending stress applied by the contracting muscle to the exposed 0.5-mm-long segment of the Ta stem, a ductile wire of 0.25 mm in diameter that is sealed into the relatively large glass capsule at one end and the Ta electrode at the other, as shown in Fig. 5. This was confirmed in a series of in vitro repetitive stress tests. Modifications were made to enhance the mechanical integrity of the glass package. This experience suggested the importance of appropriate mechanical tests to reveal the integrity of the device package prior to any application.

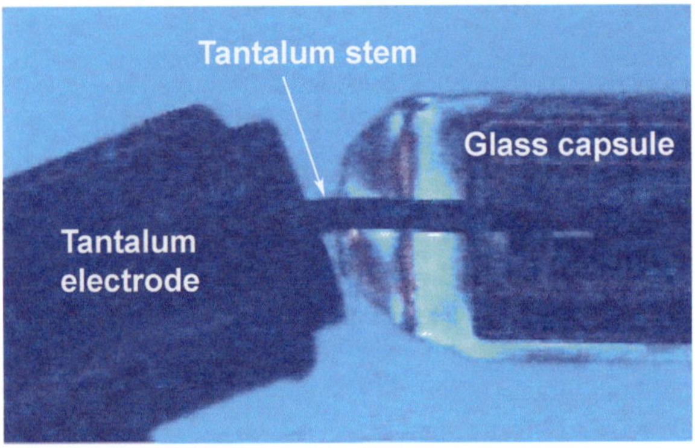

Fig. 5 A bent tantalum stem between the tantalum electrode and glass capsule. (Reprinted from [71] with permission of Springer)

4.2 Hermeticity Testing Methods and Their Limitations

High-quality hermetic seals place implantable medical devices among the most reliable assemblies [4]. Rigorous tests ensure the quality of the hermetic seals in such products. MIL-STD-883, Method 1014.10 provided details of the various hermeticity test procedures that have been adapted by the biomedical device industry. Generally speaking, a package with a standard air equivalent leak rate greater than 1×10^{-5} atm-cc/sec is considered a gross leaker. Packages with leakage below this rate are considered fine leakers. There are generally four methods of ascertaining a gross leak, including [41]:

1. Forcing a liquid, such as fluorocarbon liquid, into the package through the gross leak channel, vaporizing or decomposing this liquid in the package, thereby forcing the resultant gas out through the same leak channel, and then detecting the gas by the bubble test or the vapor detection test;

2. Forcing a liquid into the package, and then detecting its presence by a change in weight (the weight gain test) or by the deflection of the lid due to an increase in pressure by an interferometer;
3. Performing a dye penetrant test; and
4. Performing a residual gas analysis (RGA).

Helium-leak detectors, radioactive decay detectors, and interferometer (optical) leak detectors are all capable of fine leak detection. The most commonly used in the biomedical device industry is the helium-leak detector.

A helium-leak detector is a mass spectrometer tuned to analyze the helium gas. The detection limit of a helium-leak tester is generally 1×10^{-9} atm-cc/sec or better. Prior to the helium-leak test, the hermetic package is either subjected to high-pressure pure helium for a period of time ("bombed") or sealed in a helium-containing environment. Calibration of the helium-leak detector is presently accomplished using a calibrated helium-leak standard consisting of a small cylinder charged with helium at atmospheric pressure. The cylinder contains a filter through which helium exits at a fixed calibrated rate when the cylinder valve is opened, and the temperature at which the leak was calibrated is marked on the cylinder (typically 22–23°C). The calibrated helium-leak cylinder should be at this temperature when calibrating the system. Otherwise, an appropriate temperature compensation factor should be used in calculating the test object leak rate. When using the calibrated leak to set the sensitivity of the helium-leak detector, the detector meter is set for direct readout at the leak rate figure marked on the calibrated leak cylinder.

Radioactive decay is a technique in which a radioactive gas is placed inside the cavity of the device during device manufacture or by bombing after manufacture is complete. This technique is similar to the helium fine leak test method except that radioactive gas, such as krypton-85/dry N_2 mixture, is used. Krypton 85 is a radioactive inert noble gas that emits very weak gamma rays and beta particles. Parts are submerged in the radioactive gas for some time. After the part surfaces are free of residual radioactive gas, it is placed in a chamber connected to a scintillation crystal detection system that actually counts the number of Kr-85 particles inside the package. This is different from the helium fine leak test, which measures the rate of helium leaking out of the device. The leak rate of the device is calculated by a formula based on the concentration of Kr-85/N_2 tracer gas used, the bombing time and pressure, and the measured reading on the device. An Atomic Energy Commission license is necessary for possession and use of radioisotope test equipment and manufacturers are reluctant to use this method.

Optical leak testing is based on the deflection of the package lid when the pressure outside the package differs from that inside the package, either by evacuation or pressurization. The amount of deflection is measured by an interferometer, often a laser interferometer. Optical leak testing is not applicable to many devices where no suitable package configuration is available.

Fine and gross leak tests should be conducted in accordance with the requirements and procedures for the specific test conditions for the device. Combinations of fine and/or gross leak testing can be conducted in sequence or at the same time.

Cumulative Helium Leak Detection (CHLD) is a variation on conventional leak detection that allows for gross and fine leak testing in the same pass and the potential for helium-leak detection at leak rates several orders of magnitude lower than with conventional leak-detection methods.

For an implanted device with circuitry inside, moisture level is often considered the most critical piece of information as many electronic failures are directly related to moisture accumulation and condensation [36, 37, 40].

To some degree, the functional lifetime of a device can be estimated based on the moisture level accumulated inside the device. Lifetime estimation is commonly done with equation 1 based on when the moisture level inside the device reaches the dew point at body temperature, or the consensus [41] among scientists and engineers that the amount of liquid water necessary to promote corrosion is when three monolayers of liquid water form on the internal surface of the packaged device.

The dew point is a key parameter in controlling the ability of moisture condensation. The condensation process inside a package to form water droplets is a function of device temperature, internal pressure, and more importantly moisture level. With a known temperature and pressure, the dew-point level can be determined from the dew-point nomograph shown in Fig. 6 [72]. From the nomograph in Fig. 6, it can be seen that at 1.0 atm and 0°C, the moisture concentration needed for forming water droplets is 6,000 ppm. At levels below this percentage of water vapor, liquid drops will not be able to form. Hence, most materials and sealing processes are selected to keep the internal package environment at or below 5,000 ppm of moisture for the lifetime of the device. The rationale being that, at 5,000 ppm, the water vapor dew point is below the freezing mark, and therefore any moisture that would condense inside the package would be in the form of ice crystals and not be available for corrosion processes. Of course, one could argue that for implantable medical devices, the body temperature is about 37°C, so a higher moisture level should be allowed. But, some contaminants could promote moisture condensation considerably before the moisture reaches the saturation level of the dew point. A humidity test applicable to a specific medical device often has to be run to determine a safe moisture level as the baseline threshold.

$$t = -\frac{V}{L_{H2O}}\left[\ln\left(1 - \frac{Q_{H2O}}{\Delta pi_{H2O}}\right)\right] \tag{1}$$

Where: Q_{H2O} = the water that has leaked in the device in atm [41]

V = the available internal volume of the package (volume of the parts inside the package should be subtracted) in cc
t = the time in seconds
L_{H2O} = the true water leak rate = $0.471 * L_{\mathrm{He}}$ in atm-cc/sec
Δp_{iH2O} = the initial difference in the water partial pressure on the outside less the partial pressure on the inside the package (water vapor partial pressure in human body is 0.061 atm)

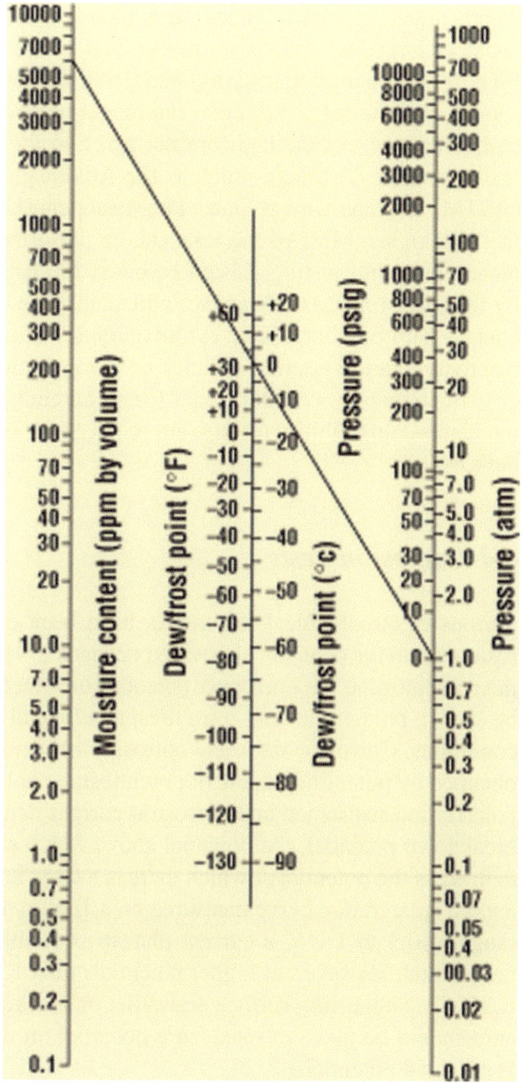

Fig. 6 Dew-point nomograph that correlates pressure, temperature, and water content (As shown, at 1.0 atm and 0°C, the dew point is 6,000 ppm of water). (Reprinted from [72] with permission of ASM International)

4.3 Biocompatibility Tests

The international standard organization ISO 10993 standard plays an important role in the assessment of the biocompatibility of a medical device. In principle, a great number of tests have to be undertaken depending on the intended use of the medical device. The standard describes tests on toxicity, carcinogenicity, and hemocompatibility, among others. Some of these tests are simple in vitro tests, while others require extensive animal experiments. For implanted medical devices, where direct contact is made with blood, tissue, or bone (such as implantable cochlear implants), extensive in vitro and in vivo tests are required [73]. Refer to ISO 10993 for more discussion and detail on the selection of individual tests that should be done for a particular biological interaction or biological effect. In general, details of test methods are not given in the ISO documents and reference is made to other documents such as the American Society of Testing and Materials (ASTM) and The United States Pharmacopeia (USP) standards for procedures and methodologies. Most of the tests can be performed by laboratories specializing in biocompatibility testing. Listed below is the battery of tests often recommended by the FDA for Class III active implantable medical devices that make long-term contact with bone or tissue: cytotoxicity, sensitization, irritation or intracutaneous reactivity, acute systemic toxicity or pyrogenicity, subchronic toxicity, genotoxicity, implantation, chronic toxicity, and carcinogenicity. Additional tests, such as the hemo-compatibility testing, are required for blood contact implantable medical devices.

4.4 Corrosion Tests

Various electrochemical techniques have been employed to characterize the corrosion behavior of medical device packages [74]. Open-circuit potential measurements determine the corrosion potential of a metal in an electrolyte. Its value can be used to predict the long-term lifespan of metal packages under passive corrosion conditions. Corrosion rates and corrosion behavior of passive coating layers can be obtained by potentiodynamic or potentiostatic polarization methods, with which the polarization resistance and corrosion current density can be determined [75]. The breakdown potential, the potential above which surface pits are initiated, is usually defined as the potential at which there is a large increase in the response current. An anodic polarization curve measured on a Ti alloy is shown in Fig. 7. In the potential range of 0.3 to 1.1 V, a current plateau is visible, which indicates the Ti surface passivation. However, at higher potentials than 1.2 V, anodic current increases dramatically, suggesting surface activation or breakdown. When used as a metal case, care should be taken to make sure potential on the Ti surface does not exceed the breakdown potential.

Electrochemical Impedance Spectroscopy (EIS) is a powerful nondestructive method to characterize biomaterials. Electrode materials, solution resistance, electrode/electrolyte interface impedance, charge transfer resistance, and surface roughness/double layer capacitance can be measured and their frequency response properties can be determined in a fast frequency scan.

Cyclic voltammetry (CV) has been employed to determine the operational potential window (the water window) limited by the H_2 and O_2 evolution potentials due to electrolysis of water on the cyclic voltammogram.

There are several ASTM standards that describe electrochemical testing techniques for the evaluation of corrosion behaviors of metal materials. Two useful methods are ASTM G5 – 94: Standard Reference Test Method for Making

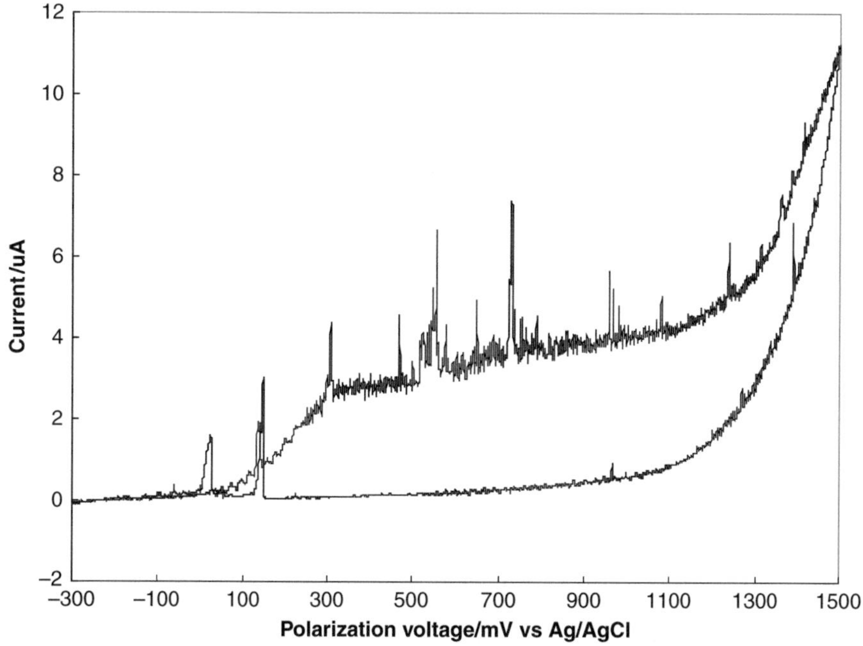

Fig. 7 An anodic polarization curve measured on a Ti alloy in saline solution at 37°C. The scan rate is 1 mV/second. See text for additional details

Potentiostatic and Potentiodynamic Anodic Polarization Measurements and ASTM G106 – 89: Standard Practice for Verification of Algorithm and Equipment for Electrochemical Impedance Measurements. Although most ASTM electrochemical testing techniques are developed for stainless steels, the test methods and procedures can be adapted for noble metals used in implantable medical devices.

Long-term stability of the metal package of medical devices is assessed in vitro through active soak tests under constant pulse stimulation. The packages are tested in buffered saline solutions at body temperature, or elevated temperatures for accelerated tests [76].

4.5 Morphological and Microstructural Characterization

Both light and scanning-electron microscopy (SEM) can provide valuable information about the surface of materials. The smoothness or roughness of surfaces can influence how materials interact with tissues and body fluids. Smoothness or roughness may also affect the binding of protein and biochemical intermediates (lymphokines and cytokines), which may also help determine a material's biocompatibility.

Electron microscopes create magnified images by using a beam of electrons as an imaging source. They resolve images at much higher magnifications than light microscopes can, often at magnifications up to 300,000 times. SEM can yield topographic images and elemental information when used in conjunction with energy-dispersive x-ray analysis (EDX) or wavelength-dispersive x-ray spectrometry (WDS). Elemental analysis using SEM/EDX or SEM/WDS is useful for qualitative and semiquantitative determination of elemental content and for obtaining correlation between microstructures and elemental composition.

Atomic force microscopy (AFM) is another powerful tool for examining the topography of a surface. It works much the same way as a profilometer does, only on a much smaller scale: a very sharp tip, often a silicon tip, is scanned across a sample surface at very short distance and the change in the vertical position reflects the topography of the surface. By collecting the amplified height data for a succession of lines it is possible to generate a three-dimensional map of the surface features with nanometer resolution. This instrument can also be used to evaluate crack formation and growth in both plastics and metals [77].

Other surface analytical techniques, such as x-ray photoelectron spectroscopy (XPS), Auger electron spectroscopy (AES) and secondary ion mass spectroscopy (SIMS) have been utilized to show that the elements of the titanium alloys are present in their surface oxides [78]. Transmission electron microscopy (TEM) and scanning transmission electron microscopy (STEM) studies showed that the oxides of the Ti-6A1-4 V alloy have a more complex microstructure and a different crystallinity, which are properties that could affect the biocompatibility of these titanium alloy implants.

X-ray diffraction enabled researchers to understand the microstructure of crystalline materials. To reveal the bonding mechanism of hermetic titanium alloys to yttria-stabilized tetragonal zirconia polycrystal (Y-TZP) ceramic-brazed joints, X-ray diffraction analysis on the fractured braze joints was conducted and revealed that the nickel titanium oxide (Ni_2Ti_4O) formed at the zirconia ceramic to titanium metal interface is responsible for the bonding [15].

4.6 Accelerated Life Test

For faster product development or improvement, accelerated life testing (ALT) can be used to determine the reliability of implants in accelerated use conditions [79]. Accelerated life testing helps to identify failures and failure modes qualitatively or predicts package lifetime quantitatively at normal use conditions (Accelerated Life Testing Online Reference, ReliaSoft's eTextbook for accelerated life testing data analysis [80]).

Among various stresses used to accelerate failures, temperature is widely accepted in accelerated life tests. The Arrhenius life-temperature model has been widely used in temperature-accelerated life testing [81]. The Arrhenius reaction rate equation proposed by the Swedish physical chemist Svante Arrhenius in 1887, is given by equation 2 [82]:

$$r = A \times \exp\left(-\frac{E_a}{kT}\right) \tag{2}$$

Where

r is the reaction rate
A is a constant with the unit sec^{-1} for first-order reactions
E_a is the activation energy (eV)
k is the Boltzman's constant (8.62×10^{-5} eV K^{-1})
T is the absolute temperature (Kelvin).

Practically, a modified equation derived from the above Arrhenius reaction rate equation is used to determine acceleration factors (K) in an accelerated life test for a package:

$$K = \exp\left[\frac{E_a}{k\left(\frac{1}{T_u} - \frac{1}{T_t}\right)}\right] \tag{3}$$

Where T_u is the intended use temperature of the device, i.e., body temperature in Kelvin and T_t is the elevated test temperature. It is clear from Eq 3 that the acceleration factor is sensitive to E_a at given test temperatures. The activation energy for the specific failure mode should be used in determining acceleration factors.

4.7 X-Ray Microscopy

X-ray microscopy permits nondestructive assessment of internal damage, defects, and degradation of a hermetic package. Illuminating a sample with X-ray energy provides images based on material density that allow for characterization of cracking, breakage, de-lamination, and defects in components. Figure 8 shows a void (about 75-μm wide) at the ceramic-to-metal joint interface of a brazed case package, which cannot be seen by visual inspection. X-ray microscopy has also been used as a 100% screening test for cochlear implants. However, cautions have to be

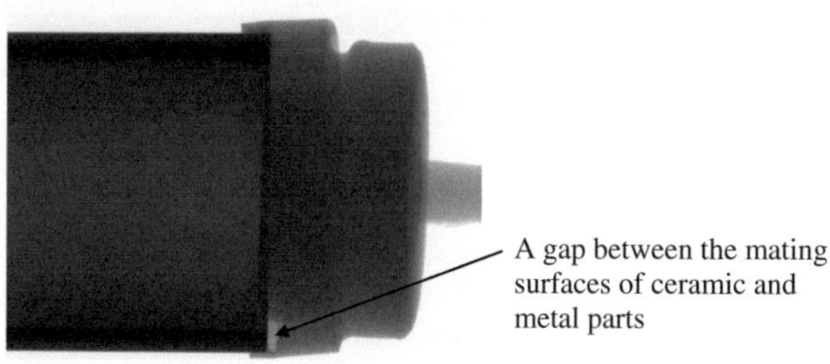

Fig. 8 A void at the ceramic-to-metal joint interface shown by X-ray imaging

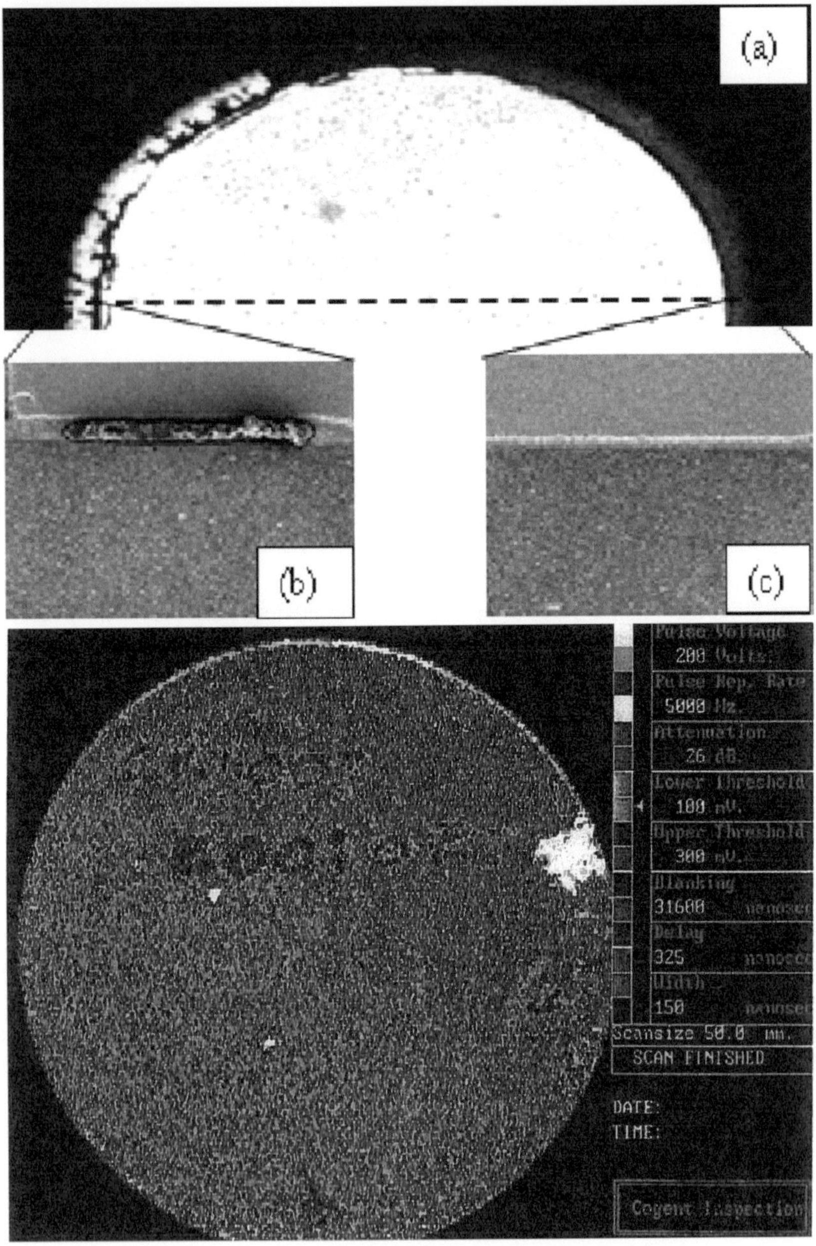

Fig. 9 SAM image (**a**) and SEM cross-section images corresponding to the left (**b**) and right (**c**) brazed areas in (**a**). Cross sectioning was done along the horizontal dotted line in (**a**). (**d**) SAM image of a 50-mm diameter brazed assembly showing a large edge-opening void at the right. (Reprinted from [83] with permission of ASM International)

taken when examining implantable medical devices containing memory chips, as an overdose of X-ray radiation can permanently erase the stored data and potentially cause device failure.

4.8 Acoustic Microscopy

Acoustic microscopy is another nondestructive testing method that uses acoustic impedance to produce high-resolution images of a sample's interior structure to detect "difficult-to-find" defects, such as interfacial separation in printed wiring boards, solder ball de-lamination in ball grid arrays, and die-attach voids, in processor element modules. Scanning acoustic microscopy (SAM) can be utilized for braze joint evaluation where the joint configuration is suitable [83]. Figure 9(a) is an SAM image of a brazed case. The large white oval area represents the body, and the gray ring on the right perimeter corresponds to a sound joint, while the white irregular area in the left perimeter corresponds to an area containing voids. Cross sectioning was made along the dotted lines and samples were prepared for SEM analysis. Figure 9(b) is the cross section of the white joint area in the SAM image where a huge void in the braze joint can be seen. Figure 9(c) is the cross section of the gray joint area in the SAM image where a good braze is shown. Figure 9(d) shows an SAM image of a 50-mm diameter assembly formed by vacuum brazing. Three voids are evident in the joint gap, including a large edge-opening void at the right [83]. For suitable joint configurations, SAM can be used for a 100% nondestructive screening test.

5 Challenges of Hermetic Packaging for Implantable Medical Devices

5.1 Long-Term Stability of Ceramic Materials

For many applications it is desirable that the implantable medical devices remain benign in the subject for the rest of the subject's life to avoid a secondary removal surgery. The intent is to leave the implanted neuromuscular microstimulator in the subject's body for the rest of his/her lifetime, which could be up to 80 or more years [15]. The cochlear implant is now the treatment of choice for children with profound and severe congenital and neonatal hearing loss [84, 85]. The long-term stability of the package for both of these devices is very important.

3 mole % Y-TZP (3Y-TZP) has much higher flexural strength and fracture toughness than the polycrystalline α–Al_2O_3. It has been adapted as the packaging material for both cochlear implants and neuromuscular microstimulators, because of an incident of fracture of an alumina-cased cochlear implant implanted in a young child. However, 3Y-TZP can suffer from low-temperature degradation (LTD) in moist environments. This aging phenomenon is caused by the transformation of

the crystalline structure from the tetragonal (T) phase to the monoclinic (M) phase, resulting in a decrease in strength and toughness, along with micro and macrocracking [86–88], which limits 3Y-TZP's long-term applications. Searching for new bioceramic materials that have improved antidegradation properties is currently one of the most active research fields. Alumina-toughened zirconia [89–91], zirconia-toughened alumina [92,93], Y-TZP with slight alumina doping [94], and zirconia ceramic with other oxide additions [95] all showed better hydrothermal stability than 3Y-TZP and have been proposed for potential use for long-term implantable packages. However, no report on their biocompatibility has been published and the manufacturing process of these materials is still maturing.

5.2 Metals and Alloys Corrosion

An implantable medical device is not only exposed to the harsh environment of the human body, but also to the electrical potential or current that is generated by the implantable medical device itself [7]. Corrosion is one of the major degradation mechanisms affecting the lifetime of metal packages used in medical implants in the body. The corrosion process will induce adverse biological reactions in the body and can lead to mechanical failure of the implants. The packaging materials or construction must eliminate the corrosion risk in the body in both passive and active conditions.

The driving force for passive electrochemical corrosion in a biological environment is the potential variation between the different materials. A typical example of passive corrosion processes is galvanic corrosion. When two dissimilar metals are in contact with each other and exposed to an electrolyte, a potential is set up between the two metals and a galvanic couple is formed. In the presence of an electrolyte, this galvanic couple acts as an electrochemical corrosion cell. In this galvanic couple, one less noble metal will become the anode while the more noble material will act as the cathode in the corrosion cell. The potential difference will result in electrochemical reactions and generate current flow in the corrosion cell, and the oxidation reaction in the cell will cause the anode to corrode. Moreover, the corrosion rate for the metals will be altered when they form a galvanic couple, and the corrosion of anodic material in the corrosion cell will be accelerated by the cathode and subsequently will corrode faster than it would have all by itself.

Crevice corrosion is another passive corrosion process which will cause localized corrosive attack. Crevice corrosion may occur in small occluded areas of a stagnant solution or in crevices where the metals are shielded from full exposure to the surrounding environment. The occluded portion of metal surface has a lower oxygen concentration than the surrounding medium due to restricted oxygen diffusion into the shielded areas. Such localized oxygen concentration difference in crevices will form a potential difference and initiate galvanic corrosion on the anode. In the crevice corrosion cell, the lower oxygen portion in the crevice acts as an anode while the exposed portion with higher oxygen acts as a cathode.

Active electrochemical corrosion on a metal package is driven by the potential or current, which is generated during neural stimulation by an implant. Most neurostimulation applications use a charge-balanced, biphasic, cathodic-first current pulse. When the metal package of an implantable device is used as a return electrode, active electrochemical corrosion will cause metal dissolution. Depending on the current density or charge density applied, the resulting voltage on the metal package may exceed the safe electrochemical window to induce irreversible Faradaic reactions, including anodic dissolution and oxide formation. In an extreme condition, the electrode voltage on the metal package may exceed water window potentials to cause water electrolysis and gas evolution. Hydrogen or oxygen evolution due to water electrolysis that is induced by a stimulus will alter pH [96]. Oxygen evolution during anodic phase will decrease pH, while hydrogen evolution in cathodic phase will increase pH. Such changes in pH will cause metal corrosion and possible tissue damage in the electrode/tissue interface [97]. High pH produced by the cathodic reaction of water analysis reduction leads to dissolution of the passive oxidation layer. A recent study on hydrogen gas evolution induced by neural stimulus revealed that free chlorine (in the forms of HOCl, ClO^-, and Cl_2) was also detected along with the hydrogen evolution [98]. It appeared that the hydrogen and chlorine evolution reactions proceeded simultaneously at rates directly related to the charge injected.

Alloys used in the brazing process of the metal package are susceptible to dealloying corrosion. In a dealloying corrosion process, one or more elements are selectively dissolved, leaving behind a porous residue of the remaining elements [99]. Under a high stimulus, the resulting electrode voltage may exceed a critical potential that indicates the transition from passive and stable alloy to rapid dealloying. Dealloying of metal packages will lead to stress corrosion cracking and will eventually compromise hermeticity.

5.3 Challenges in Accelerated Life Test

Accelerated life tests usually take too long to be conducted online, as part of any product development cycle. Therefore, they must be conducted offline, well before the components, materials, or processes are needed for a given application. For these reasons, ALT is usually conducted generically, using generic samples which represent the materials, components, and processes used for a variety of products [100].

The most significant potential problem with quantifiable accelerated testing is that failure modes produced might not be modes occurring under normal operating conditions [101]. Thus care should be taken in an accelerated life test to keep failure modes unchanged from a normal use condition. Often, multiple failure modes are associated with packages under tests. In order to achieve a reliable lifetime prediction, the acceleration conditions and the accelerated factors should be identified for each failure mechanism. In the case of the ceramic-to-metal package for the second-generation microstimulator, there are two possible failure modes: one mode is the loss of the hermeticity of the package due to corrosion occurred in the ceramic-to-metal seal. The second is 3Y-TZP ceramic surface flaking or self-transformation. The second mode is more likely when a qualified ceramic-to-metal joint is provided. A reliable lifetime prediction of the microstimulator package has been done by determining the Arrhenius factor of the ceramic degradation based on quantitative information obtained from an accelerated aging test with a dummy microstimulator ceramic-to-metal package at a series of temperatures and in vivo studies carried out with microstimulator implants in sheep and rats [15]. It is concluded that a neuromuscular microstimulator packaged with 3Y-TZP ceramic can remain hermetic and retain the ability to withstand a minimum of 15 pounds of cracking load in three-point bending tests after what is equivalent to 70 years of implantation in a human body.

To ensure the acceleration test conditions and factors truly correlate the failure modes produced by ALT with those that occur in normal use conditions, real-time tests should be carried out in parallel with the accelerated tests [102].

5.4 Hermeticity Test Reliability for Miniature Devices

Miniaturization of implantable medical devices continuously poses challenges to hermetic packaging practices. For example, the reliability of hermetic tests decreases with the further reduction of device size. Assuming no moisture is present within a sealed package at the final seal, and no moisture outgassing from the internal material after final sealing, at 1×10^{-10} atm-cc/sec helium-leak rate (4.71×10^{-11} atm-cc/sec moisture leak rate), a calculation based on equation 1 suggests that it will take less than 2 years to reach the 5,000 ppm moisture level in a 0.05 cc package. To guarantee a 10-year functional life (based on a final moisture level $\leq$5,000 ppm) of a medical device with a smaller internal free volume of 0.005 cc, a true helium-leak rate of 6.05×10^{-15} atm-cc/sec (an equivalent H_2O leak rate of 2.85×10^{-15} atm-cc/sec) is required. This is certainly beyond the capacity of any helium-leak detector currently available. The current state-of-the-art for helium-leak testing is claimed by a manufacturer that states that its cumulative helium-leak detector has a 1×10^{-13} atm-cc/sec true helium-leak rate capability [103]. This equipment utilizes metallic seals to eliminate any polymer/plastic seals that could absorb helium. The real operational leak-rate detection capability is about 5×10^{-12} atm-cc/sec and the productivity is limited. A separate room with good ventilation is needed to maintain a low-helium background level for this equipment, and a single use metal o-ring is required for each part to be tested.

The requirement of a helium-leak test for a MEMS device package with an internal volume of around 0.001 cc or below for chronic implantation (>10 years) is definitely beyond the capability of any current helium-leak detector. That is where getter materials come into play. A getter material can absorb various gaseous species and can be used to extend the effective lifetime of a medical device by absorbing moisture and other detrimental gas species, such as hydrogen [104–106] and oxygen [40].

5.5 Design challenges for Miniature Devices

The reduced size of implantable medical devices means the amount of water necessary to increase the humidity to corrosive levels in the interfacial environment becomes exceedingly small. It takes a shorter period of time for moisture or other ions to go through a narrower sealing wall or an interfacial pathway. Higher quality bulk materials and void-free interfacial sealing are essential to achieve the same degree of reliability for a smaller package than a bigger one. The challenges associated with hermetic seal design and material processing also increase with further reduction in device size [107].

5.6 Hermetic Packaging of MEMS for Implantable Medical Devices

There is tremendous interest in the development of MEMS for medical applications. In the most general sense, MEMS refers to miniature components or systems that are fabricated using techniques that were originally developed by the microelectronics fabrication industry, and then modified for the production of microstructures, micromachines, or microsystems such as sensors and actuators [108–110]. Currently, there are numerous research, development, and commercialization efforts underway to create high-performance clinical devices by exploiting the potential for size miniaturization and integration with microelectronics afforded by microfabrication and micromachining techniques [108, 111–114].

Materials commonly used in the fabrication and packaging of standard MEMS devices, including silicon, silicon dioxide, silicon nitride, polycrystalline silicon, silicon carbide, titanium, and SU-8 epoxy photoresist, were evaluated for cytotoxicity using the ISO 10993-5 standard [115, 116]. The data from this evaluation indicated that all above-mentioned MEMS materials are suitable candidates for the development of implantable medical devices. The deployment of implantable MEMS devices based on the silicon and related microelectronics materials has generally relied on protective coatings, such as biocompatible silicone gels, to isolate the MEMS components from the hostile body environment. Two primary drawbacks can result from this protective packaging approach: attenuation of signal/stimulus that must be communicated between the physiological environment and the device and an increased size that detracts from the benefits of miniaturization particularly when working in constrained spaces or at the cellular level.

Several hermetic packaging technologies could potentially lead to successful deployment of MEMS for implantable medical devices [107–111]:

Najafi et al. have developed a biocompatible hermetic package with high-density multifeedthroughs designed to withstand corrosive environments [107, 108]. This technology utilizes electrostatic bonding of a custom-made glass capsule to a silicon substrate to form a hermetically sealed cavity. Even though biocompatibility and long-term stability have been demonstrated, the required high voltage (2,000 V) and high temperature (320 to 350°C) during the process limits its applications.

"Epi-seal" encapsulation developed by Kenny et al. at Stanford University consists of a 20~50-μm thick epitaxially grown polysilicon encapsulation layer covered by 4 μm passivation oxide. Aluminum partially covers the encapsulation to form electrical interconnects [109]. Investigation on the hermeticity and diffusion behaviors of "epi-seal" wafer-scale polysilicon thin-film encapsulation revealed that hermeticity of the encapsulation is gas species specific: hydrogen and helium easily travel in and out of the encapsulation, but nitrogen and argon do not [110].

Chiao and Lin reported that a glass-silicon package formed by rapid thermal processing aluminum-to-silicon nitride bonding can be used for MEMS packaging applications; a Pyrex® (Corning 7740) glass is used [111]. Accelerated hermeticity testing showed that for packages with a sealing ring width of 200 μm and an average sealing area of 1,000 × 1,000 μm^2, the lower bound of the 90% confidence interval of mean time to failure is estimated as 270 years under "tropical" conditions (35°C, 1 atmosphere and 95% relative humidity).

6 Conclusions

Advances in hermetic packaging technology have helped in the successful commercialization of many implantable medical devices, including implantable pacemakers, cardioverter defibrillators, implantable neuromuscular stimulators, and cochlear implants. The continued success of such devices is very much dependent on the reliability of the hermetic package. The packaging methods discussed in this chapter will continue to play important roles in the realm of hermetic packaging for implantable medical devices.

Many issues associated with hermetic packaging have yet to be completely understood, let alone overcome. The continued miniaturization of future implantable medical devices provides both opportunities and challenges for packaging/materials engineers to improve the current packaging methods and to develop new methods. Reliable hermetic micropackaging technologies are the key to a wide utilization of MEMS in miniaturized implantable medical devices.

Acknowledgements The authors would like to thank the Alfred Mann Foundation and Second Sight Medical Product Inc. for their support while they were working on this manuscript. They also thank Dr. Schnittgrund G, Dr. Duttaahmed S, and Grannis S for their detailed review of the manuscript.

References

1. FDA Consumer (2000) 34(2):7
2. Thwaites T (1995) Total recall for medical implants: New Scientist, p. 1212
3. Tummala R, Rymaszewski E (1989) Microelectronics packaging handbook, New York, Van Nostrand Reinhold
4. Ely K (2000) Manufacturing issues in hermetic sealing of medical products, http://www.devicelink.com/mddi/archive/00/01/015.html. Accessed 20 Jan 2008
5. Bhadra N, Kilgore KL and Peckham PH (2001) Implanted stimulators for restoration of function in spinal cord injury, Medical Engineering and Physics. 23:19–28

6. Strojnik P, Peckham PH (2006) Implantable stimulators for neuromuscular control. In: Bronzino JD (ed) Medical devices and systems, The biomedical engineering handbook, 3rd edn. CRC Press, Taylor and Francis Group, Boca Raton, FL
7. Nichols M F (1994) The challenges for hermetic encapsulation of implanted devices- A review, Biomed. Eng. 22(1): 39–67
8. Lussignea RW (1997) Liquid crystal polymers: new barrier materials for packaging, Packaging Technology, October 1997
9. Farrell B, Jaynes P, Johnson W et al. (2003) The liquid crystal polymer packaging solution, Proc. IMAPS 2003 International Symposium, Boston, MA pp 18–23
10. Loeb GE, Byers CL, Rebscher SJ et al. (1983) Design and fabrication of an experimental cochlear prosthesis. Med & Biol Eng Comput, 21:241–254
11. Forde M, Ridgely P (2006) Implantable cardiac pacemakers. In: Bronzino JD (ed) Medical Devices and Systems, The Biomed Eng handbook, 3rd edn. CRC Press, Taylor and Francis Group, Boca Raton, FL
12. Duffin EG (2006) Implantable defibrillator. In: Bronzino JD (ed) Medical Devices and Systems, The Biomed Eng handbook, 3rd edn. CRC Press, Taylor and Francis Group, Boca Raton, FL
13. McDermott H (1989) An advanced multiple channel cochlear implant. Biomedical Engineering, IEEE Transactions on. 36:789–797
14. Loeb GE, Richmond FJR (2001) BION^{TM} implants for therapeutic and functional electrical stimulation. In: Chapin JK, Moxon KA (ed) Neural prostheses for restoration of sensory and motor function. CRC Press, Boca Raton, FL
15. Jiang G (2005) Development of ceramic-to-metal package for BION microstimulator, Ph.D. dissertation, University of Southern California
16. Hochmair I, Nopp P, Jolly C et al. (2006) Trends in Amplification, 10 (4):201–220
17. Weiland JD, Liu W, Humayun MS (2005) Retinal Prosthesis, Annu. Rev. Biomed. Eng. 7:361–401
18. McKinney JRV, Lemons J (1987) The dental implant, PSG Publ., Littleton, MA
19. Hulbert SF, Bokros JC, Hench LL et al. (1987) Ceramics in clinical applications: Past, present, and future, in High Tech Ceramics, Vincenzini P ed, Elsevier, Amsterdam, pp189–213
20. Christel P, Meunier A, Dorlot JM et al. (1988) Biomechanical compatibility and design of ceramic implants for orthopedic surgery, in Bioceramics: materials characteristics versus in-vivo behavior. Ducheyne P and Lemons J, eds. Ann New York Acad Sci 523:234
21. Hulbert S (1993) The use of alumina and zirconia in surgical implants. In: An introduction to bioceramics. Hench LL and Wilson J, eds. World Scientific, Singapore, P25–40
22. Miller JA, Talton JD, Bhatia S (1996) in Clinical performance of skeletal prostheses, Hench LL, Wilson J eds, Chapman and hall, London, p41–56
23. Piconi C, Maccauro G (1999) Review: zirconia as a ceramic biomaterial. Biomatls 20:1–25
24. Tsukuma K, Shimada M (1985) Strength, fracture toughness and vickers hardness of CeO_2-stabilized tetragonal ZrO_2 polycrystals (Ce-TZP). J of Matls Sci 20:1178–1184
25. Schneider SJ (ed) (1991) Ceramics and glasses In: Engineered Materials Handbook, Volume 4, ASM International
26. Schwartz MM (1992) Handbook of structural ceramics, McGraw-Hill Publishers, USA
27. Guillou MO, Henshall JL, Hooper RM et al. (1992) Indentation fracture testing and analysis, and its application to zirconia, silicon carbide and silicon nitride ceramics. J of Hard Matls 3:421–434
28. Whitney ED (1994) Ceramic cutting tools – materials, development and performance, Noyes Publications, Park Ridge, NJ
29. Griffin EA, Mumm DR, Marshall DB (1996) Rapid prototyping of functional ceramic composites. Amer Ceram Soc Bull 75:65–68
30. Jiang G, Mishler D, Davis R et al. (2005) Zirconia to Ti-6Al-4 V braze joint for implantable biomedical device. J of Biomed Mater Res: Part B – Applied Biomatls 72B:316–321

31. Daulton J (2006) Self-centering braze assembly, US patent: 7,132,173 B2
32. Haller MI, He TX, Daulton J (2006) Electrode assembly for a microstimulator, US patent: 7,103,408 B2
33. Loeb GE, Richmond FJR, Baker LL (2006) The BION devices: injectable interfaces with peripheral nerves and muscles, Neurosurg Focus 20 (5):E2
34. Webster JG (1978) Medical instrumentation application and design, Houghton Mifflin, Boston
35. Williams DF (1981) Biocompatibility of clinical implant materials, Vols 1 and 2, CRC press, Boca Raton, FL
36. Thomas RW (1976) Moisture, myths, and microcircuits, in IEEE Trans. On Parts, Hybrids, and Packaging, p167–171
37. DerMarderosian A (1978) Electrochemical migration of metals, Proc. Int'l Microelectronics Symp., pp134–141
38. Grunthaner FJ, Griswold TW, Clendening PJ (1975) Migratory gold resistive shorts: chemical aspects of a failure mechanism, Proc. 13th annual proceedings. International reliability physics symposium, pp 99–106
39. DerMarderosian A, Murphy C (1977) Humidity threshold variations for dendritic growth on hybrid substrate, Proc Int Reliability Phys Symp, Las Vegas, NV
40. Roswell AE, Clymer GK (1971) Thermal fatigue lead-soldered semiconductor device, US Patent 3,735,208
41. Greenhouse H (1999) Hermeticity of electronic packages, Noyes Publication / William Andrew Publishing LLC, Norwich, New York, USA
42. Wong CP (1998) Polymers for encapsulation: materials processes and reliability, Chip scale review, Vol. 2, No. 1, 30
43. Donaldson PEK (1983) The cooper cable: an implantable multiconductor cable for neurological prostheses. J of medical and biological engineering and computing 21:371–374
44. Lovely DF, Olive MB, Scott RN (1986) Epoxy moulding system for the encapsulation of microelectronic devices suitable for implantation, J. of medical and biological engineering and computing, Vol24, No. 2:206–208
45. Loeb GE, Bak MJ, Salcman M et al. (1977) Parylene as a chronically stable, reproducible microelectrode insulator, IEEE Trans. Biomed. Eng., 24 (2):121–128
46. Yuen TG, Agnew WF, Bullara LA (1987) Tissue response to potential neuro-prosthetic material implanted subdurally. Biomaterials, 8 (2):138–141
47. Stieglitz T (2005) Methods to determine the stability of polymer encapsulations. The 10th annual conference of the international functional electrical stimulation society, Montréal, Canada
48. Donaldson PEK (1976) The encapsulation of microelectronic devices for long service life. IEEE trans. Biomed. Eng., 23:281–285
49. Petersen ME, Sergent J (1979) Metal hermetic package selection, Electronic Packaging and Production, p. 150–156
50. Graeme C (2003) Cochlear implants, fundamentals and applications, AIP series in modern acoustics and signal processing by Beyer RT (editor in chief) Springer, New York
51. Bealka JD, Da Costa PH (2003) Feedthrough devices, US patent: 6,586,675 B1
52. Mastrogiacomo J (2007) New ceramic technology contributes to advances in medical implants, http://www.morgantechnicalceramics.com/articles/medical_implants.htm. Accessed 20 Jan 2008
53. Peytour C, Berthet P, Barbier F et al. (1990) Interface microstructure and mechanical behavior of brazed Ti6Al4V/zirconia joints, J. Mater. Sci. Lett, 9:1129–31
54. Santella ML, Pak JJ (1993) Brazing titanium-vapor-coated zirconia. Welding Res Supplement, 165–172
55. Agathopoulos S, Moretto P, Peteves SD et al. (1997) Brazing of zirconia to Ti and Ti6Al4V. In 1996 Amer Ceram. Soc. Meeting, Indianapolis, USA 1996, Ceram Joining, Ceram Trans. Indianapolis 77:75–82

56. Lasater BJ (2001) Methods for hermetically sealing ceramic to metallic surfaces and assemblies incorporating such seal, US Patent: 6,221,513 B1
57. Fey K and Jiang G (2003) Application and manufacturing method for a ceramic to metal seal; US Patent: 6,521,350 B2
58. Messler RW (2004) Joining of materials and structures-–from pragmatic process to enabling technology, Elsevier Butterworth Heinemann, Burlington, MA
59. Correia RN, Emiliano JV, Moretto P (1998) Microstructure of diffusional zirconia-titanium and zirconia-Ti6Al4V alloy joints. J Matl Sci 33:215–221
60. Agathopoulos S, Correia RN, Joanni E et al. (2002) Interactions at zirconia-Au-Ti interfaces at high temperatures, Key Eng Matls 206–213:487–90.
61. Messler RW (1993) Joining of advanced materials, Elsevier Butterworth Heinemann Science, Burlington, MA
62. Agathopoulos S, Pina S, Correia RN (2002) A review of recent investigations on zirconia joining for biomedical applications. Ceram Trans 138:35–147
63. Falvo A, Furgiuele FM, Maletta C (2005) Laser welding of a NiTi alloy: Mechanical and shape memory behavior. Materials Science and Engineering: A. 412:235–240
64. Wu MH (2001) Fabrication of Nitinol materials and components, Proceedings of the international conference on shape memory and super-elastic technologies, Kunming, China, 285–292
65. Schetky LM, Wu MH (2003) Issues in the further development of Nitinol properties and processing for medical device applications, Proceedings from the Materials & Processes for Medical Devices Conference, Anaheim, California, pp 271–276
66. Korinko PS, Malene SH (2001) Considerations for the weldability of types 304 L and 316 L stainless steel. J of failure analysis and prevention 1:61–68.
67. Greenberg RJ, Mann AE, Talbot N et al. (2007) Biocompatible bonding method and electronics package suitable for implantation, US Patent: 7,211,103
68. Loeb GE, Zamin CJ, Schulman JH et al. (1991) Injectable microstimulator for functional electrical stimulation, North Sea Conference on Biomedical Engineering, Antwerp, Belgium
69. Singh J, Peck RA, Loeb GE (2001) Development of BION Technology for functional electrical stimulation: Hermetic Packaging, Proc. IEEE-EMBS Istanbul, Turkey
70. Dupont AC, Bagg SD, Chun S et al. (2002) Clinical Trials of BION^{TM} Microstimulators, Proc. IFESS, Ljubljana, Slovenia
71. Loeb GE, Peck RA, Singh J et al. (2006) Mechanical loading of rigid intramuscular implants, Biomed microdevices, Vol . 9, No. 6:901–910
72. ASM International (1989) Electronic material handbook, Packaging Vol. 1, CRC press
73. Ligtvoet KM, Wijcherson A, Bakker EJ (2005) Biocompatibility of medical devices, In: DI Sens symposium-book
74. Mansfeld F (2003) The use of electrochemical techniques for the investigation and monitoring of microbiologically influenced corrosion and its inhibition – a review. Materials and Corrosion 54:489–502
75. Chohayeb AA, Fraker AC, Eichmiller FC et al. (1996) Corrosion Behavior of Dental Casting Alloys Coupled with Titanium, in Medical Applications of Titanium and Its Alloys: The Material and Biological Issues, ASTM STP 1272, S. A. Brown and J. E. Lemons, eds., American Society for Testing and Materials, West Conshohocken, PA
76. Zhou D, Mech B, Greenberg R (2000) Accelerated corrosion tests on Silicon wafers for implantable medical devices. Proc., of 198th Electrochemical Society Meeting, p363
77. Goken M (1999) Atomic Force Microscopy of Metallic Surfaces. Adv Matls & Processes 155:35–37
78. Lausmaa J, Ask M, Rolander U et al. (1989) Preparation and analysis of Ti and alloyed Ti surfaces used in the evaluation of biological response. Mater. Res. Soc. Symp. Proc. 110:647–653

79. Meeker and Hahn (1985) How to plan an accelerated life test: some practical guidelines, The ASQC basic references in quality control, Vol. 10
80. http://www.weibull.com/acceltestwebcontents.htm, accessed on 10 March, 2008.
81. Nelson W (1990) Accelerated testing, statistical models, test plans, and data analysis, John Wiley & Sons, New York
82. Parker SP (editor-in-chief) (1994) McGraw-Hill Dictionary of Scientific and technical Terms, 5th edition, McGraw-Hill
83. Jacobson DM, Humpston G (2005) Principles of brazing, ASM International, p165.
84. Osberger MJ (1997) Current issues in cochlear implants in children. The hearing review, Vol 4, p 29
85. Severens JL, Brokx JPL and van den Broek (1997) Cost analysis of cochlear implants in deaf children in the Netherlands. Amer J of Otology 18:714
86. Tsukuma K, Kubota Y, Tsukidate T (1984) Advances in ceramics, Vol. 12, Science and technology of zirconia II. Edited by N. Claussen, M. Ruhle, and A. H. Heuer. American Ceramic Society, Columbus, OH, p. 382
87. Sato T, Shimada M (1985) Transformation of Yttria-doped Tetragonal ZrO_2 Polycrystals by Annealing in Water. J Am Cer Soc 68 (6): 356–369
88. Somiya S, Yoshimura M (1987) Zirconia ceramics, Uchida Rokakuho Publishing Co., Ltd, Tokyo
89. Li JF and Watanabe R (1999) Mechanical properties of PSZ-matrix ceramic composites containing Al2O3 particles with various sizes. Key Eng Matls 161–163:299–302
90. Begand S, Oberbach T, Glien W (2005) ATZ – a new material with a high potential in joint replacement. Key Eng Matls 284–286:983–986
91. Begand S, Oberbach T, Glien W et al. (2008) Kinetic of the phase transformation of ATZ compared to biograde Y-TZP. Key Eng Matls 361–336:763–766
92. Ikedal J, Pezzotti G, Nakanishi T (2006) Phase stability of zirconia toughened alumina composite for artificial joints. Key Eng Matls 309–311:1243–1246
93. Begand S, Oberbach T, Glien W (2007) Corrosion behavior of ATZ and ZTA ceramics. Key Eng Matls 330–332:1227–1230
94. Zhang B, Isobe T, Satani S et al. (1999), The effect of alumina addition on phase transformation and mechanical properties in partial stabilized zirconia. Key Eng Matls 161–163:307–310
95. Hirano M, Inada H (1991) Hydrothermal stability of yttria- and ceria-doped tetragonal zirconia-alumina composites. J of Matls Sci 26:5047–5052
96. Zhou D, Chu A, Agazaryan A et al. (2004) Towards an implantable micro pH electrode array for visual prostheses, in Nanoscale Devices, Materials, and Biological Systems: Fundamentals and Applications (Cahay M ed.) pp. 563–576. Electrochemical Society
97. Huang CQ, Carter PM, Shepherd RK (2001) Stimulus induced pH changes in cochlear implants: An *in vitro* and *in vivo* Study. Annals of Biomedical Engineering 29:791–802
98. Sanders C, Nagler E, Zhou D et al. (2007) Dynamic Interactions of Retinal Prosthesis Electrodes with Neural Tissue and Materials Science in Electrode Design, in Artificial Sight, Basic Research, Biomedical Engineering, and Clinical Advances, Humayun MS et al. (Eds.) Ch 11:209–226, Springer
99. Dursun A, Pugh DV, Corcoran SG (2003) A Steady-State Method for Determining the Dealloying Critical Potential. Electrochemical and Solid-State Letters, 6 (8) B32–B34
100. http://www.tutorialsweb.com/reliability/reliability4.htm#r4.1, accessed on 10 March, 2008.
101. Utter RE (2005) Accelerated life test – your key to new product success. http://www.innovativethermal.com/articles/, accessed 10 March, 2008
102. Edell DJ (2004) Insulating biomaterials in Neuroprosthetics, Theory and Practice, edited by Horch KW & Dhillon GS, pp. 517–579

103. Pernicka JC (2006) Pernicka unveils world's first CHLD hermeticity test system for medical and space applications, available online at http://www.pernicka.com/Press_CHLD.htm, Accessed 10 March, 2008
104. Bredendiek-Kämper S, Klewe-Nebenius H, Pfennig G et al. (1989) Surface analytical characterization of the hydrogen getter material ZrCo. Fresenius' Journal of Analytical Chemistry 335:669–674
105. Lee SM, Park YJ, Lee HY et al. (2000) Hydrogen absorption properties of a Zr–Al alloy ball-milled with Ni powder, Intermetallics 8:781–784
106. Liu CZ, Shi LQ, Xu SL et al. (2004) Kinetics of hydrogen uptake for getter materials. Vacuum 75:71–78
107. Ramesham R, Ghaffarian R (2000) Challenges in interconnection and packaging of microelectromechanical systems (MEMS) Electronic Components and Technology Conference Proceedings. 50th Volume , Issue 2000:666–675
108. Roy S, Ferrara LA, Fleischman AJ et al. (2001) Micro-electromechanical systems and neurosurgery: a new era in millennium. Neurosurgery 49:779–797
109. Roy S, Fleischman AJ (2003) Cytotoxicity evaluation for microsystems materials using human cells. Sens and Mat., 15:335–340
110. Roy S, Mehregany M (1999) Introduction to MEMS. In Helbajian H (ed.), Microengineering aerospace systems. The aerospace press, El Segundo, CA pp 1–28
111. Ferrara LA, Fleischman AJ, Togawa D et al. (2003) An in vivo biocompatibility assessment of MEMS materials for spinal fusion monitoring. Biomed Microdev 5:297–302
112. Fleischman AJ (2003) Miniature high frequency focused ultrasonic transducers for minimally invasive imaging procedures. Sens Actu A: Phys 103:76–82
113. McAllister DV, Allen MG, Prausnitz MR (2000) Micro-fabricated micro-needles for gene and drug delivery. Ann Rev Biomed Eng 2:289–313
114. Polla DL, Erdman AG, Robbins WP et al. (2000) Microdevices in medicine. Ann Rev Biomed Eng, 2:551–576
115. Kotzar G, Freas M, Abel P et al. (2002) Evaluation of MEMS materials of construction for implantable medical devices. Biomaterials 23:2737–2750
116. Roy S, Ferrara LA, Fleischman AJ et al. (2007) MEMS and neurosurgery, In Ferrari M (editor in chief) BioMEMS and biomedical nanotechnology, Vol 3: Therapeutic Micro/nanotechnology (ed. Desai T and Bhatia S) pp 95–123, Springer US.

Science and Technology of Bio-Inert Thin Films as Hermetic-Encapsulating Coatings for Implantable Biomedical Devices: Application to Implantable Microchip in the Eye for the Artificial Retina

Orlando Auciello and Bing Shi

Abstract Extensive research has been devoted to the development of neuron prostheses and hybrid bionic systems to establish links between the nervous system and electronic or robotic prostheses with the main focus of restoring motor and sensory functions in blind patients. Artificial retinas, one type of neural prostheses we are currently working on, aim to restore some vision in blind patients caused by retinitis picmentosa or macular degeneration, and in the future to restore vision at the level of face recognition, if not more. Currently there is no hermetic microchip-size coating that provides a reliable, long-term (years) performance as encapsulating coating for the artificial retina Si microchip to be implanted inside the eye. This chapter focuses on the critical topics relevant to the development of a robust, long-term artificial retina device, namely the science and technology of hermetic bio-inert encapsulating coatings to protect a Si microchip implanted in the human eye from being attacked by chemicals existing in the eye's saline environment. The work discussed in this chapter is related to the development of a novel ultrananocrystalline diamond (UNCD) hermetic coating, which exhibited no degradation in rabbit eyes. The material synthesis, characterization, and electrochemical properties of these hermetic coatings are reviewed for application as encapsulating coating for the artificial retinal microchips implantable inside the human eye. Our work has shown that UNCD coatings may provide a reliable hermetic bio-inert coating technology for encapsulation of Si microchips implantable in the eye specifically and in the human body in general. Electrochemical tests of the UNCD films grown under $CH_4/Ar/H_2$ (1%) plasma exhibit the lowest leakage currents ($\sim 7 \times 10^{-7}$ A/cm^2) in a saline solution simulating the eye environment. This leakage is incompatible with the functionality of the first-generation artificial retinal microchip. However, the growth of UNCD on top of the Si microchip passivated by a silicon nitride layer or the oxide layers is also under investigation in our group as introduced in this chapter. The electrochemically

O. Auciello (✉)
Materials Science Division, Center for Nanoscale Materials, Argonne National Laboratory, Argonne, IL 60439, USA
e-mail: auciello@anl.gov

D.D. Zhou, E. Greenbaum (eds.), *Implantable Neural Prostheses 2*,
Biological and Medical Physics, Biomedical Engineering, DOI 10.1007/978-0-387-98120-8_3,

induced leakage will be reduced by at least one to three orders of magnitude to the range of 10^{-10} A/cm^2, which is compatible with reliable, long-term implants.

Contents

1 Scientific and Technological State-of-the-Art of Bio-inert Coatings for Encapsulation of Implantable Microchips

Implantable medical devices and neural prostheses are being used to address a growing number of medical problems. This rapidly growing field integrates physiology and microfabrication technology to produce devices implantable in the human body for diagnosis and monitoring, therapeutic treatment, and restoration of sense and motor functionalities in persons affected by degradation or loss of those functionalities, either through genetically induced degenerative conditions or through traumatic events. Examples of prostheses for restoration of functionality include (a) artificial joints currently used to replace natural human joints degraded due to age-related problems and/or genetics; (b) neural prostheses such as the cochlear, brain, or retinal electrode implants to produce electrical stimulation of nerves or brain or ganglion retinal cells; and (c) implantable microchips, such as those used as part of the artificial cochlear and retinal implants to restore hearing to deaf people and sight to blind people due to retina degeneration, and many other implantable devices.

Neural prostheses have been used increasingly in treatments for a variety of neurological disorders such as the early cochlear implants and an early artificial vision device [1–4]. Extensive research has been devoted to the development of neuron prostheses and hybrid bionic systems to establish links between the nervous system and electronic or robotic prostheses with the main focus of restoring motor and sensory functions in those who are suffering of dysfunctional conditions. Scientists all over the world have been developing semiconductor-based implants for the light-sensitive retina for more than a decade. Artificial retinas are currently aimed at restoring some vision in blind patients with eye conditions such as retinitis picmentosa or macular degeneration, and in the future to restore vision to the level of face recognition and more.

In 2002, the U.S. Department of Energy (DOE) set up the Artificial Retina Research Project. The goal is to design a device with hundreds to a thousand microelectrodes. This resolution will help restore limited vision that enables reading, unaided mobility, and facial recognition [5]. The basic principle of the retinal implant under development by the DOE consortium is to translate the visual information captured by an externally worn camera into a pattern or electrical stimulation pulses that are applied to the retina at the ganglion cell level for transmission to the brain to restore image formation. Clinical trials of a prototype retinal prosthesis show that completely blind individuals interpret this stimulus as a visual sensation. The current prototype used in clinical trials is a low-resolution device, which involves an array of 16 electrodes (Pt dots in a polymer matrix) connected through thin wires to the microchip located outside the eye. In order to improve this device, it will be necessary to develop an implantable microchip that can stimulate the ganglion cells in the retina at hundreds to thousands of individual locations, in the same way a display uses thousands of pixels to create an image [6–7].

This chapter focuses on the critical topics relevant to the development of a robust, long-term artificial retina device, namely the science and technology of hermetic bio-inert encapsulating coatings to protect a Si microchip implanted in the human eye from attack by chemicals present in the eye's saline environment.

Currently there is no hermetic microchip-size coating that provides a reliable, long-term (years) performance as encapsulating coating for the artificial retina Si microchip to be implanted inside the eye. The work discussed in this chapter is related to the development of a novel ultrananocrystalline diamond (UNCD) hermetic coating, which looks promising in relation to the performance in passive tests where the coating was exposed to the environment in rabbit eyes and exhibited no degradation. However, the most stringent test of having a UNCD-coated Si chip performing with low leakage current ($< 10^{-8}$ A/cm^2) to the saline for an equivalent time of several years, with applied voltages to the chip, as required to excite the ganglion cells, has not been demonstrated yet, since this program is at an early stage of research and development. In addition, it remains to be demonstrated that the interface between the UNCD coating and the electrode layer, whatever it may be (e.g., Pt or Au), which will provide the excitation of the ganglion cells in the retina, will not exhibit delamination, which will render the device useless.

The current implantable microchip, which is a critical component of the artificial retina, is based on silicon (Si). Silicon cannot be inserted inside the eye because it is chemically attacked by the saline solution under the operational conditions of the microchip. Therefore, for the Si-based microchip to be implanted in the eye, it must be protected from interaction with the saline solution by a hermetic, biocompatible, and bio-inert encapsulating coating. The package must simultaneously protect the chip from corrosive eye fluids, protect the eye from the chip materials, and allow the circuit to interface with the retina through microelectrodes. Although the package is designed to protect the implantable microchip, it must also protect the biological tissue surrounding the microchip. The microelectronics technology has made significant progress in the past decades and found widespread application in many areas, but packaging and assembly of these devices have not been able to keep pace, particularly for biomedical implants [8].

2 Process and Design Considerations for Hermetic Bio-inert Coatings for Implantable Artificial Retina

Hermetic coatings for encapsulation of implantable biomedical devices should have a double functionality of protecting the implantable microdevice and the surrounding tissues, in order to yield devices with high performance, long service time, and free of electronic failure. Currently, researchers are exploring two types of packaging technologies, namely (a) hard case and (b) thin-film encapsulating coatings. The hard-case technology uses metal, ceramic, or glass cases to protect electronics of implants. A hard case is bulky and presents difficulties for miniaturization. For implantable microchips, especially for neural prostheses devices, thin-film encapsulating coatings provide a better platform choice because of size considerations. Thus, this review focuses on a discussion of the science and technology of inorganic thin films, specifically, a novel material named ultrananocrystalline diamond (UNCD) and oxide coatings as potentially strong candidates for application as hermetic-encapsulating bio-inert/biocompatible coatings. In this chapter, we will mainly focus on the UNCD thin films. The materials synthesis, characterization, and electrochemical properties of these hermetic coatings are reviewed for application as encapsulating coating for the artificial retinal microchips implantable inside the human eye.

2.1 Materials for Hermetic-Encapsulating Coatings

The technology for packaging implantable microchips is very challenging because of the size, material properties, mechanical structure and rigidity, biocompatibility, required lifetime, and maximum allowable temperature. Reliability, long-term stability, and acceptable cost are essential requirements for the packaging [9].

Implantable devices require different packaging systems depending on the shape, size, and the locations where they are implanted. The surface of the retina microchip will be exposed to the physiological environment of the eye. Therefore, the permanently implanted microchip must fulfill special requirements in terms of biocompatability and bio-inertness without adverse reactions to the surrounding retina tissue that may influence its physiological function and without electrochemical reactions with the physiological environment that might generate gas (i.e., oxygen or hydrogen) and lead to chemical attack of the microchip surface affecting its functionality. The microchip used in the artificial retinal implant of the DOE consortium and most other implantable electronic medical devices uses Si as the structural and functional material. However, the Si CMOS device performance can be affected by sodium ion incorporation from the saline. Also, Si is soluble in human fluids, specifically saline over a long time [10]. In addition, hermetic coatings with good biocompatibility and bio-inertness are critically needed for encapsulation of bioimplantable devices.

Coating materials currently being evaluated for encapsulating artificial retinal prototype implants include SiO_2 [7], SiNx [11], SiC [12], silicone elastomer, polytetrafluoroethylene, polyimide [13–15], and parylene [16]. However, it has been shown that SiO_2 coatings exhibit dissolution and decay in some retinal implants when implanted in animals up to 6 months [2, 10]. SiNx coatings often exhibit structural defects (pin holes) plus chemical reaction with the saline components, which result in compositional changes of the coating, thus loss of insulating properties and therefore loss of protection of the microphotodiode underneath [11]. Polyimide and other polymers are inexpensive and flexible, but tend to absorb significant quantities of water, which in turn leads to electrical leakage after some time. Implantable medical devices should last decades, however, and polymers are not considered hermetic for that period of time [17]. Synthesis and characterization of oxide films as hermetic-encapsulating coatings are described in detail in the chapter *Microchip-Embedded Capacitors for Implantable Neural Stimulators*, in relation to oxide films for capacitors, because of their double functionalities for microchip-embedded capacitor application and as components of hermetic-encapsulating coatings.

2.2 Carbon-Based Ultrananocrystalline Diamond (UNCD) Coatings and Film

In the 1950s to 1980s, two main techniques were developed for the synthesis of artificial diamond: (a) high-pressure/high-temperature synthesis that yielded artificial diamond gems [18], and (b) chemical vapor deposition (CVD) [19] that made possible the synthesis of diamond in thin-film form on various substrates. Different CVD techniques have been adopted in the synthesis of diamond thin films. Three main CVD techniques include hot-filament CVD [20], plasma-assisted CVD [21], and combustion CVD [22]. There are many types of plasma-assisted CVD techniques according to the energy sources, such as the microwave plasma-assisted CVD (also known as microwave plasma-enhanced CVD) [23], direct current plasma-assisted CVD [24], direct current thermal plasma CVD [25], radio frequency plasma-assisted CVD [26], and radio frequency thermal plasma CVD [26–28].

Research on diamond thin films focused on the CVD technique to produce diamond films with microcrystalline structure. In order to grow diamond, which in the bulk is thermodynamically less stable than graphite, early CVD methods [19, 20] involved volatile hydrocarbons, an energy source, and an abundant source of atomic hydrogen [21, 22]. The growth species is CH_3^*, typically produced by a methane-containing plasma. Atomic hydrogen drives the hydrogen abstraction reactions that (1) prepare the CH_3^* adsorption site by removing a hydrogen atom from the hydrogen-terminated diamond surface, and (2) remove the hydrogen atoms from the adsorbed CH_3^*, thereby permitting the carbon atom to move into the position corresponding to an extension of the diamond lattice. Additionally, atomic hydrogen preferentially etches the graphitic phase. By using a plasma containing H_2 (99%) and CH_4 (1%), it is possible to grow diamond films that are largely free of non-diamond secondary phases. However, atomic hydrogen also etches the diamond phase, resulting in the formation of intergranular voids and a columnar morphology with grain size and rms surface roughness typically ~10% of the film thickness (grains size $\leq 1\ \mu m$) [29–32]. These conventional CVD polycrystalline diamond thin films have rougher surfaces than expected and polishing on these synthesis diamond thin films is very time consuming. Studies in recent years focused on the growth of nanocrystalline diamond (NCD) (grain size of a few hundred nanometers) and microcrystalline diamond (MCD) (grain size from 30 nm to hundred of nanometers) [33, 34].

Diamond films possess exceptional mechanical properties, chemical inertness, biocompatibility, and are ideal candidates as protective coatings for bioimplants. However, the high substrate temperatures (> 700°C) required for the synthesis of MCD and NCD films, using conventional CVD techniques described above, involving hydrogen-rich plasma chemistries, limits the selection range of the substrate materials. Current silicon microchips based on the complementary metal oxide semiconductor technology (CMOS), which is the technique that is also used for bioimplantable microchips, cannot be subjected to processing temperatures above ~400°C, otherwise the devices fail. To use diamond films as the hermetic coating of the artificial retina chips, the deposition/processing temperature should be ≤ 400°C. The large grain size and low nucleation density of MCD coatings make it difficult to achieve conformal deposition as hermetic coating on components that have irregular surfaces. On the other hand, conventional NCD film synthesis techniques involving the H_2/CH_4 chemistry require relatively high synthesis temperatures (600–800°C), which are incompatible with the thermal budget of Si-based CMOS devices to avoid destruction of the device.

Ultrananocrystalline diamond (UNCD) thin films developed and patented at Argonne National Laboratory [35, 36] exhibit a unique nanostructure characterized by grain sizes of 3–5 nm in diameter and 0.4 nm wide grain boundaries, which yields a combination of smooth surface, low friction, and high wear resistance [37], mechanical properties similar to single crystal diamond [38], biocompatibility and bio-inertness [39, 40], and chemical inertness [38]. The atomically abrupt grain boundaries consist of mixture of sp^3 (diamond) and sp^2 (graphite) bonding. Most importantly, UNCD thin films are synthesized using microwave plasma-enhanced chemical vapor deposition (MPECVD) with an argon-rich Ar-rich(99%)/CH_4(1%) plasma chemistry, which is different from the H_2/CH_4 chemistry used in the synthesis of MCD and NCD films. The novel Ar-rich/CH_4 chemistry involves C2 dimers as the main nucleation species and contributors, together with CH_3 radicals, to the UNCD film growth. Because the C2 dimers can insert directly onto the substrate surface without H abstraction, the activation energy for this process is only about 6 kcal/mol as opposed to 20–30 kcal/mol for MCD and NCD. Thus, because of the very low activation energy for nucleation and growth of the UNCD films, they can be synthesized at 400°C [41] with growth rates similar to those observed for high temperature growth. Because of these unique properties, UNCD could be a suitable candidate for the hermetic coating for artificial retina microchips.

The hermetic UNCD-encapsulating coatings for artificial retina microchips are produced using $Ar/CH_4/H_2$ gas mixtures injected into a chamber initially evacuated to a relatively high vacuum ($\sim 10 \times 10^{-7}$ Torr), with small amounts of hydrogen gas added in order to decrease the electrical conductivity of the film and render it electrochemically inert. During film growth, hydrogen is added to the gas mixture – 1 to 20% – in order to optimize the resistance of the films. Figure 1 shows a schematic of the MPCVD system used to grow UNCD films.

The synthesis temperature for UNCD films is 400°C to be within the thermal budget of the Si microchip. In order to compare the property change with the synthesis temperature, UNCD films have also been synthesized at 600°C and 800°C. Conductive double-polished Si (N-type, R 0.001~0.005 Ω-cm) strips (3 mm × 25 mm × 0.55 mm) are used as the substrate materials. The substrates are seeded via immersion in a suspension of nanometer-sized diamond powder (~5 nm grains) in methanol solution and ultrasonication for 30 min prior to the deposition to gain high diamond nucleation density. One end of the coated Si strips is covered with a Si cover slide (Fig. 2a) to avoid growth of the insulating UNCD coating and provide a place for electrical connection to the impedance system for measuring the leakage current as a part of electrochemical tests. SEM observation shows (Fig. 2b) that the UNCD layer is dense and conformally coats the Si strip entirely, which is a requirement for an encapsulating coating.

Uniformity of the hermetic coatings is one of the key properties to insure the quality of the coatings. Using the MPCVD technique to grow the UNCD films, the arrangement of Si strips on the substrate holder is critical to obtain good film-thickness uniformity. There is a hot spot in the center area of the plasma, which results in temperature changes in closely related positions. On the other hand, temperature changes are minimal in positions outside the central area. Figure 3 shows different sample position arrangements explored to achieve optimum thickness uniformity.

The film-thickness uniformity has been more difficult to optimize when UNCD films are grown using MPCVD systems powered by microwaves of 2.45 GHz.

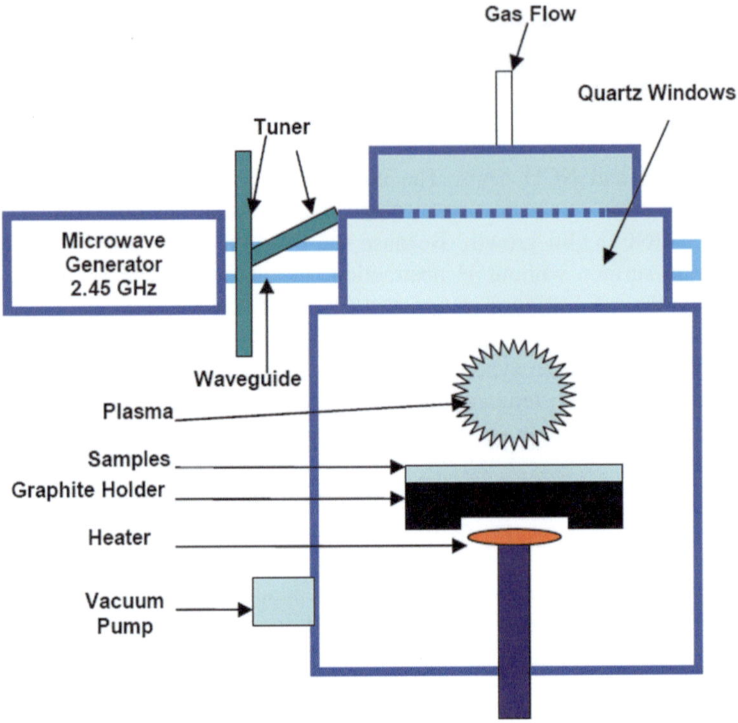

Fig. 1 Schematic diagram of MPCVD IPLAS system (Innovative Plasma Systems) for UNCD synthesis

However, a new MPCVD system installed in the Center for Nanoscale Materials at Argonne, which features 915 MHz microwave power, has already demonstrated that film-thickness uniformity with $\leq \pm 5\%$ is achievable, which is needed for commercial implementation of the UNCD hermetic coatings.

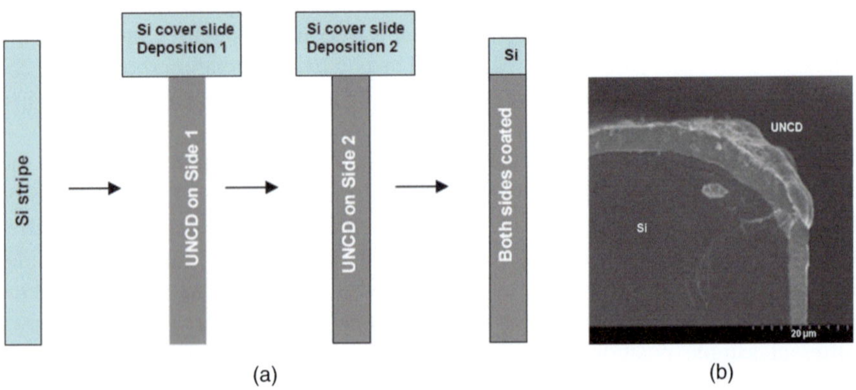

Fig. 2 (**a**) Schematic procedures for growth of UNCD films on Si-simulating microchips; (**b**) Cross-section SEM picture of encapsulating UNCD film grown with 1% H_2 in Ar/CH_4 plasma showing the UNCD conformal hermetic coating covering the Si substrate

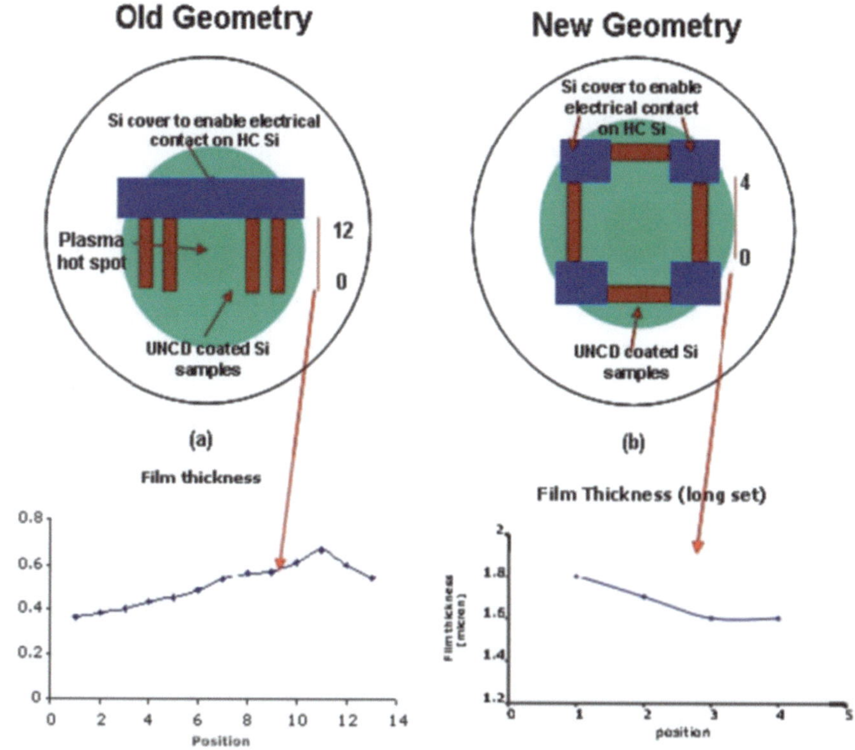

Fig. 3 Optimization of UNCD film-thickness uniformity via position on the substrate holder in the MPCVD system

2.3 *Oxide Films Alone or as Component of Hybrid UNCD/Oxide for Hermetic-Encapsulating Coatings*

The investigation of UNCD films has demonstrated that UNCD films are intrinsically good hermetic coatings, although they may need further chemical optimization to increase the insulating properties or may need to be combined with highly insulating oxide layers. These strategies are part of the appropriate materials integration with the Si microchip that may be needed to eliminate edge-induced field effects and/or particulate-induced pinholes that may increase leakage current in the saline environment of the eye. Therefore, it is relevant to investigate oxide films as alternative hermetic-encapsulating coatings or as components of hybrid UNCD/oxide coatings. Oxide films can be synthesized by magnetron sputter-deposition or more recently using atomic layer deposition (ALD), which provides a unique growth process capable of growing films one atomic layer at the time.

Single or multiple layers of TiAlOx, Al_2O_3, TiAlOx/Al_2O_3, and TiO_2/Al_2O_3 were synthesized by magnetron sputter-deposition techniques, and more recently by atomic layer (ALD) deposition (see details of oxide layer synthesis in the chapter *Microchip-Embedded Capacitors for Implantable Neural Stimulators*).

3 Characterization of Bio-inert Hermetic-Encapsulating Coatings

3.1 Characterization of Chemical, Microstructural, and Morphological Properties of UNCD Coatings

The chemical, microstructural, and morphological properties of UNCD thin films have been characterized, using a variety of complementary analytical techniques, as discussed below.

Visible Raman spectroscopy analysis was performed to characterize the chemical bonds in the UNCD films. The Raman analysis was performed using a Renishaw Raman Microscope, with a He–Ne laser at a wavelength of 632.8 nm (energy 1.96 eV). Due to the resonant Raman Effect, there is a much larger Raman scattering cross section for sp^2 (graphite)-bonded carbon than for sp^3 (diamond)-bonded carbon, when a laser with a wavelength in the visible region is used because the energy of the incident photons is much lower than the energy of the band gap [42].

Two peaks, D band (1330 cm^{-1}) (the activated AIg mode due to the finite crystal size) and G band (1580 cm^{-1}) (E2g modes of single-crystal graphite), are observed in the UNCD visible Raman spectra, which represent the sp^2-bonded carbon component accumulated along the grain boundary. The characteristic Raman peak of a diamond grain (1332 cm^{-1}), which usually overlaps with the D peak, is observed when sufficiently large amounts of hydrogen are added to the plasma, because the introduction of more hydrogen leads to the growth of larger diamond grains. UNCD has been synthesized with plasma containing 2, 10, and 20% hydrogen and the results have been compared (Fig. 4). Slight increases of the peak intensities around 1150 cm^{-1} and 1440 cm^{-1} are found, which reveal that more hydrogen was incorporated into the grain boundaries. Raman spectra show that when the hydrogen flow is larger than 10%, the typical UNCD structures are not retained. The strong and sharp peak of 1332 cm^{-1} appears over the D peak due to a large fraction of microcrystalline diamond (MCD) included in the films.

Besides the comparison of the hydrogen contents, the comparisons of Raman spectra on the growth at different temperatures are shown in Figs. 5 and 6 [40]. Results show that when the synthesis temperature decreases, the peak positions gradually shift from 1330 and 1560 cm^{-1} (800°C) to 1360 and 1540 cm^{-1} (at 400°C). The height ratio of 1560 and 1330 cm^{-1} peaks reverses between 700°C and 600°C. Visible Raman is more sensitive to the sp^2-bonded carbon as mentioned above; the changes of the peaks reflect the changes in the grain boundaries.

To observe the sp^3 information of UNCD, UV Raman spectroscopy (4.66 eV at 266 nm) is used on the analysis of UNCD synthesized at different temperatures. This energy can excite both π and σ states and therefore can probe both sp^2- and sp^3-bonded carbon. Results show that the diamond peak at 1332 cm^{-1} increases in intensity as the deposition temperature decreases (Fig. 5). The broad feature center at 1600 cm^{-1} is due to sp^2-bonded carbon.

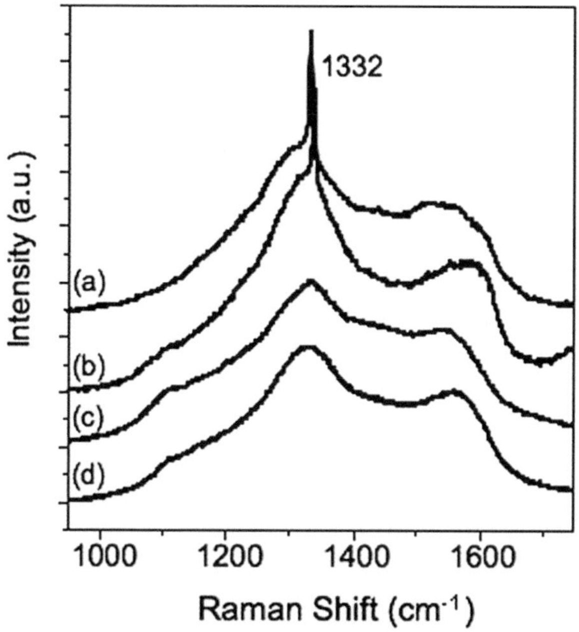

Fig. 4 Raman spectra of UNCD films grown with different H_2 percentage in the Ar/CH_4 gas mixture used to grow UNCD films: (**a**) 20%, (**b**) 10%, (**c**) 2%, and (**d**) 0%. All curves have been normalized against the intensity of the peak around 1590 cm^{-1}. UNCD films were grown in the ANL group (Reproduced from Ref. [40] with permission from Wiley)

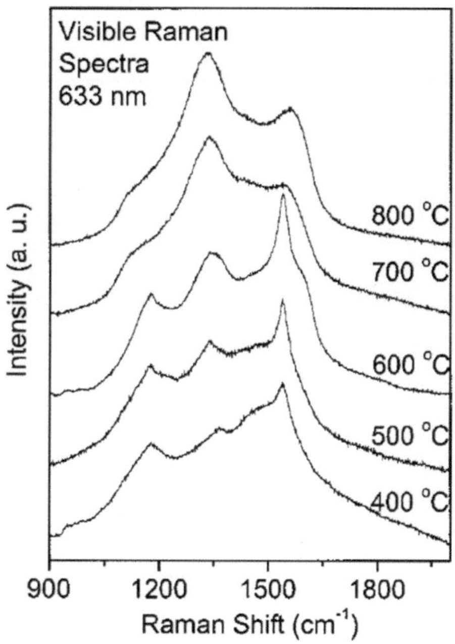

Fig. 5 Visible Raman spectra of UNCD films deposited at different temperatures (Reproduced from Ref. [41] with permission from American Institute of Physics)

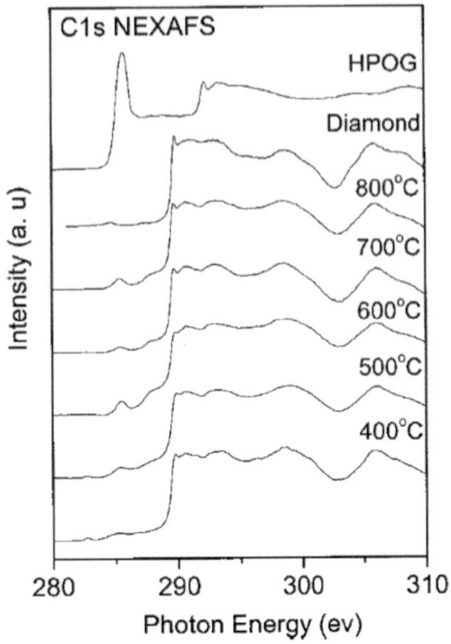

Fig. 6 UV Raman spectra of UNCD films deposited at different temperatures (Reproduced from Ref. [41] with permission from American Institute of Physics)

The microstructure information of surface and cross-section morphologies of the UNCD thin films was observed using a Hitachi S-4700 field emission scanning electron microscope (FE-SEM). Transmission electron microscopy (TEM) observation was carried out using a Philips CM30 microscope operated at 300 kV and a high-resolution JEOL 4000EXII system.

SEM analyses on UNCD films deposited with different hydrogen percentages are shown in Fig. 7. Results showed that the incorporation of hydrogen leads to dramatic changes to the UNCD film microstructure. The more hydrogen incorporated, the

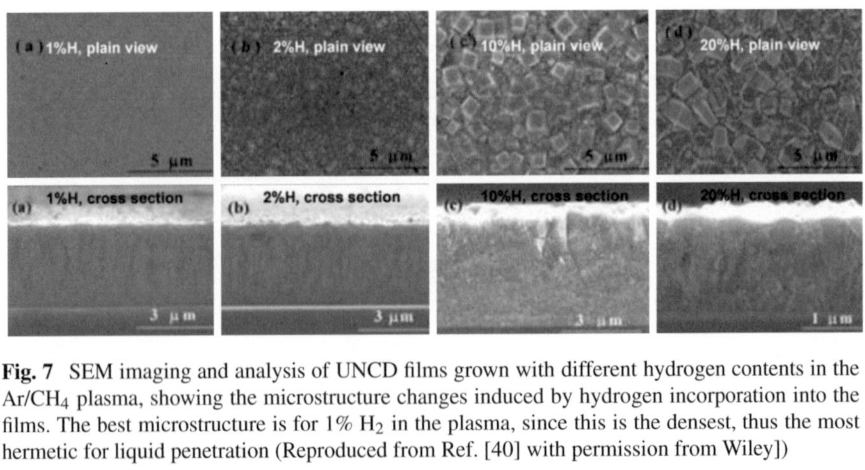

Fig. 7 SEM imaging and analysis of UNCD films grown with different hydrogen contents in the Ar/CH_4 plasma, showing the microstructure changes induced by hydrogen incorporation into the films. The best microstructure is for 1% H_2 in the plasma, since this is the densest, thus the most hermetic for liquid penetration (Reproduced from Ref. [40] with permission from Wiley])

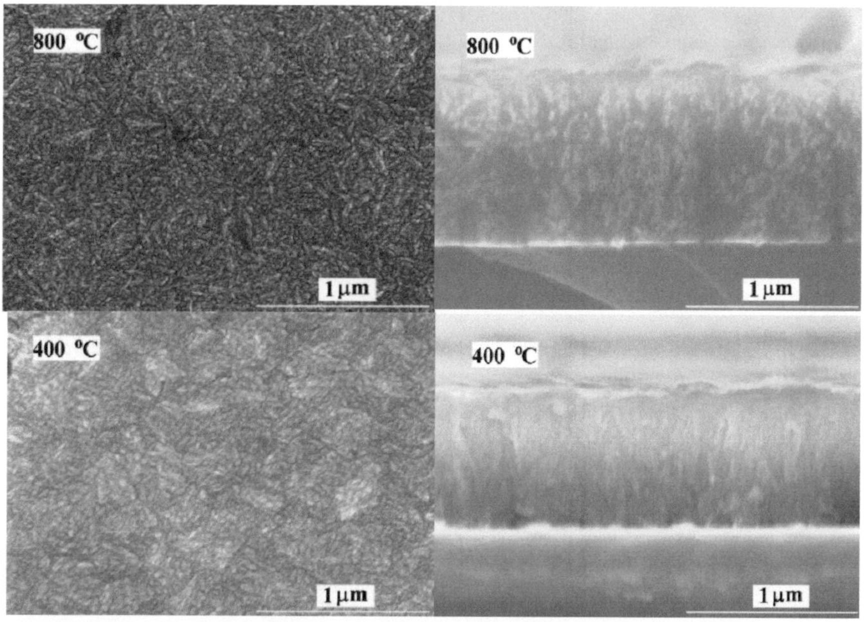

Fig. 8 Plain view (*top*) and cross-section (*bottom*) SEM images of surface morphologies and microstructures of UNCD films grown at 400°C and 800°C, respectively (Reproduced from Ref. [41] with permission from American Institute of Physics)

rougher the surface and the larger the grain size for the UNCD films. The UNCD films grown with 10% and 20% hydrogen in the gas phase showed much larger diamond grains, which is the symbolic microstructure for microcrystalline diamond. These results match the results from Raman analysis, which showed that higher incorporation of hydrogen leads to larger grain size.

The comparisons of the microstructures of UNCD thin films deposited at different temperatures are shown in Fig. 8. The growth rates estimated from the corresponding SEM cross-section images shown in Fig. 8 are 0.2 μm/h at 400°C and 0.25 μm/h at 800°C, which indicates that the growth of UNCD films with the Ar rich/CH_4 plasma chemistry is less temperature dependent.

The results shown in Fig. 8 indicate that the UNCD films deposited at 400°C have similar microstructures to those of the films deposited at higher temperatures. These films are very dense and pinhole-free (Fig. 9). Three different scales are given to provide thorough views of the films from microstructure to macrostructure. The SEM images of the very dense UNCD films indicate that they are good candidate materials as hermetic coatings.

High-resolution TEM studies of UNCD thin films grown at low and high temperatures show similar nanostructures (Fig. 10). As shown in the bright field TEM images, the density of microcrystals with stacking faults and twinning planes is slightly higher for UNCD films grown at 400°C than for UNCD films grown

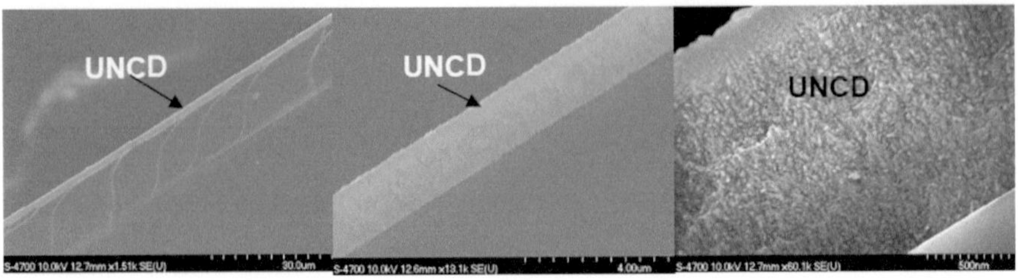

Fig. 9 Cross-section SEM images of a UNCD-coated high-conductivity Si substrate, using 1% H_2/98% Ar/1% CH_4 gas mixture at nominal 400°C [43]

at 800°C. The grain sizes of the films grown at different temperatures are similar, with no significant differences. UNCD films deposited at 400°C maintain the atomically abrupt grain boundaries observed for the UNCD films deposited at 800°C.

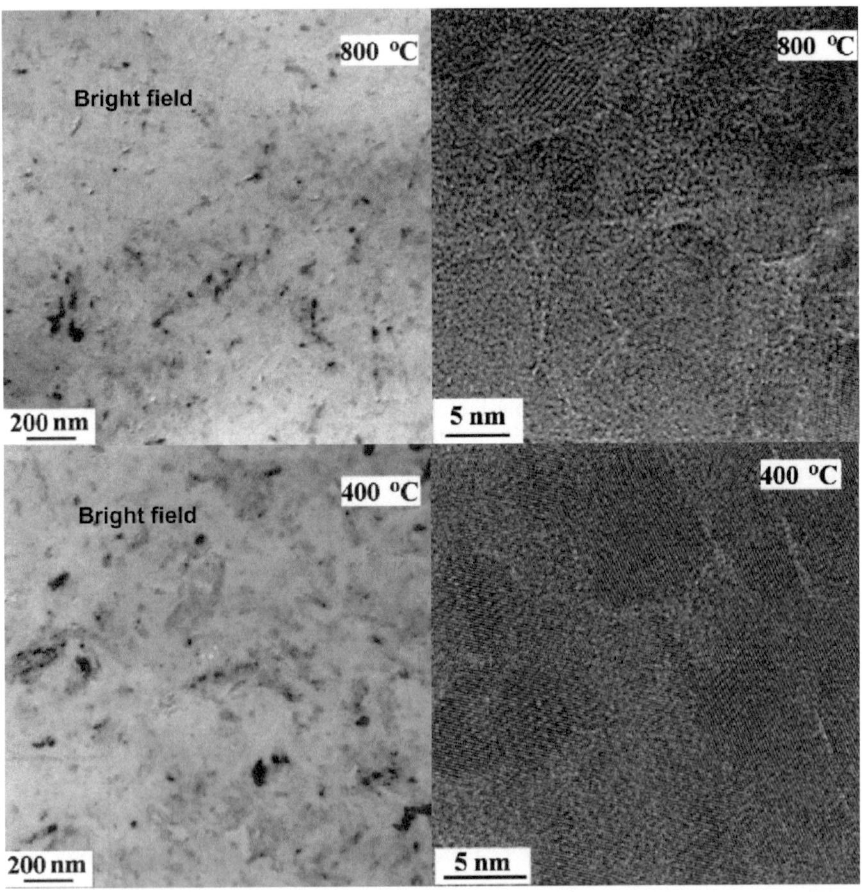

Fig. 10 Plain-view high-resolution TEM images of UNCD films grown on Si at 400°C and 800°C, using 1% H_2/98% Ar/1% CH_4 gas mixture at nominal 400 (Reproduced from Ref. [41] with permission from American Institute of Physics)

3.2 Characterization of Microstructural and Morphological Properties of Oxide Films for Hybrid Hermetic-Encapsulating Coatings

Al_2O_3, TiO_2, and Al_2O_3/TiO_2 multilayer films were synthesized using ALD deposition technique. Surface morphology and the microstructure of these oxides were observed using scanning electron microscopy (SEM) (Fig. 11), which shows that the TiAlOx oxide has formed a dense hermetic layer on a Si surface, even around sharp corners.

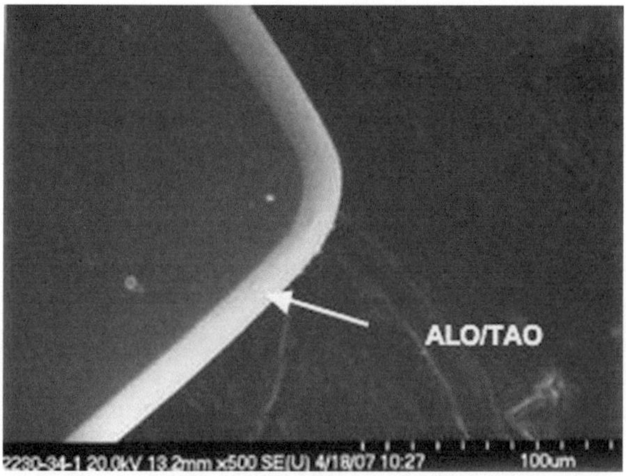

Fig. 11 Cross-section SEM picture of a TiAlOx layer grow on Si, using ALD

3.3 Characterization of Electrochemical Performance in Saline Solution for Hermetic UNCD Coatings

To characterize the electrochemical performance of the hermetic UNCD films, cyclic voltammetry tests were performed, using three electrodes in a potentiostat (Solartron 1287A; Solartron Analytical). Phosphate-buffered saline (PBS) was used as the electrolyte. A platinum rod was used as the counter electrode. An Ag/AgCl electrode was used as the reference electrode.

The results of testing UNCD films grown at different temperatures and with different amounts of incorporated hydrogen are shown in Fig. 12. Results showed that the leakage current densities are on the same order of magnitude for all films, with the films grown at 400°C having the lowest leakage current. UNCD thin films grown with 2% hydrogen in the plasma have a higher leakage current than UNCD films grown with 1% hydrogen in the plasma (Fig. 12(b)), due to the larger grain size resulting from a higher hydrogen content during the synthesis.

Samples that have the maximum leakage current at 8×10^{-6}A/cm^2 using cyclic voltammetry tests have been tested in a different mode: leakage current versus DC voltage. Leakage current vs. DC voltage measurements were performed at +5 V for

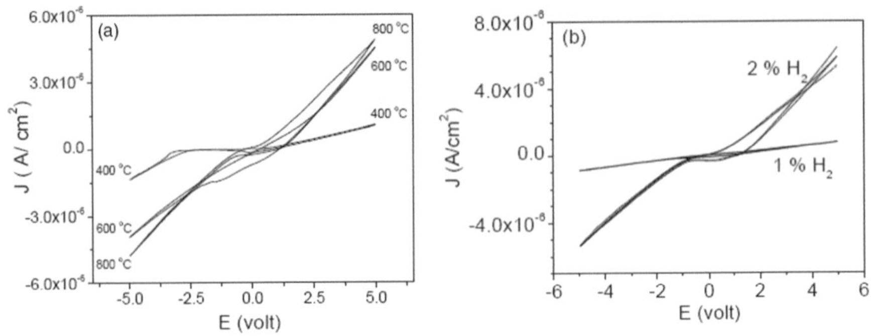

Fig. 12 Cyclic voltammogram tests on the leakage current of UNCD films; (**a**) different temperatures; (**b**) with different hydrogen content in plasma (Reproduced from Ref. [40] with permission from Wiley)

three consecutive days. The leakage current was initially $\sim 2 \times 10^{-7}$ A/cm^2, and then decayed to $\sim 1 \times 10^{-7}$ A/cm^2. After the 3-day test, the samples were exposed to -5 V for three consecutive days. The leakage current was initially 5×10^{-7} A/cm^2, and then increased slowly up to $\sim 10^{-6}$ A/cm^2. It is not totally understood yet why the leakage current density is higher while using the cyclic voltammetry test mode. It appears that there is some surface conditioning at +5 V that might produce a surface layer that results in a decrease of the leakage current, while that surface layer might be deteriorated during the negative voltage excursion. More work is needed, particularly, using biphasic voltage pulses, more similar to the real working conditions of the retina microchip.

3.4 Characterization of Electrochemical Performance in Saline Solution for Hermetic Oxide Films

This work is being done to explore bio-inert/biocompatible oxides as alternative hermetic coatings for encapsulation of the retina microchip. Cyclic voltammetry investigation of different oxide materials was performed using the three-electrode mode on the potentiostat (Solartron 1287A; Solartron Analytical). Phosphate-buffered saline (PBS) was used as the electrolyte. A platinum rod was used as the counter electrode. An Ag/AgCl electrode was used as the reference electrode. Figure 13 shows the summary of the oxide materials that have been investigated. All complex oxide or heterostructure oxide coatings show promising low leakage currents up to -5 V, except for TiO_2.

Further studies are currently under way to optimize the insulating property of the oxides as the hermetic coatings. Figure 14 shows the most recent research on Al_2O_3/TiAlOx as the hermetic coatings for the artificial retina project. Figure 14(a) shows that a Si chip coated with an Al_2O_3/TiAlOx layer exhibits a leakage current as low as 7×10^{-10} A/cm^2, which is well below the required leakage current for

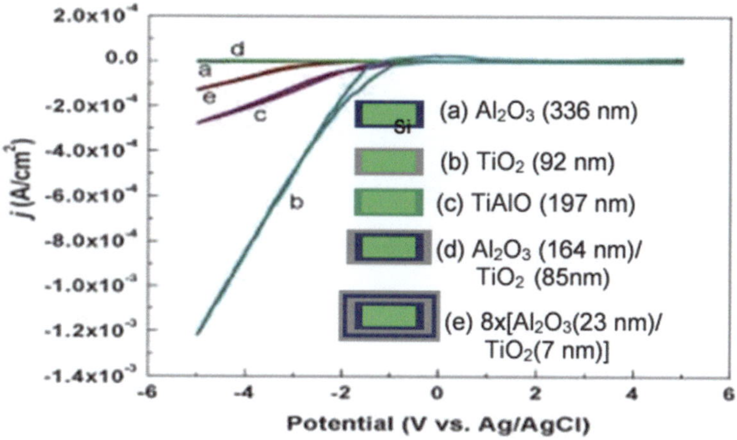

Fig. 13 Leakage current vs. voltage curves from electrochemical tests in saline for various oxide coatings produced by ALD

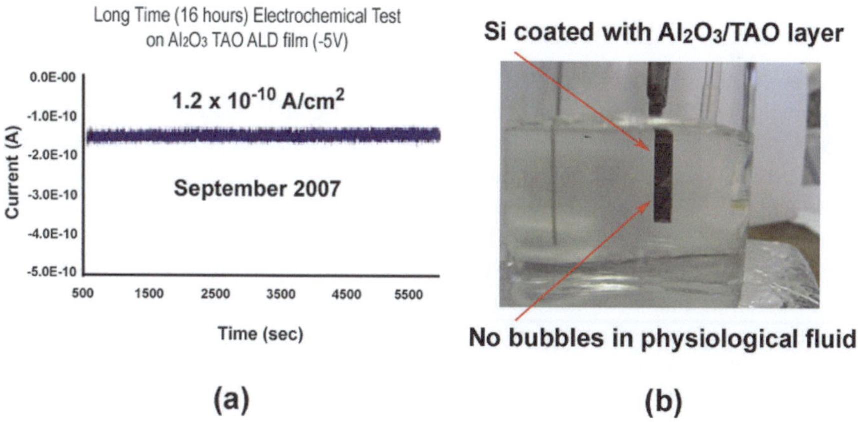

Fig. 14 (**a**) Leakage current vs. time measurement for an Al_2O_3 (1 μm)/TiAlOx (300 nm)-coated Si substrate; (**b**) picture taken during testing showing no bubbling of the coating in physiological fluid under a constant DC voltage bias

an implantable device inside the human eye. Figure 14(b) confirms the result shown in Fig. 14(a) by showing no bubbling from the coated Si chip under a 5 V DC bias voltage.

3.5 Characterization of UNCD/CMOS Integration

A critical component of the development of UNCD as a hermetic bio-inert coating for encapsulation of Si-based CMOS implantable chips is the demonstration that the CMOS devices perform to specification before and after being coated with UNCD

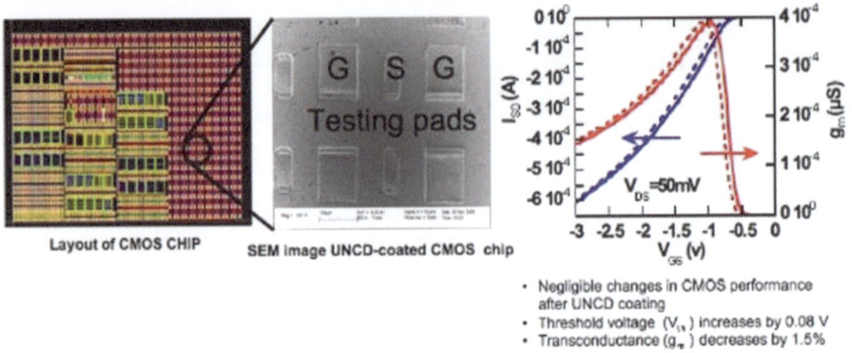

Fig. 15 Measurement of CMOS-device performance before and after coating with UNCD film, for four hours at 400°C. There is negligible change in the electrical parameters of the CMOS device after UNCD deposition, demonstrating that the Argonne-patented UNCD growth process is compatible with CMOS devices (200 mm CMOS wafers were supplied by S. Pacheco (Freescale) and the electrical characterization was performed by Prof. Z. Ma's group at University of Wisconsin-Madison)

films at 400°C. We demonstrated the integration of UNCD with CMOS devices by growing UNCD layers on 200 mm commercial CMOS wafers kindly provided by our colleague S. Pacheco from Freescale. All CMOS devices performed with practically no degradation in output electrical parameters after being coated with UNCD films about 1 μm thick produced during four hours deposition at 400°C, using the Argonne-patented MPCVD growth technique (see Fig. 15).

All CMOS devices tested from different areas of the CMOS wafer performed to specification, thus demonstrating the required integration of UNCD with CMOS devices.

3.6 In Vivo Animal Tests of Hermetic-Encapsulating Coatings for Artificial Retina

Several UNCD-coated Si samples (∼5 mm× 5 mm× 0.5 mm), where the Si was totally encapsulated by the UNCD layer, as well as two with defects (UNCD coatings with several pinholes due to the low nucleation density) grown at low temperature were implanted in rabbit eyes to evaluate the biocompatibility and long-term biostability of UNCD in the physiological environment of animal eyes. These tests were passive in the sense that the implants were just Si substrates without any CMOS circuits, so there was no power applied to the Si substrate. After six-months implantation, no acute surface damage or evidence for biomaterial attachment was noted, as determined by SEM (Fig. 16(a) and (b)) and XPS analysis (Fig. 16(c)). Reactions were observed in two samples with defective coatings, presumably due to reaction with the rabbit eye's physiological saline through pinholes.

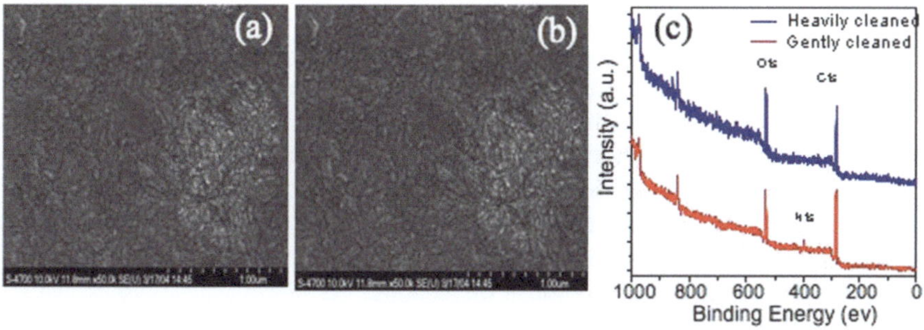

Fig. 16 SEM picture of the surface of UNCD coating before (**a**) and after (**b**) exposure to the saline solution in a rabbit's eye; (**c**) XPS analysis showing no chemical reactions with the saline solution

In addition, no deterioration of the retina was elicited by the UNCD coating. These results indicate that the UNCD coatings exhibit promising bio-inert performance in animal eyes, and tests in human eyes are thus warranted.

4 Challenges for Bio-inert Microchip Encapsulation Hermetic Coatings

There are three issues that should be taken into account when developing hermetic coatings for encapsulation of implantable medical devices, and specifically in the case of this paper, for microchip retinal implant, namely, (a) bio-inertness and biocompatibility; (b) mechanical robustness; and (c) chemical inertness and electrical insulation. Chemical inertness and electrical insulation are critical for encapsulation of a retinal microchip in order to avoid electrical shorts from the independent connector lines to electrodes that are covered by the UNCD and to avoid electrochemical reactions at the surface of the UNCD when exposed to the eye's saline solution. Prior work by our group showed that the grain boundaries in the UNCD films consisted largely of sp^2-bonded carbon atoms and that electron transport occurred through those grain boundaries [40]. In the case of using UNCD coatings as an electron-emitting surface in field electron emitters for cold cathodes, this is a desirable property. However, in the case of application of UNCD as hermetic coating for encapsulation of retinal implant microchips, electrical conductivity, even if small, can result in undesirable electrochemical reactions. The data presented in this study indicate that the insulation characteristics of UNCD layers can be greatly improved by engineering the grain boundaries of UNCD, via elimination of dangling bonds in sp^2-coordinated carbon atoms, which results in elimination of sites that contribute to grain boundary-based electrical conduction in UNCD. The approach we have chosen to eliminate dangling bonds at the grain boundaries is to saturate them with hydrogen atoms. Theoretical calculations also reveal that hydrogen exists in the grain boundaries in the form of C-H bonds. Moreover, the hydrogen addition increases the coordination of carbon and decreases the density of state in the diamond band gap. It can also decrease the whole concentration in p-type semiconductors or neutralize dopant in n-type materials. Our study confirmed that hydrogen incorporation into UNCD grain boundaries made these coatings better insulators with low enough leakage currents to be compatible with the functionality required for a hermetic coating for the retinal microchip implant. However, a critical issue under investigation is the possibility that sharp edges or corners of the coated microchip could concentrate electric fields and/or dust particles, resulting in pinholes that can compromise the coating. To address these issues, we are presently exploring hybrid oxide/UNCD coatings, where a highly insulating oxide such as Al_2O_3 is grown first on the Si chip, followed by a very thin metallic layer (this is used also to enhance the growth of the UNCD layer) on top of the oxide, but electrically isolated from the underlying Si substrate, followed by a UNCD layer to provide the bio-interface to the retina.

5 Conclusions and a Future Outlook

Our work has shown that UNCD coatings grown on Si substrates at temperatures $\leq 400°C$ range, using PECVD with CH_4/Ar plasmas, may provide a reliable hermetic bio-inert coating technology for encapsulation of Si microchips implantable in the eye specifically and in the human body in general. SEM analysis showed that films grown in this temperature range exhibit surface morphology and microstructure characteristics similar to the best UNCD films demonstrated in the past in our laboratory. Electrochemical tests of the plain UNCD coatings yielded relatively high leakage currents ($\sim 10^{-4}$ A/cm^2), which are not compatible with the needed functionality in the eye environment. Introducing hydrogen into Ar/CH_4 mixture resulted in the incorporation of hydrogen into the grain boundaries and saturated the dangling bonds, thus leading to a greatly decreased leakage current. UNCD films grown using CH_4/Ar/H_2 (1%) plasma exhibit the lowest leakage currents ($\sim 7 \times 10^{-7}$ A/cm^2) in a saline solution simulating the eye environment. This leakage is incompatible with the functionality of the first-generation artificial retinal microchip. However, it is expected that when growing UNCD on top of the Si microchip passivated by a silicon nitride layer or the oxide layers also under investigation in our group, the electrochemically induced leakage will be reduced by at least one to three orders of magnitude to the range of 10^{-10} A/cm^2, which is compatible with reliable, long-term implants.

Acknowledgments We wish to acknowledge support from the U. S. Department of Energy, BES-Materials Science for work in the Materials Science Division, under contract W-31-109-ENG-38. The work at the Center for Nanoscale Materials and at the Electron Microscopy Center for Materials Research at Argonne National Laboratory was supported by the U.S. Department of Energy-Office of Science under Contract No. DE-AC02-06CH11357 by UChicago Argonne, LLC. We also acknowledge the many colleagues and postdoctorals who have made substantial contributions to the work discussed in this chapter over the years, namely: J. Birrell, J.A. Carlisle, L. Chen,

W. Fan, R. Greenberg, D.M. Gruen, M. Humayun, B. Kabius, W. Li, Q. Lin, C. Liu, B. Mech, A. V. Sumant, J. Wang, J. Weiland, and X. Xiao.

References

1. Weiland JD, Liu W, Humayun MS (2005) Retinal prosthesis. Annu Rev Biomed Eng 7: 361.
2. Djourno A, Eyries C (1957) Prothese auditive par excitation electrique a distance du nerf sensorial a l'aide d'un bobinage inclus a demeure. Presse Med 35: 14.
3. House WF (1976), Cochlear implants. Ann Otol Rhinol Laryngol 85(Suppl.27) Pt 2: 1.
4. Brindley G, Rushton D (1974) Implanted stimulators of the visual cortex as visual prosthetic devices. Trans Am Acad Ophthalmol Otolaryngol 78: Op 741.
5. http://artificialretina.energy.gov/about.shtml
6. Humanyun MS (1974) Intraocular retinal prosthesis. Trans Am Ophthamol Soc 99: 271.
7. Meyer J (2001) Retina implant-a bio MEMS challenge. Sens Actuator A 97–98: 1.
8. Najafi K (2003) Micropackaging technologies for intergrated microsystems: applications to MEMS and MOEMS: Micromach Microfabr Process Technol VIII, Proc. SPIE 4979: 1.
9. Najafi K (2007) Packaging of implantable microsystems. IEEE Sens Conf 58.
10. Hammerle H, Kobuch K, Kohler K, Nisch W, Sachs H, Stelzle M (2002) Biostability of micro-photodiode arrays for subretinal implantation. Biomaterials 23: 797.
11. Rojahn M (2003) Encapsulation of a retina implant. Ph D dissertation, University of Stuggart.
12. Stuart F, Edell D, Guzelian A, Liu Y, Edell R (2003) Plasma-enhanced chemical vapor deposited silicon carbide as an implantable dielectric coating. J Biomed Mater Res A 67: 856.
13. Seo J, Kim S, Chung H, Kim H, Yu H, Yu Y (2004) Biocompatibility of polyimide microelectrode array for retinal stimulation. Mater Sci Eng C 24: 185.
14. Naples GG, et al. (1990) In: Agnew WF, McCreery DB (eds) Neural Prostheses: Fundamental Studies, Biophysics and Bioengineering Series. Prentice Hall, New Jersey, pp 107.
15. Heiduschka P, Thanos S (1998) Implantable bioelectronic interfaces for lost nerve functions. Prog Neurobiol 55: 433.
16. Stieglitz T, Haberer W, Lau C, Goertz M (2004) Development of an inductively coupled epiretinal visual prosthesis. Proc Ann Int Conf IEEE Eng Med Biol Soc, 26th, San Francisco, CA.
17. Weiland JD, Liu W, Humayun MS (2005) Retinal Prothesis Annu Rev Biomed Eng. 7: 361.
18. Field JE (1992) Properties of Diamond. Academic Press, London.
19. Spitsyn BV, Bouilov LL, Derjaguin BV (1981) Vapor growth of diamond on diamond and other surfaces. J Cryst Growth 52: 219.
20. Matsumoto S, Sato Y, Tsutsumi M, Setaka N (1982) Growth of Diamond particles from Methane-Hydrogen Gas. J Mater Sci 17: 3106.
21. Matsumoto S (1989) Development of CVD Diamond Synthesis Techniques. Proc. of 1st Symp on Diamond and Diamond-like Films. Electrochem Soc Proc 89–12: 50.
22. Hirose Y, Kondo N (1988) Program and Book of Abstracts. Jpn Appl Phys Spring Meeting, 434.
23. Kamo M, Sato Y, Matsumoto S, Setaka N (1983) Diamond synthesis from gas phase in microwave plasma, J Cryst Growth 62(3): 642.
24. Suzuki K, Sawabe S, Yasuda H, Inzuka T (1987) Growth of diamond thin films by DC plasma chemical vapor deposition. Appl Phys Lett 50 (12): 728.
25. Kurihara K, Sasaki K, Kawaradi M, Koshino N (1988) High rate synthesis of diamond by DC plasma jet chemical vapor deposition. Appl Phys Lett 52:437.
26. Matsumoto S (1985) Chemical Vapor deposition of diamond in RF glow discharge. J Mater Sci Lett 4(5): 600.
27. Matsumoto M, Hino M, Kobayashi T (1987) Synthesis of diamond films in a RF induction thermal plasma. Appl Phys Lett 51: 737.

28. Backmann PK and Messier R (1989) Emerging technology of diamond thin films. C&EN 67(20): 24.
29. Harris SJ, Goodwin DG (1993) Growth on the reconstructed diamond (100) surface J Phys Chem 97: 23.
30. Dischler B, Wild C (1998) Low-Pressure Synthetic Diamond: Manufacturing and Applications. Springer, Heidelberg.
31. Status and Applications of Diamond and Diamond-like Materials: an Emerging Technology (1990), National Materials Advisory Board, Commission on Engineering and Technical Systems. National Research Council NMAB-445, National Academy Press, p. 28.
32. Bachmann PK, Hagemann HJ, Lade H, Leers D, Weichert DU, Wilson H, Fournier D, Plamann K (1995) Diamond chemical vapor deposition: gas compositions and film properties. Diam Relat Mater 4: 820.
33. Sharda T, Bhattacharyya S (2004) Advances in Nanocrystalline Diamond, Nalwa, HS (Ed), Encyclopedia of Nanoscience and Nanotechnology vol. 2, American Scientific Publishers, California p. 337.
34. Butler JE, Sumant AV (2008) The CVD of Nanodiamond Materials. Chem Vap Deposition 14: 145.
35. Gruen DM (1999) Nanocrystalline Diamond Films. Annu Rev Mater Sci 29: 211.
36. Zhou D, Gruen DM, Qin LC, McCauley TG, Krauss AR (1998) Control of diamond film microstructure by Ar additions to CH4/H2 microwave plasmas. J Appl Phys 84: 1981.
37. Sumant AV, Krauss AR, Gruen DM, Auciello O, Erdemir A, Williams M, Artiles AF Adams W (2005) Ultrananocrystalline Diamond Film as a Wear-Resistant and Protective Coating for Mechanical Seal Applications. Tribol Trans 48: 24.
38. Auciello O, Birrell J, Carlisle JA, Gerbi JE, and Xiao X, Peng B, Espinosa HD (2004) Diamond Thin Films. J Phys Condens Matter 16(16): R539.
39. Wang J, Firestone MA, Auciello O, and Carlisle JA (2004) Surface Functionalization of Ultrananocrystalline Diamond Films by Electrochemical Reduction of Aryldiazonium Salts. Langmuir 20: 11450.
40. Xiao X, Wang J, Carlisle JA, Merch B, Greenberg R, Freda R, Humayun MS, Weiland J, Auciello O (2006) In Vitro and in Vivo evaluation of ultrananocrystalline diamond for coating of implantable retinal microchips. J Biomed Mater 77B: 273.
41. Xiao X, Birrell J, Gerbi JE, Auciello O Carlisle JA (2004) Low temperature growth of ultrananocrystalline diamond. J Appl Phys 96:2232.
42. Birrell J, Gerbi JE, Auciello O, Gibson JM, Johnson J, Carlisle JA (2005) Interpretation of the Raman spectra of ultrananocrystalline diamond. Diam Relat Mater 14: 86.
43. Shi B, Jin Q, Chen L, Auciello O (2006) Study of ultrananocrystalline diamond films as implantable biomedical devices: Assessment of their biocompatibility to cell attachment and growth. MRS Fall Meeting, Boston.

The Electrochemistry of Charge Injection at the Electrode/Tissue Interface

Daniel R. Merrill

Abstract The physical basis for electrical stimulation of excitable tissue is presented with emphasis on the fundamental mechanisms of charge injection at the electrode/tissue interface. Faradaic and non-Faradaic charge-transfer mechanisms are presented and contrasted. An electrical model of the electrode/tissue interface is given. The physical basis for the origin of electrode potentials is given. Electrochemical reversibility is discussed. Two-electrode and three-electrode systems are compared. Various methods of controlling charge delivery during pulsing are presented. Commonly used electrode materials and stimulation protocols are reviewed in terms of stimulation efficacy and safety. Principles of stimulation of excitable tissue are reviewed. Mechanisms of damage to tissue and the electrode are reviewed.

Contents

D.R. Merrill (✉)
Alfred E. Mann Foundation for Scientific Research, Santa Clarita, CA 91355, USA
e-mail: danm@aemf.org

This chapter is adapted from a previously published review [1].

D.D. Zhou, E. Greenbaum (eds.), *Implantable Neural Prostheses 2*,

Biological and Medical Physics, Biomedical Engineering, DOI 10.1007/978-0-387-98120-8_4,

1 Physical Basis of the Electrode/Electrolyte Interface

Electrical stimulation of excitable tissue is the basis of clinical therapeutic electrical stimulation and functional electrical stimulation, including deep brain stimulation and stimulation of muscles, peripheral nerves, or sensory systems. When a metal electrode is placed inside a physiological medium such as extracellular fluid (ECF), an interface is formed between the two phases. In the metal electrode phase and in attached electrical circuits, charge is carried by electrons. In the physiological medium, or in more general electrochemical terms the electrolyte, charge is carried by ions, including sodium, potassium, and chloride in the ECF. The central process that occurs at the electrode/electrolyte interface is a transduction of charge carriers from electrons in the metal electrode to ions in the electrolyte.

In the simplest system, two electrodes are placed in an electrolyte, and electrical current may pass between the electrodes through the electrolyte. One of the two electrodes is termed a working electrode (WE), and the second is termed a counter electrode (CE). The working electrode is defined as the electrode that one is interested in studying, with the counter electrode being necessary to complete the circuit for charge conduction. In electrophysiology experiments, it is common to use a third electrode termed the reference electrode (RE), which defines a reference for electrical potential measurements.

A change in electrical potential occurs upon crossing from one conducting phase to another (from the metal electrode to the electrolyte) at the interface itself, in a very narrow interphase region (at most a few hundred angstroms in width). The basis for this is described in more detail in Section 1.4. The change or gradient in electrical potential corresponds to an electric field, measured in volts/meter, at the interface. This gradient exists even in the equilibrium condition when there is no current flow. Electrochemical reactions may occur in this interphase region if the electrical potential profile is forced away from the equilibrium condition. In the absence of current, the electrical potential is constant throughout the electrolyte beyond the narrow interphase region. During current flow, a potential gradient exists in the electrolyte, generally many orders of magnitude smaller than at the interface.

There are two primary mechanisms of charge transfer at the electrode-electrolyte interface, illustrated in Fig. 1. One is a non-Faradaic reaction, where no electrons are

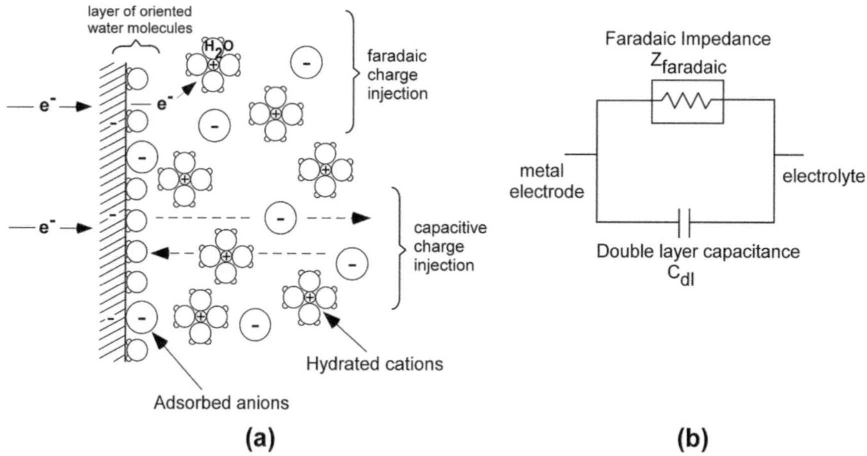

Fig. 1 The electrode/electrolyte interface, illustrating Faradaic charge transfer (*top*) and capacitive redistribution of charge (*bottom*) as the electrode is driven negative. (**a**) Physical representation (**b**) Two-element electrical circuit model for mechanisms of charge transfer at the interface. The capacitive process involves reversible redistribution of charge. The Faradaic process involves transfer of electrons from the metal electrode, reducing hydrated cations in solution (symbolically $O + e^- \rightarrow R$, where the cation O is the oxidized form of the redox couple O/R). An example reaction is the reduction of silver ions in solution to form a silver plating on the electrode, reaction (8a). Faradaic charge injection may or may not be reversible

transferred between the electrode and electrolyte. Non-Faradaic reactions include redistribution of charged chemical species in the electrolyte. The second mechanism is a Faradaic reaction in which electrons are transferred between the electrode and electrolyte, resulting in reduction or oxidation of chemical species in the electrolyte. Faradaic reactions are further divided into reversible and nonreversible Faradaic reactions, which are detailed in Section 1.3. Reversible Faradaic reactions include those where the products either remain bound to the electrode surface or do not diffuse far away from the electrode. In an irreversible Faradaic reaction, the products diffuse away from the electrode.

1.1 Capacitive/Non-Faradaic Charge Transfer

If only non-Faradaic redistribution of charge occurs, the electrode/electrolyte interface may be modeled as a simple electrical capacitor called the double-layer capacitor C_{dl}. This capacitor is formed due to several physical phenomena [2, 3, 4, 5, 6]. First, when a metal electrode is placed in an electrolyte, charge redistribution occurs as metal ions in the electrolyte combine with the electrode. This involves a transient transfer of electrons between the two phases, resulting in a plane of charge at the surface of the metal electrode, opposed by a plane of opposite charge, as counterions, in the electrolyte. The excess charge on the electrode surface, symbolized by q^M or σ^M, takes the form of an excess or deficiency of electrons and is present on a very thin layer (< 0.1 angstrom thick) at the surface. In the electrolyte, counterions take the form of excess cations or anions, symbolized by q^S. If q^M is an excess of electrons, then q^S is an excess of cations, and if q^M is a deficiency of electrons, then q^S is an excess of anions, i.e., net electroneutrality is maintained and $q^M = -q^S$. A second reason for formation of the double layer is that some chemical species such as halide anions may specifically adsorb to the solid electrode, acting to separate charge. A third reason is that polar molecules such as water may have a preferential orientation at the interface, and the net orientation of polar molecules separates charge.

If the net charge on the metal electrode is forced to vary (as occurs with charge injection during stimulation), a redistribution of charge occurs in the solution. Consider two metal electrodes immersed in an electrolytic salt solution. A voltage source is applied across the two electrodes so that one electrode is driven to a relatively negative potential and the other to a relatively positive potential. At the interface that is driven negative, the metal electrode has an excess of negative charge (Fig. 1). This will attract positive charge (cations) in solution toward the electrode and repel negative charge (anions). In the interfacial region, there will be net electroneutrality, because the negative charge excess on the electrode surface will equal the positive charge in solution near the interface. The bulk solution will also have net electroneutrality. At the second electrode, the opposite processes occur, i.e., the repulsion of anions by the negative electrode is countered by attraction of anions at the positive electrode. If the total amount of charge delivered is sufficiently small, only charge redistribution occurs, there is no transfer of electrons across the interface, and the interface is well modeled as a simple capacitor. *If the polarity of the applied voltage source is then reversed, the direction of current is reversed, the charge redistribution is reversed, and charge that was injected from the electrode into the electrolyte and stored by the capacitor may be recovered.*

1.2 Faradaic Charge Transfer and the Electrical Model of the Electrode/Electrolyte Interface

Charge may also be injected from the electrode to the electrolyte by Faradaic processes of reduction and oxidation, whereby electrons are transferred between the two phases. Reduction, which requires the addition of an electron, occurs at the electrode that is driven negative, while oxidation, requiring the removal of an electron, occurs at the electrode that is driven positive. Faradaic charge injection results in the creation of chemical species, which may either go into the solution or remain bound to the electrode surface. Unlike the capacitive charge injection mechanism, if these Faradaic reaction products diffuse sufficiently far away from the electrode, they cannot be recovered upon reversing the direction of current. Fig. 1(b) illustrates a simple electrical circuit model of the electrode/electrolyte interface consisting of two elements [7, 8, 9]. C_{dl} is the double-layer capacitance representing the ability of the electrode to cause charge flow in the electrolyte without electron transfer. Z_{faradaic} is the Faradaic impedance representing the Faradaic processes of reduction and oxidation where electron transfer occurs between the electrode and electrolyte. One may generally think of the capacitance as representing charge storage, and the Faradaic impedance as representing charge dissipation.

The following are illustrative examples of Faradaic electrode reactions that may occur. Cathodic processes, defined as those where reduction of species in the electrolyte occur as electrons are transferred from the electrode to the electrolyte, include such reactions as

$$2\,H_2O + 2e^- \rightarrow H_2 \uparrow 2\,OH^- \quad \text{reduction of water} \tag{1}$$

$$Fe^{3+} + e^- \longleftrightarrow Fe^{2+} \quad \text{simple electron transfer} \tag{2}$$

$$Cu^{2+} + 2\,e^- \longleftrightarrow Cu \quad \text{metal deposition} \tag{3}$$

$$PtO + 2H^+ + 2e^- \longleftrightarrow Pt + H_2O \quad \text{oxide formation and reduction} \tag{4}$$

$$IrO + 2H^+ + 2\,e^- \longleftrightarrow Ir + H_2O \quad \text{oxide formation and reduction} \tag{5a}$$

$$IrO_2 + 4H^+ + 4\,e^- \longleftrightarrow Ir + 2H_2O \quad \text{oxide formation and reduction} \tag{5b}$$

$$2IrO_2 + 2H^+ + 2\,e^- \longleftrightarrow Ir_2O_3 + H_2O \quad \text{oxide formation and reduction} \tag{5c}$$

$$Pt + H^+ + e^- \longleftrightarrow Pt - H \quad \text{hydrogen atom plating} \tag{6}$$

$$M^{(n+1)+}\,(OH)_{(n+1)} + H^+ + e^- \longleftrightarrow M^{n+}\,(OH)_n + H_2O \quad \text{valency changes within an oxide} \tag{7}$$

$$Ag^+ + e^- \longleftrightarrow Ag \quad \text{reduction of silver ions} \tag{8a}$$

$$AgCl \longleftrightarrow Ag^+ + Cl^- \quad \text{dissolution of silver chloride} \tag{8b}$$

Anodic processes, defined as those where oxidation of species in the electrolyte occur as electrons are transferred to the electrode, include:

$$2H_2O \rightarrow O_2 \uparrow + 4\,H^+ + 4\,e^- \quad \text{oxidation of water} \tag{9}$$

$$Pt + 4Cl^- \rightarrow [PtCl_4]^{2-} + 2\,e^- \quad \text{corrosion} \tag{10}$$

$$2Cl^- \rightarrow Cl_2 \uparrow + 2\,e^- \quad \text{gas evolution} \tag{11}$$

$$Fe \rightarrow Fe^{2+} + 2\,e^- \qquad \text{anodic dissolution} \qquad (12)$$

$$2\,Ag + 2OH^- \longleftrightarrow Ag_2O + H_2O + 2\,e^- \qquad \text{oxide formation} \qquad (13)$$

Reaction (1) is the irreversible reduction of water (which is typically abundant as a solvent at 55.5 M) forming hydrogen gas and hydroxyl ions. The formation of hydroxyl raises the pH of the solution. Reversible reactions, where species remain bound or close to the electrode surface, are demonstrated by reactions (2), (3), (4), (5), (6), (7), and (8). In reaction (2), the electrolyte consists of ferric and ferrous ions. By driving the metal electrode to more negative potentials, electrons are transferred to the ferric ions forming ferrous ions. In reaction (3) a copper metal electrode is immersed in a solution of cuprous ions. The cuprous ions in the solution are reduced, building up the copper electrode. Reactions (4) and (5a), (5b), (5c) are the reversible formation and subsequent reduction of an oxide layer on platinum and iridium, respectively. Reaction (6) is reversible adsorption of hydrogen onto a platinum surface responsible for the so-called pseudocapacity of platinum. Reaction (7) is the general form of reversible valency changes that occur in a multilayer oxide film of iridium, ruthenium, or rhodium, with associated proton or hydroxyl ion transfer [10, 11, 12, 13]. Reactions (8a) and (8b) are the reversible reactions of a silver chloride electrode driven cathodically. Silver ions in solution are reduced to solid silver on the electrode (reaction 8a). To maintain the solubility constant $K_S \equiv (a_{Ag+})(a_{Cl-})$, where a is the ionic activity, as silver ions in solution are reduced the AgCl salt covering the electrode dissolves to form silver and chloride ions in solution (reaction 8b) (these reactions are discussed in more detail at the end of this section). In reaction (9), water molecules are irreversibly oxidized forming oxygen gas and hydrogen ions, and thus lowering the pH. Reaction (10) is the corrosion of a platinum electrode in a chloride-containing media. In reaction (11), chloride ions in solution are oxidized forming chlorine gas. In reaction (12), an iron metal electrode is dissolved forming ferrous ions that go into solution. Reaction (13) represents a reversible oxide formation on a silver electrode. As electrons are removed from the silver metal, Ag^+ ions are formed. These Ag^+ ions then combine with hydroxyl (OH^-) ions from solution forming an oxide layer (Ag_2O) on the surface of the silver electrode. Note the transfer of charge that occurs. As electrons are transferred to the electrode and then the external electrical circuit, the silver electrode is oxidized ($Ag \rightarrow Ag^+$). Because hydroxyl ions associate with the silver ions, the silver oxide is electroneutral. However, since hydroxyl has been removed from the solution, there is a net movement of negative charge from the electrolyte (loss of hydroxyl) to the electrode (electrons transferred to the electrode and then to the electrical circuit). The loss of hydroxyl lowers the solution pH.

1.3 Reversible and Irreversible Faradaic Reactions

There are two limiting cases that may define the net rate of a Faradaic reaction [9, 14, 15]. At one extreme, the reaction rate is under kinetic control; at the other extreme, the reaction rate is under mass transport control. For a given metal electrode and electrolyte, there is an electrical potential (voltage) called the equilibrium potential where no net current passes between the two phases. At electrical potentials sufficiently close to equilibrium, the reaction rate is under kinetic control. Under kinetic control, the rate of electron transfer at the interface is determined by the electrode potential and is *not* limited by the rate at which reactant is delivered to the electrode surface (the reaction site). When the electrode potential is sufficiently different from equilibrium, the reaction rate is under mass transport control. In this case, all reactant that is delivered to the surface reacts immediately, and the reaction rate is limited by the rate of delivery of reactant to the electrode surface.

Faradaic reactions are divided into reversible and irreversible reactions [9]. The degree of reversibility depends on the relative rates of kinetics (electron transfer at the interface) and mass transport. A Faradaic reaction with very fast kinetics relative to the rate of mass transport is reversible. With fast kinetics, large currents occur with small potential excursions away from equilibrium. Since the electrochemical product does not move away from the surface extremely fast (relative to the kinetic rate), there is an effective storage of charge near the electrode surface, and *if the direction of current is reversed then some product that has been recently formed may be reversed back into its initial (reactant) form.*

FAST KINETICS RELATIVE TO MASS TRANSPORT →
CHARGE - STORAGE CAPACITY = REVERSIBLE

In a Faradaic reaction with slow kinetics, large potential excursions away from equilibrium are required for significant currents to flow. In such a reaction, the potential must be forced very far from equilibrium before the mass transport rate limits the net reaction rate. In the lengthy time frame imposed by the slow electron-transfer kinetics, chemical reactant is able to diffuse to the surface to support the kinetic rate, and product diffuses away quickly relative to the kinetic rate. Because the product diffuses away, there is no effective storage of charge near the electrode surface, in contrast to reversible reactions. *If the direction of current is reversed, product will not be reversed back into its initial (reactant) form, since it has diffused away within the slow time frame of the reaction kinetics.* Irreversible products may include species that are soluble in the electrolyte (e.g., reaction 12), precipitate in the electrolyte, or evolve as a gas (e.g., reactions 1, 9, and 11). Irreversible Faradaic reactions result in a net change in the chemical environment, potentially creating chemical species that are damaging to tissue or the electrode. Thus, as a general principle, an objective of electrical stimulation design is to avoid irreversible Faradaic reactions.

SLOW KINETICS → NO CHARGE - STORAGE CAPACITY
(PRODUCT DIFFUSES AWAY) = IRREVERSIBLE

In certain Faradaic reactions, the product remains bound to the electrode surface. Examples include hydrogen atom plating on platinum (reaction 6) and oxide formation (reaction 13 as an example). These can be considered a logical extreme of slow mass transport. Since the product remains next to the electrode, such reactions are a basis for reversible charge injection.

1.4 The Origin of Electrode Potentials and the Three-Electrode Electrical Model

Electrochemical potential is a parameter that defines the driving force for all chemical processes and is the sum of a chemical potential term and an electrical potential term [16]. It is defined as

$$\overline{\mu}_i^{\beta} \equiv \mu_i^{\beta} + z_i\, e\, \phi^{\beta} \qquad (14)$$

where $\overline{\mu}_i^{\beta}$ is the electrochemical potential of particle i in phase β, μ_i^{β} is the chemical potential of particle i in phase β, and ϕ^{β} is the inner potential of the particle in phase β (the electrical potential in the bulk).

Two phases in contact are defined to be in electrochemical equilibrium when the electrochemical potential of any given chemical species is the same in each phase. If the electrochemical potentials of some species are unequal, there is a driving force for the net transfer of such species between the phases. For a metal electrode and a solution of metal ions in contact to be in equilibrium, the electrochemical potential of an electron must be the same in each phase. When two isolated phases are brought into contact, electron transfer may occur if the electrochemical potentials are unequal. Consider immersing a metal electrode into an electrolyte with an electrochemical reduction/oxidation (redox) couple, for example ferric and ferrous ions ($Fe^{3+}$and Fe^{2+}). Assume that while in isolation, the metal electrode has a higher chemical potential for electrons than the redox couple. Upon bringing the electrode into contact with the electrolyte, electrons will transfer from the metal to the redox couple, driving the reaction $Fe^{3+} + e^{-} \rightarrow Fe^{2+}$ to the right as ferric ions are reduced to ferrous ions. Upon transferring electrons, an electrical potential difference develops between the phases that repels further transfer. Equilibrium is reached when the electrostatic force cancels the driving force due to a difference in chemical potentials for an electron. At equilibrium, there is no further transfer of electrons, and a distinctive difference in inner potentials $\Delta\phi$ exists between the two phases (the inner potential ϕ is the electrical potential inside the bulk of the phase). The difference in inner potentials between a metal phase and solution phase in contact, $\Delta\phi_{\text{metal - solution}}$, defines the electrode interfacial potential.

It is an experimental limitation that a single interfacial potential cannot be measured. Whenever a measuring instrument is introduced, a new interface is created, and one is unable to separate the effects of the two interfaces. It is tempting to wonder why one cannot simply place one voltmeter probe on a metal electrode and a second voltmeter probe into the electrolyte and measure an electrode potential, as shown in Fig. 2. The electrode potential of interest is $\Delta\phi_{\text{metal - solution}}$. By introducing the measuring device (a metal voltmeter probe) into the electrolyte solution, a new interface is created with its own

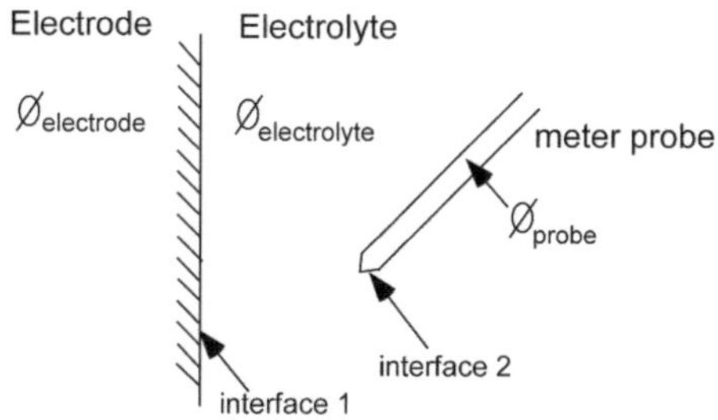

Fig. 2 Voltmeter Probe in an Electrolyte Introduction of the metal probe creates a second electrode/electrolyte interface

difference in potentials$\Delta\phi_{\text{solution - probe}}$. It is impossible to separate the components $\Delta\phi_{\text{metal - solution}}$ and $\Delta\phi_{\text{solution - probe}}$ from the measured potential. Note that if the other voltmeter probe (touching the metal electrode, not shown in Fig. 2) consists of a different material than the electrode, a third interface is formed, with a third difference in inner potentials.

Evaluation must be of a complete electrochemical cell, which is generally considered as two electrodes separated by an electrolyte. Practically, potentials are measured as complete cell potentials between two electrodes, either from the working electrode to the counter electrode or from the working electrode to a reference electrode. A cell potential is the sum of two interfacial potentials (electrode_1 to electrolyte plus electrolyte to electrode_2), as well as any potential difference occurring across the electrolyte as current flows. In the absence of current, the cell potential between the working electrode and second (counter or reference) electrode is called the open-circuit potential and is the sum of two equilibrium interfacial potentials from the working electrode to the electrolyte and from the electrolyte to the second electrode.

The term "electrode potential" is not defined consistently in the electrochemistry literature. Some authors define the electrode potential as the potential between an electrode and a reference electrode, and others define it as the (immeasurable) interfacial potential. For clarity and accuracy, when the term "electrode potential" is used it should be specified what this potential is with respect to, e.g., the electrolyte, a reference electrode, or another electrode.

Consider the electron-transfer reaction between a metal electrode and a reduction/oxidation (redox) couple O and R in solution:

$$\text{O}+n\,e^{-} \longleftrightarrow \text{R} \tag{15}$$

where O is the oxidized species of the couple, R is the reduced species, and n is the number of electrons transferred

If the concentrations of both O and R in solution are equal, then the electrical potential of the redox couple equilibrates at $E^{\Theta'}$, defined as the formal potential. More generally, if the concentrations of O and R are unequal, the equilibrium potential or Nernst potential, E_{eq}, may be calculated by the Nernst equation [9,16]:

$$E_{\text{eq}} = E^{\Theta\prime} + (RT/nF)\ln\{[\text{O}]\,/\,[\text{R}]\} \tag{16}$$

where [O] and [R] are concentrations in the bulk solution, R is the gas constant $\sim$ 8.314 J/mol-°K, T is the absolute temperature, and F is Faraday's constant $\sim$96,485 C/mol of electrons

The Nernst equation (16) relates the equilibrium electrode potential E_{eq} (the electrical potential of the working electrode with respect to any convenient reference electrode) to the bulk solution concentrations [O] and [R] when the system is in equilibrium. As the bulk concentration [O] increases or the bulk concentration [R] decreases, the equilibrium potential becomes more positive.

In a system containing only one redox couple that has fairly fast kinetics, the measured open-circuit potential equals the equilibrium potential of the redox couple. If the kinetics of the redox couple are slow, the open-circuit potential (an empirical parameter) may not quickly attain the equilibrium potential after a perturbation, and if other contaminating redox couples (affectionately known as "dirt") are present that affect the equilibrium state, the measured open-circuit potential does not readily correlate with any single redox equilibrium potential.

If one begins with a system that is in equilibrium and then forces the potential of an electrode away from its equilibrium value, for example by connecting a current source between the working and counter electrodes, the electrode is said to become polarized. Polarization is measured by the overpotential η, which is the difference between an electrode's potential and its equilibrium potential (both measured with respect to some reference electrode):

$$\eta \equiv E - E_{eq} \tag{17}$$

The electrode interface model of Fig. 1(b) demonstrates the mechanisms of charge injection from an electrode; however, it neglects the equilibrium interfacial potential $\Delta\phi$ that exists across the interface at equilibrium. This is modeled as shown in Fig. 3(a). In addition to the electrode interface, the solution resistance R_S (alternatively referred to as the access resistance R_A or the ohmic resistance R_Ω) that exists between two electrodes in solution is modeled.

An electrical potential difference, or voltage, is always defined between two points in space. During electrical stimulation, the potentials of both the working and counter electrodes may vary with respect to some third reference point. A third electrode whose potential does not change over time, the reference electrode, may be employed for making potential measurements. Potentials of the working electrode and counter electrode may then be given with respect to the reference electrode. An electrical circuit model of a three-electrode system, including the working electrode, counter electrode, and reference electrode immersed into an electrolyte, is shown in Fig. 3(b). The reference electrode is used for potential measurements and is not required to pass current for stimulation; a two-electrode system (working and counter electrodes) is sufficient for stimulation. As current is passed between the working and counter electrodes through the electrolytic solution, the interfacial potentials, $V_{\text{WE-solution}}$ and $V_{\text{CE-solution}}$, will vary from their equilibrium values, i.e., there are overpotentials associated with both interfaces. Also, as current flows there is a voltage drop across the resistive solution equal to the product of current

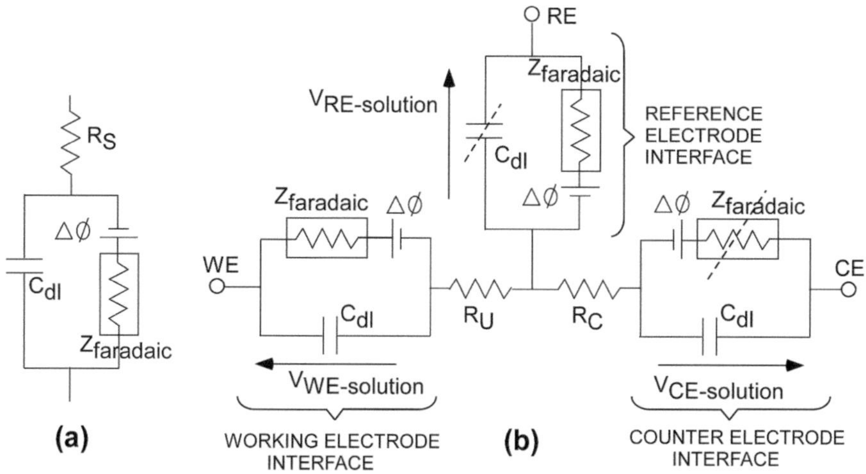

Fig. 3 Electrical Circuit Models (a) Single-Electrode/Electrolyte Interface (b) Three-Electrode System External access to the system is at three points labeled "WE", "CE", and "RE". If the counter electrode has a large surface area, it may be considered as strictly a capacitance as shown. A reference electrode with very low valued Faradaic resistance will maintain the interfacial potential $V_{\text{RE-solution}}$ constant

and solution resistance: $v = iR_S$. Thus, if current flows and there is a change in the measured potential $V_{\text{WE-CE}}$, this change may be from any of three sources: (1) an overpotential at the working electrode as the interfacial potential $V_{\text{WE-solution}}$ varies, (2) an overpotential at the counter electrode as the interfacial potential $V_{\text{CE-solution}}$ varies, and (3) the voltage drop iR_S in solution. In the two-electrode system, one may only measure $V_{\text{WE-CE}}$, and the individual components of the two overpotentials and iR_S cannot be resolved. A third (reference) electrode may be used for potential measurements. An ideal reference electrode has a Faradaic reaction with very fast kinetics, which appears in the electrical model as a very low resistance for the Faradaic impedance Z_{faradaic}. In this case, no significant overpotential occurs at the reference electrode during current flow, and the interfacial potential $V_{\text{RE-solution}}$is considered constant. Examples of common reference electrodes are the reversible hydrogen electrode (RHE), the saturated calomel electrode (SCE), and the silver-silver chloride electrode [17]. In the three-electrode system, if current flows through the working and counter electrodes and a change is noted in the measured potential $V_{\text{WE-RE}}$, this change may be from either of two sources: (1) an overpotential at the working electrode as the interfacial potential $V_{\text{WE-solution}}$ varies, and (2) the voltage drop iR_S in solution. Unlike the two-electrode system, only one overpotential contributes to the measured potential change. Furthermore, the overpotential at the working electrode can be estimated using the process of correction. This involves estimating the value of the solution resistance between the working electrode interface and the reference electrode interface, called the uncorrected solution resistance R_U, and multiplying R_U by the measured current. This product $V_{\text{corr}} = iR_{U\ \text{estimated}}$ is then subtracted from the measured $V_{\text{WE-RE}}$ to yield the two interfacial potentials $V_{\text{WE-solution}}$ and $V_{\text{RE-solution}}$. Since $V_{\text{RE-solution}}$ is constant, any change in $V_{\text{WE-RE}}$ is attributed to an overpotential at the working electrode interface. Figure 4 illustrates the electrical potential profiles of a two-electrode system and a three-electrode system, under conditions of no current flow and with current.

1.5 Faradaic Processes: Quantitative Description

Equation (18) below, the current–overpotential equation [9], relates the overpotential to net current density through an electrode going into a Faradaic reaction and defines the full characteristics of the Faradaic impedance.

$$i_{\text{net}} = i_0\{\frac{[\text{O}]\,(0,t)}{[\text{O}]_\infty} \exp\left(-\alpha_{\text{c}} n f\,\eta\right) - \frac{[\text{R}]\,(0,t)}{[\text{R}]_\infty} \exp\left(+\left(1-\alpha_c\right) n f \eta\right)\} \tag{18}$$

where i_{net} is the net Faradaic current density across the electrode/electrolyte interface, i_0 is the exchange current density, $[\text{O}](0,t)$ and $[\text{R}](0,t)$ are concentrations at the electrode surface ($x = 0$) as a function of time, $[\text{O}]_\infty$ and $[\text{R}]_\infty$ are bulk concentrations, $\alpha_{\text{c}}$is the cathodic transfer coefficient and equals ~0.5, n is the number of moles of electrons per mole of reactant oxidized (equation 15), $f \equiv F/R\,T$, F is Faraday's constant ~96,485 C/mol of electrons, R is the gas constant ~ 8.314 J/mol-°K, and T is the absolute temperature

This equation relates the net current of a Faradaic reaction to three factors of interest: (1) the exchange current density i_0, which is a measure of the kinetic rate of the reaction, (2) an exponential function of the overpotential, and (3) the concentration of reactant at the electrode interface. The exponential dependence of Faradaic current on overpotential indicates that for a sufficiently small overpotential, there is little Faradaic current, i.e., for small potential excursions away from equilibrium, current flows primarily through the capacitive branch of Fig. 1, charging the electrode capacitance, not through the Faradaic branch. As more charge is delivered through an electrode interface, the electrode capacitance continues to charge, the overpotential increases, and the Faradaic current (proportional to exp (η)) begins to be a significant fraction of the total injected current. For substantial cathodic overpotentials, the left term of equation (18) dominates; for substantial anodic overpotentials, the right term dominates.

The two exponential terms represent the reduction and oxidation rates, respectively. The net current is the sum of the reduction and oxidation currents, as shown in Fig. 5. At the equilibrium potential E_{eq}, when $\eta = 0$, the rates are equal and opposite and may be relatively small (compared to when driven away from equilibrium), and the net current is zero. As the electrode potential moves away from equilibrium, one or the other term will begin to dominate. At overpotentials near equilibrium, the current increases rapidly with changes in overpotential due to the exponential form of equation (18).

A large value for i_0 represents a reaction with rapid electron exchange between the electrode and electrolyte (called the heterogeneous reaction); a small value for i_0 represents a reaction with slow electron transfer in the heterogeneous reaction. The

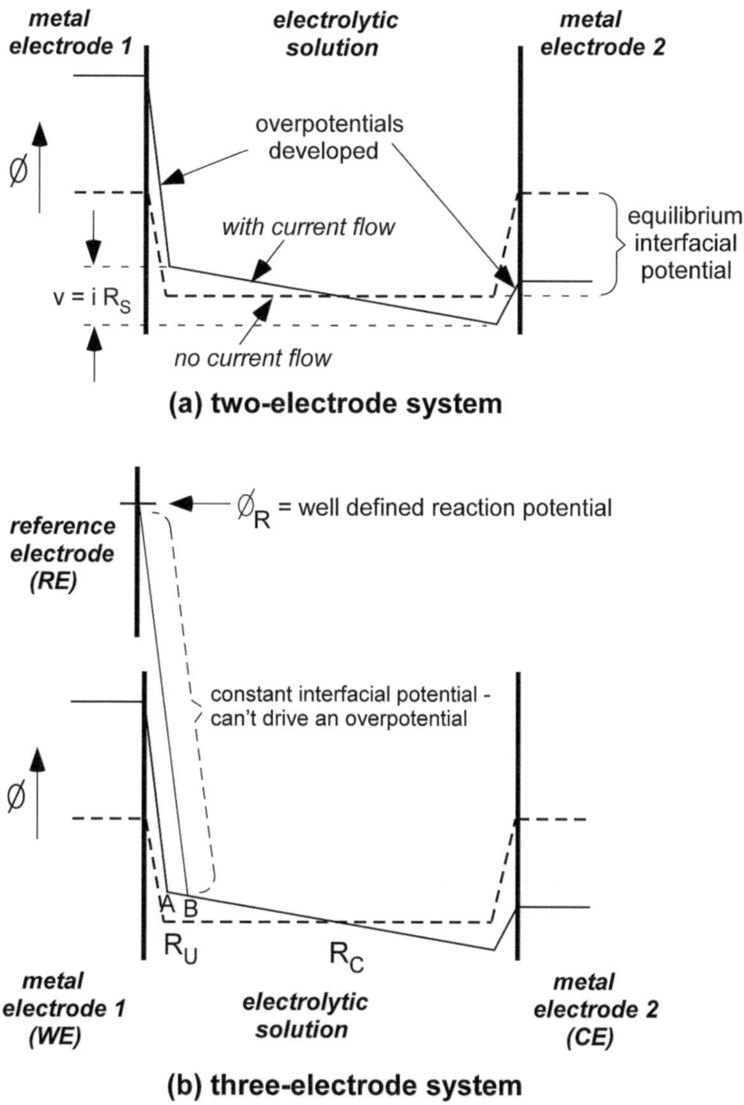

Fig. 4 Electrical Potential Profiles (**a**) Two-electrode system. In the absence of current, two equilibrium interfacial potentials exist, and the cell potential measured between the two electrodes is the difference between these equilibrium potentials. As shown the equilibrium potentials are the same (as would be the case if the same metal was used for both electrodes), and the cell potential would be zero. Upon passing current, overpotentials develop at both interfaces (one interfacial potential becomes greater, one smaller). The net change in measured cell potential is due to three sources: the voltage drop in solution $i\,R_S$ and two overpotentials η_1 and η_2. (**b**) Three-electrode system. The measured potential is between the working electrode and reference electrode. Since no substantial overpotential can be developed at the reference electrode, any change in measured potential upon passing current is due to two sources: the overpotential at the working electrode-solution interface, and the solution drop $i\,R_U$, where the uncorrected resistance R_U is the solution resistance between the WE interface and RE interface

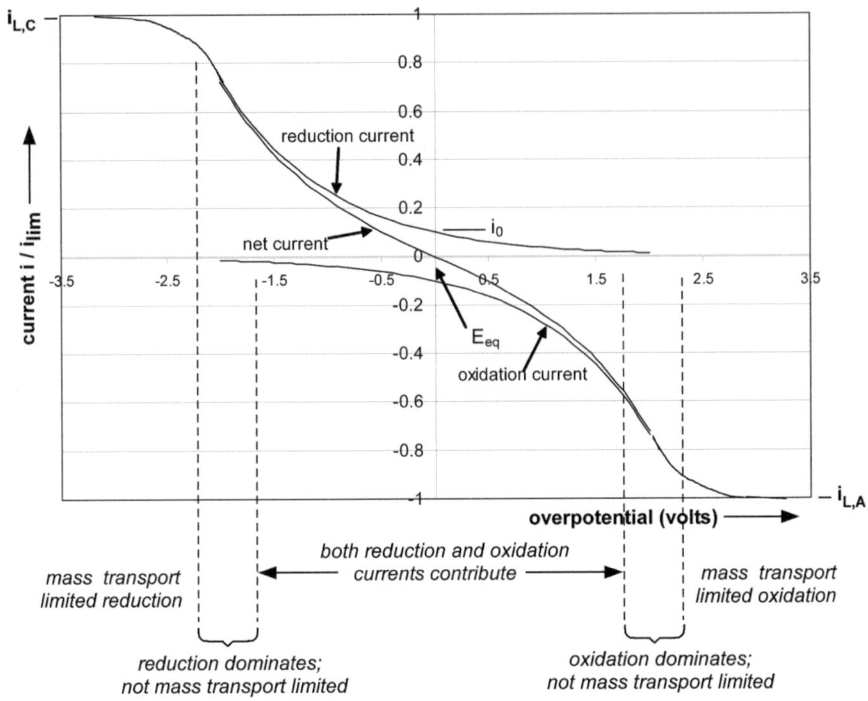

Fig. 5 Net Current vs. Overpotential, Oxidation and Reduction Curves Three characteristic regions are displayed: (1) near equilibrium, both reduction and oxidation currents contribute, then as the overpotential increases, (2) either reduction or oxidation dominate, initially in the absence of mass-transport limitation, and then (3) with mass-transport limitation

values for exchange current density i_0 may range over several orders of magnitude, e.g., from 10^{+1} to 10^{-12}A/cm^2. In the example shown in Fig. 5, i_0 is 0.1 of i_L, the limiting current. Figure 6 illustrates how the current–overpotential relation is highly dependent upon the exchange current. Three values of exchange current density are plotted. In each case α_c= 0.5, so the plot is symmetric about $\eta = 0$. For a kinetically fast system with a large exchange current density, such as $i_0 = 10^{-3}$A/cm^2, no significant overpotential may be achieved before a large current ensues. As the exchange current density decreases, one must go to higher overpotentials (further from the equilibrium value of $\eta = 0$) before a given current is noted. For a finite detection level of current (a real instrument), a reaction with low exchange current density will not manifest until relatively high overpotentials are achieved.

If currents are low or if the electrolytic solution is well stirred, so that the surface concentrations [O](0,t) and [R](0,t) are essentially equal to the bulk concentrations, then equation (18) reduces to

$$i_{\text{net}} = i_0 \left\{ \exp\left(-\alpha_c\, n f\, \eta\right) - \exp\left(+\left(1-\alpha_c\right)\ n f\, \eta\right) \right\} \tag{19}$$

This is the Butler-Volmer equation, which describes the current–overpotential relationship when mass-transfer effects are negligible. This may be a useful approximation of (18) when the current is less than 10% of the limiting current.

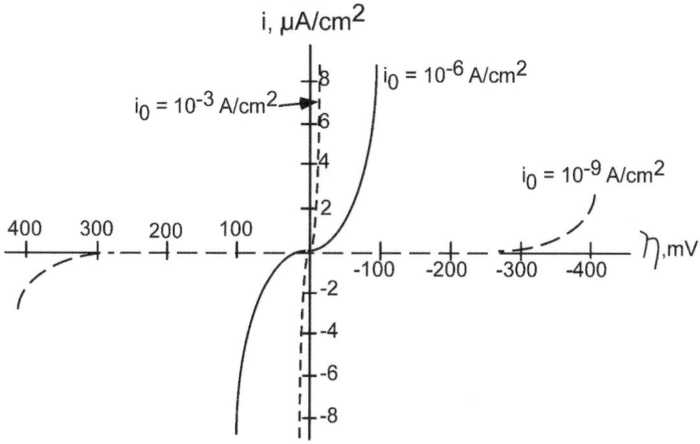

Fig. 6 Current–Overpotential Dependence on Exchange Current Density Three example exchange current densities are shown. Large densities correspond to kinetically fast reactions. At large exchange current densities, little overpotential is required for substantial current density

As η increases away from zero, one of the two terms of the current–overpotential relationship (representing either reduction or oxidation) will dominate:

$$\text{for negative overpotentilas} \qquad i_{net} = i_0 \exp\left(-\alpha_c\, nf\, \eta\right) \tag{20a}$$

$$\text{for positive overpotentials} \qquad i_{\text{net}} = i_0 \left\{-\exp\left(+\left(1-\alpha_c\right) nf\, \eta\right)\right\} \tag{20b}$$

Near equilibrium, the surface concentrations of O and R are approximately equal to the bulk concentrations. As more charge is delivered and the overpotential continues to increase (in either direction), the surface concentration of reactant may decrease. The Faradaic current will then begin to level off, corresponding to the current becoming limited by mass transport of reactant, not electron transfer kinetics. At the limiting currents, $i_{L,c}$(cathodic, for negative overpotentials) or $i_{L,a}$(anodic, for positive overpotentials), the reactant concentration at the electrode surface approaches zero, and the terms $[O](0,t)/[O]_\infty$ or $[R](0,t)/[R]_\infty$ counteract the exponential terms in equation (18), dominating the solution for net reaction rate. At the limiting currents, the slope of the reactant concentration gradient between the electrode surface and the bulk electrolyte determines the rate of reactant delivery, and thus the current. At overpotentials where mass transport limitation effects occur (but prior to $i_{L,c}$ or $i_{L,a}$), Faradaic current takes the form:

$$\text{for negative overpotentials} i_{\text{net}} = \frac{[O]\,[0,t]}{[O]_\infty} i_0 \exp\left(-\alpha_c\, nf\, \eta\right) \tag{21a}$$

$$\text{for positve overpotentials} i_{\text{net}} = -\frac{[R]\,[0.t]}{R_\infty} i_0 \exp\left(+\left(1-\alpha_c\right) nf\, \eta\right) \tag{21b}$$

Equations (19), (20), and (21) are illustrated as three regions on the current–overpotential plot, shown in Fig. 5.

The mass transport limited currents $i_{L,c}$ and $i_{L,a}$ are given by equations (22a) and (22b) below.

$$i_{L,c} = -n\,F\,A\,k_{d,O}\,[O]_\infty \tag{22a}$$

$$i_{L,a} = n\,F\,A\,k_{d,R}\,[R]_\infty \tag{22b}$$

where A is the electrode area and k_d is the mass transport rate, given by $k_d = D/\delta$, where D is the diffusion coefficient and δ is the diffusion layer thickness

For very small overpotentials, the Butler-Volmer equation (19) can be approximated by

$$i_{net} = i_0\,(-nf\,\eta) \tag{23}$$

since $e^x \sim 1 + x$ for small x. Thus at small overpotentials, the current is a linear function of overpotential. The ratio $-\eta/i$ is called the charge transfer resistance R_{ct}, which is given by

$$R_{ct} = RT/nF\,i_0 \tag{24}$$

A small value for R_{ct} corresponds to a kinetically fast reaction.

When the overpotential is relatively large, only one of the two terms in (18) is significant, and either the reduction current or the oxidation current becomes negligible. Figure 7 is a Tafel plot, which is a plot of log i vs. η. The straight-line approximations of Fig. 7, with slopes of $-\alpha\,n\,F/2.3\,R\,T$ for the cathodic reaction and $(1-\alpha)\,n\,F/2.3\,R\,T$ for the anodic reaction, are good approximations when the reverse reaction supplies less than 1% of the total current. Note that the intercept of the straight lines on the $\eta = 0$ axis is at log i_0. If the kinetics are fairly

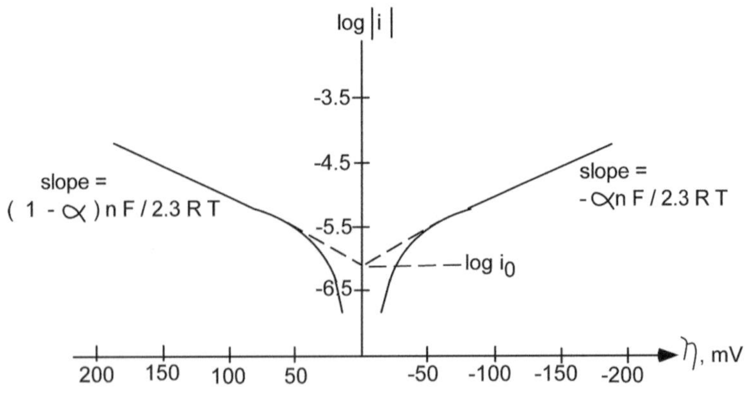

Fig. 7 Tafel plot

fast, no Tafel straight line will be noted, because the mass-transfer-limited current will be reached before the reverse reaction is negligible. Existence of a Tafel region requires that there are no mass-transfer effects. Slow kinetics allow a Tafel region to be observed. Tafel behavior may be used as an indicator of irreversible kinetics, requiring large overpotentials, yielding an essentially unidirectional reaction that is irreversible. The Tafel equation for negative overpotentials is given by:

$$i_{\text{net}} \sim i_0 \exp\left(-\alpha_{\text{c}}\, n f\, \eta\right) \tag{25}$$

As current is passed between a working electrode and reference electrode through an electrolyte, both the working and counter electrodes' potentials move away from their equilibrium values, with one moving positive of its equilibrium value and the other moving negative of its equilibrium value. Total capacitance is proportional to area, with capacitance C_{dl} = (capacitance/area) × area. Capacitance/area is an intrinsic material property. Capacitance is defined as the ability to store charge, and is given by

$$C_{\text{dl}} \equiv \mathrm{d}q/dV \tag{26}$$

where q = charge and V = the electrode potential with respect to some reference electrode

Thus an electrode with a relatively large area and total capacity (as is often the case for a counter electrode) can store a large amount of charge (dq) with a small overpotential (dV). During stimulation, the use of a large counter electrode keeps the potential of the counter electrode fairly constant during charge injection (near its equilibrium value), and there is little Faradaic current (equation 18). Significant overpotentials may be realized at a small working electrode. A typical reason for using a small electrode area is to achieve high spatial resolution during recording or stimulation. It is common to neglect the counter electrode in analysis, and while this is often a fair assumption it is not always the case.

1.6 Ideally Polarizable Electrodes and Ideally Nonpolarizable Electrodes

Two limiting cases for the description of an electrode are the ideally polarizable electrode and the ideally nonpolarizable electrode [8, 9, 14]. The ideally polarizable electrode corresponds to an electrode for which the Z_{faradaic} element has infinite resistance (i.e., this element is absent). Such an electrode is modeled as a pure capacitor, with C_{dl} = dq/dV (equation 26), in series with the solution resistance. In an ideally polarizable electrode, no electron transfer occurs across the electrode/electrolyte interface at any potential when current is passed; rather all current is through capacitive action. No sustained current flow is required to support a large voltage change across the electrode interface. An ideally polarizable electrode is not used as a reference electrode, since the electrode potential is easily perturbed

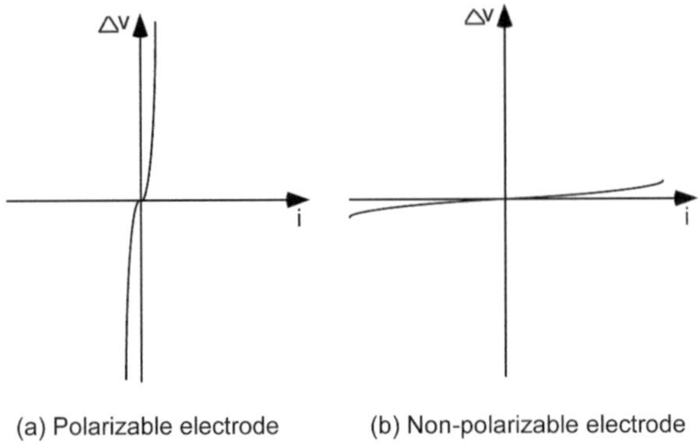

Fig. 8 Current–voltage relationships of highly polarizable and nonpolarizable electrodes

away from the equilibrium potential. A highly polarizable (real) electrode is one that can accommodate a large amount of injected charge on the double layer prior to initiating Faradaic reactions, corresponding to a relatively small exchange current density, e.g., $i_0 = 10^{-9}$ A/cm^2.

The ideally nonpolarizable electrode corresponds to an electrode for which the Z_{faradaic} element has zero resistance; thus only the solution resistance appears in the model. In the ideally nonpolarizable electrode, current flows readily in Faradaic reactions and injected charge is accommodated by these reactions. No change in voltage across the interface occurs upon the passage of current. This is the desired situation for a reference electrode, so that the electrode potential remains near equilibrium even upon current flow. A highly nonpolarizable (real) electrode, for which the Z_{faradaic} element has very small resistance, has a relatively large exchange current density, e.g., $i_0 = 10^{-3}$ A/cm^2. Most real electrode interfaces are modeled by a C_{dl} in parallel with a finite Z_{faradaic}, together in series with the solution resistance (Fig. 3a). Figure 8(a) illustrates a highly polarizable electrode, which rapidly develops a potential upon the passage of current, and Fig. 8(b) illustrates a highly nonpolarizable electrode, which does not readily support a change in potential upon current flow.

Consider a metal electrode consisting of a silver wire placed inside the body, with a solution of silver ions between the wire and ECF, supporting the reaction $Ag^+ + e^- \longleftrightarrow Ag$. This is an example of an electrode of the first kind, which is defined as a metal electrode directly immersed into an electrolyte of ions of the metal's salt. As the concentration of silver ions $[Ag^+]$ decreases, the resistance of the interface increases. At very low silver ion concentrations, the Faradaic impedance Z_{faradaic} becomes very large, and the interface model shown in Fig. 3(a) reduces to a solution resistance R_S in series with the capacitance C_{dl}. Such an electrode is an ideally polarizable electrode. At very high silver concentrations, the Faradaic impedance approaches zero and the interface model of Fig. 3(a) reduces to a solution resistance in series with the Faradaic impedance Z_{faradaic}, which is approximated by the solution resistance only. Such an electrode is an ideally nonpolarizable electrode.

The example of a silver electrode placed in direct contact with the ECF, acting as an electrode of the first kind, is impractical. Silver is toxic, silver ions are not innate in the body, and any added silver ions may diffuse away. The silver wire electrode is a highly polarizable electrode since the innate silver concentration is very low and the Faradaic reaction consumes little charge; thus this configuration is not usable as a reference electrode. The equilibrium potential, given by a derivation of the Nernst equation $E_{eq} = E^{\Theta} + (RT/nF)\ \ln [Ag^+] = (59\ \text{mV/decade})\ \log [Ag^+]$ at 25°C, is poorly defined due to the low silver ion concentration. A solution to these problems is to use an electrode of the second kind, which is defined as a metal coated with a sparingly soluble metal salt. The common silver/silver chloride (Ag/AgCl) electrode, described by reactions 8a and 8b, is such an electrode. This consists of a silver electrode covered with silver chloride, which is then put in contact with the body. The Ag/AgCl electrode acts as a highly nonpolarizable electrode. The equilibrium potential $E_{eq} = E^{\Theta} + 59 \log [Ag^+]$ can be combined with the definition of the silver chloride solubility constant:

$K_s \equiv [Ag^+][Cl^-] \sim 10^{-10} M^2$, to yield $E_{eq} = E^{\Theta} + 59 \log [K_s/Cl^-] = E^{\Theta} + 59 \log K_s - 59 \log [Cl^-] = E^{\Theta\prime} - 59 \log [Cl^-]$. The equilibrium potential of this electrode of the second kind is seen to be dependent on the finite chloride concentration rather than any minimal silver concentration, and is well defined for use as a reference electrode.

2 Charge Injection Across the Electrode/Electrolyte Interface During Electrical Stimulation

2.1 Charge Injection During Pulsing: Interaction of Capacitive and Faradaic Mechanisms

As illustrated in Fig. 1, there are two primary mechanisms of charge injection from a metal electrode into an electrolyte. The first consists of charging and discharging the double-layer capacitance causing a redistribution of charge in the electrolyte, but no electron transfer from the electrode to the electrolyte. C_{dl} for a metal in aqueous solution has values on the order of 10–20 $\mu F/cm^2$ of real area (geometric area multiplied by the roughness factor). For a small enough total injected charge, all charge injection is by charging and discharging of the double layer. Above some injected charge density, a second mechanism occurs consisting of Faradaic reactions where electrons are transferred between the electrode and electrolyte, thus changing the chemical composition in the electrolyte by reduction or oxidation reactions. Figure 1 illustrates a single Faradaic impedance representing the electron transfer reaction $O + n\,e^- \longleftrightarrow R$. Generally there may be more than one Faradaic reaction possible, which is modeled by several branches of $Z_{faradaic}$ (one for each reaction), all in parallel with the double-layer capacitance. The current–overpotential equation 18 and Fick's first and second laws for diffusion give the complete description of processes occurring for any Faradaic reaction.

In addition to the double-layer capacitance, some metals have the property of pseudocapacity [8], where a Faradaic electron transfer occurs, but because the product remains bound to the electrode surface, the reactant may be recovered (the reaction may be reversed) if the direction of current is reversed. Although electron transfer occurs, in terms of the electrical model of Fig. 1, the pseudocapacitance is better modeled as a capacitor, since it is a charge-storage (not dissipative) process. Platinum is commonly used for stimulating electrodes, as it has a pseudocapacity (by reaction 6) of 210 $\mu C/cm^2$ real area [18], or equivalently 294 $\mu C/cm^2$ geometric area using a roughness factor of 4.[1]

It is a general principle when designing electrical stimulation systems that one should avoid onset of irreversible Faradaic processes, which may potentially create damaging chemical species, and keep the injected charge at a low enough level where it may be accommodated strictly by reversible charge injection processes. Unfortunately this is not always possible, because a larger injected charge may be required to cause the desired effect (e.g., initiating action potentials). Reversible processes include charging and discharging of the double-layer capacitance, reversible Faradaic processes involving products that remain bound to the surface such as plating of hydrogen atoms on platinum (reaction 6) or the reversible formation and reduction of a surface oxide (reactions 4, 5), and reversible Faradaic processes where the solution phase product remains near the electrode due to mass diffusion limitations.

When the exchange current density is very low and significant overpotentials are required for measurable Faradaic current, a relatively large total charge can be injected (and thus a relatively large overpotential achieved) through the capacitive mechanism before Faradaic reactions commence. When the exchange current density is high, little injected charge is accepted into capacitive charge, and small overpotentials are achieved, before onset of significant Faradaic reactions. The desirable paradigm for a stimulating electrode is to use either capacitive charge injection or charge injection through reversible Faradaic processes (such as reversible oxide formation), thus minimizing irreversible Faradaic reactions that lead to either electrode or tissue damage.

The net current passed by an electrode, modeled as shown in Fig. 1, is the sum of currents through the two parallel branches. The total current through the electrode is given by

$$i_{\mathrm{total}} = i_{\mathrm{C}} + i_{\mathrm{f}} \tag{27}$$

where i_{C} is the current through the capacitance and i_{f} is the current through Faradaic processes

[1] The relationship between capacitance and stored charge is given by equation 29. A one volt potential excursion applied to a double layer capacitance of 20 $\mu F/cm^2$ yields 20 $\mu C/cm^2$ stored charge, which is an order of magnitude lower than the total charge storage available from platinum pseudocapacitance.

The current through Faradaic processes is given by the current–overpotential equation 18. The current through the capacitance is given by equation (28) below.

$$i_C = C_{dl}\, dv/dt = C_{dl}\, d\eta/dt \tag{28}$$

The capacitive current depends upon the rate of potential change, but not the absolute value of the potential. The Faradaic current, however, is exponentially dependent upon the overpotential, or departure from the equilibrium potential. Thus, as an electrode is driven away from its equilibrium potential, essentially all charge initially flows through the capacitive branch since the overpotential is small near equilibrium. As the overpotential increases, the Faradaic branch begins to conduct a relatively larger fraction of the injected current. When the overpotential becomes great enough, the Faradaic impedance becomes sufficiently small that the Faradaic current equals the injected current. At this point the Faradaic process of reduction or oxidation conducts all injected charge, and the potential of the electrode does not change, corresponding to the capacitor not charging any further.

In terms of charge going into the different processes, the charge on the double-layer capacitance is proportional to the voltage across the capacitance:

$$q_C = C_{dl} \Delta V \tag{29}$$

thus if the electrode potential does not change in time, neither does the stored charge. The charge into Faradaic processes however does continue to flow for any nonzero overpotential. The Faradaic charge is the integration of Faradaic current over time, which by equation (18) is proportional to an exponential of the overpotential integrated over time:

$$q_f = \int i_f\, dt \propto \int \exp(\eta)\, dt \tag{30}$$

The charge delivered into Faradaic reactions is directly proportional to the mass of Faradaic reaction product formed, which may be potentially damaging to the tissue being stimulated or the electrode.

2.2 Methods of Controlling Charge Delivery During Pulsing

Charge injection from an electrode into an electrolyte (e.g., extracellular fluid) is commonly controlled by one of three methods. In the current-controlled (also called galvanostatic) method, a current source is attached between the working and counter electrodes and a user-defined current is passed. In the voltage-controlled (also called potentiostatic) method, current is driven between the working electrode and counter electrode as required to control the working electrode potential with respect to a third (reference) electrode. This may be used for electrochemical measurements of certain neurotransmitters [19]. This method is most often not used for stimulation and is not discussed further in this review. In the third method, $V_{WE\text{-}CE}$ control, a voltage source is applied between the working and counter electrodes. While this is the simplest method to implement, neither the potential of the working electrode nor the potential of the counter electrode (with respect to a third reference electrode) are controlled; only the net potential between the working and counter electrodes is controlled.

2.3 Charge Delivery by Current Control

The current-controlled method is commonly used for electrical stimulation of excitable tissue. This typically takes the form of pulsing. In monophasic pulsing, a constant current is passed for a period of time (generally on the order of tens to hundreds of microseconds), and then the external stimulator circuit is open-circuited (it is effectively electrically removed from the electrodes) until the next pulse. Among the different pulsing schemes, monophasic pulsing results in the greatest amount of irreversible Faradaic reaction product (detailed in the next section), which may result in tissue or electrode damage; thus it is not used in chronic stimulation. In biphasic pulsing, a constant current is passed in one direction, then the direction of current is reversed, and then the circuit is open-circuited until the next pulse. In biphasic pulsing the first phase, or stimulating phase, is used to elicit the desired physiological effect such as initiation of an action potential, and the second phase, or reversal phase, is used to reverse electrochemical processes occurring during the stimulating phase. It is common to use a cathodic pulse as the stimulating phase (the working electrode is driven negative with respect to its prepulse potential), followed by an anodic-reversal phase (the working electrode is driven positive), although anodic pulsing may also be used for stimulation (discussed in Section 4). Figure 9 illustrates definitions of the key parameters in pulsing. The frequency of stimulation is the inverse of the period or time between pulses. The interpulse interval is the period of time between pulses. Figure 9(b) illustrates charge-balanced biphasic pulsing, where the charge in the stimulation phase equals the charge in the reversal phase. Figure 9(c) illustrates charge-imbalanced biphasic pulsing (detailed in Section 4) where there are two phases, but the reversal phase has less charge than the stimulating phase. Figure 9(d) illustrates the use of an interphase delay, where an open-circuit is introduced between the stimulating and reversal phases.

Upon application of a cathodic current pulse to an electrode that starts at a potential close to the equilibrium potential, the term exp (η) is small and initially little charge goes into any Faradaic reactions, thus the initial charge delivery goes into charging the double-layer capacitance. As charge goes onto the double layer, the electrode potential moves away from equilibrium (an overpotential η develops), and the Faradaic reaction $\mathrm{O} + ne^{-} \rightarrow \mathrm{R}$ starts to consume charge, with net current density proportional to exp (η). The total injected current then goes into both capacitive current i_c, causing the electrode capacitance to continue to charge to more negative potentials, and Faradaic current i_f. At sufficiently negative potentials, another

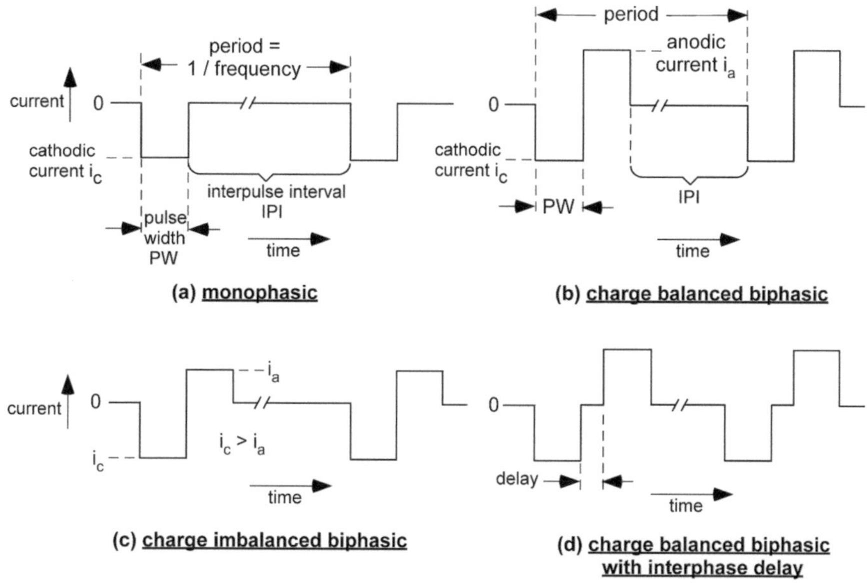

Fig. 9 Common pulse types and parameters

Faradaic reaction with a lower exchange current density (thus more irreversible) than the first may start. In the case where this second reaction is the reduction of water, the reaction will not become mass transport limited (water at 55.5 M will support substantial current), and an electrode potential will be reached where the non-mass transport limited reduction of water accepts all further injected charge. The water window is a potential range that is defined by the reduction of water in the negative direction, forming hydrogen gas, and the oxidation of water in the positive direction, forming oxygen. Because water does not become mass transport limited in an aqueous solution, the potentials where water is reduced and oxidized form lower and upper limits respectively for electrode potentials that may be attained, and any electrode driven to large enough potentials in water will evolve either hydrogen gas or oxygen gas. Upon reaching either of these limits, all further charge injection is accommodated by the reduction or oxidation of water.

2.4 Pulse train response during current control

Based on the simple electrical model of Fig. 1, one may predict different characteristics in the potential waveforms resulting from monophasic pulsing, charge-balanced biphasic pulsing, and charge-imbalanced biphasic pulsing. Consider what occurs when an electrode, starting from the open-circuit potential, is pulsed with a single cathodic pulse and then left open-circuit (illustrated in Fig. 10(a), pulse 1). Upon pulsing the electrode initially charges with injected charge being stored reversibly

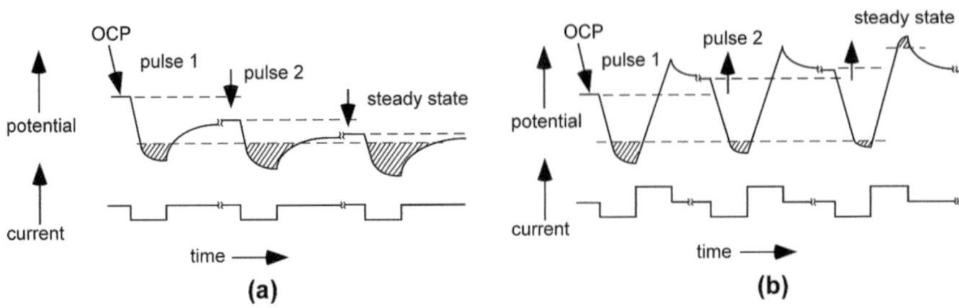

Fig. 10 Electrode Potentials in Response to Monophasic and Charge-Balanced Biphasic Pulse Trains (**a**) Ratcheting of potential during monophasic cathodic pulsing. The prepulse potential of successive pulses moves negative until all injected charge goes into irreversible processes. (**b**) Ratcheting during charge-balanced cathodic first biphasic pulsing. The prepulse potential of successive pulses moves positive until the same amount of charge is lost irreversibly during the cathodic and anodic phases. Shaded areas represent periods of irreversible reactions

on the double-layer capacitance, causing the electrode potential to move negative. As the potential continues to move negative, charge begins to be delivered into Faradaic currents (whose magnitude is an exponential function of the overpotential). At the end of the pulse when the external circuit is opened, charge on the double-layer capacitance continues to discharge through Faradaic reactions. This causes the electrode potential to move positive, and as the electrode discharges (i.e., the overpotential decreases) the Faradaic current decreases, resulting in an exponential discharge of the electrode. Given a long enough time, the electrode potential will approach the open-circuit potential. However, if the electrode is pulsed with a train whose period is short with respect to the time constant for discharge (as may occur with neural stimulation, with a period of perhaps 20 ms), i.e., if a second cathodic pulse arrives before the electrode has completely discharged, then the potential at the start of the second pulse (the prepulse potential) is more negative than the prepulse potential of the first pulse (which is the open circuit potential). Because the potential during the second pulse begins at a more negative potential than the first, a smaller fraction of the injected charge goes into reversible charging of the double-layer capacitance. The Faradaic reactions begin accepting significant charge at an earlier time than in the first pulse, and there is more charge delivered to irreversible reactions during the second pulse than during the first as the overall potential range traversed is more negative during the second pulse (Fig. 10(a), pulse 2). Upon going to open-circuit after the second pulse, the electrode discharges through Faradaic reactions. Because the potential at the end of the second current pulse is more negative than the potential at the end of the first current pulse, the potential range during discharge between pulses 2 and 3 is also more negative than between pulses 1 and 2, and likewise the prepulse potential of pulse 3 is more negative than the prepulse potential of pulse 2. This "ratcheting" of the electrode potential continues until the following condition is met:

$$\text{Unrecoverable Charge } (Q_{\text{ur}}) \text{ per pulse} = \text{Injected Charge } \left(Q_{\text{inj}}\right) \text{ per pulse} \quad (31)$$

i.e., all injected charge goes into irreversible Faradaic reactions that occur either during the pulse or during the open-circuit interpulse interval period. Charge delivered into irreversible processes is defined as unrecoverable charge Q_{ur}. Once condition (31) is met, the pulsing is in the steady state, and the potential excursions repeat themselves with each pulse cycle.

Next consider the electrode response when a charge-balanced stimulation protocol is used; cathodic then anodic, followed by open-circuit. The electrode begins from open-circuit potential. Upon applying the first cathodic pulse, the double-layer reversibly charges, and then the electrode may begin to transfer charge into Faradaic reactions as the potential moves negative. The anodic pulse then causes the electrode potential to move back positive (illustrated in Fig. 10(b), pulse 1). Unlike the exponential decay during the monophasic pulsing, the electrode potential now changes according to the anodic current and the double-layer capacitance, and there is reversal of charge from the double layer. Because not all of the injected charge during the cathodic pulse went into charging of the double layer, only some fraction of the injected cathodic charge is required in the anodic phase to bring the electrode potential back to the prepulse value. Since the anodic pulse is balanced with the cathodic pulse, the electrode potential at the end of the anodic phase of pulse 1 is positive of the prepulse potential of pulse 1 (the open circuit potential). During the anodic phase and during the open-circuit following the anodic phase, if the potential becomes sufficiently positive, anodic Faradaic reactions such as electrode corrosion may occur. During the open-circuit period, the electrode discharges exponentially through anodic Faradaic reactions back toward the open-circuit potential, moving negative with time. By the beginning of pulse 2, the potential is still positive of the prepulse potential for pulse 1 (the open-circuit potential). Thus, as long as any charge is lost irreversibly during the cathodic phase, the potential at the end of the charge-balanced anodic phase will be positive of the prepulse potential, and a ratcheting effect is seen. Unlike the monophasic case, the ratcheting of the electrode prepulse potential is now in a positive direction. Steady state occurs when one of the two following conditions is met:

(1) *There are no irreversible Faradaic reactions during either the cathodic or anodic phases, and the electrode simply charges and then discharges the double layer (the potential waveform appears as a sawtooth):*

$$Q_{\text{ur cathodic}} = Q_{\text{ur anodic}} = 0 \tag{32}$$

or

(2) *The same amount of charge is lost irreversibly during the cathodic phase and during the combined anodic phase and interpulse interval:*

$$Q_{\text{ur cathodic}} = Q_{\text{ur anodic+IPI}} \neq 0 \tag{33}$$

If irreversible processes do occur, for cathodic first charge-balanced biphasic pulsing, the electrode potential will move positive of the open-circuit potential, and during steady-state continuous pulsing there is an equal amount of unrecoverable charge delivered into cathodic- and anodic-irreversible processes.

Finally consider the electrode response when a charge-imbalanced stimulation protocol is used (not illustrated). The electrode begins from open-circuit potential. The response to the first cathodic pulse is the same as with the monophasic or charge-balanced biphasic waveforms. The anodic phase then causes the electrode potential to move back positive, but since there is less charge in the anodic phase than cathodic, the electrode potential does not move as far positive as it did with the charge-balanced biphasic waveform. The potential at the end of the anodic phase will be closer to the open-circuit potential than during charge-balanced pulsing. The maximum positive potential will be less when using the charge-imbalanced waveform than when using the charge-balanced waveform. This has the advantage that charge delivered into anodic Faradaic processes such as metal corrosion is reduced with respect to charge-balanced stimulation. The prepulse potential will move under factors as explained for the monophasic and charge-balanced biphasic waveforms until the following condition is met:

The net imbalance in injected charge is equal to the net difference in unrecoverable charge between the cathodic phase and the combined anodic phase and interpulse interval:

$$(Q_{\text{inj cathodic}} - Q_{\text{inj anodic}}) \equiv Q_{\text{imbal}} = (Q_{\text{ur cathodic}} - Q_{\text{ur anodic}+IPI}) \quad (34)$$

During charge-imbalanced stimulation, the shift in prepulse potential may be either positive or negative of the open-circuit potential depending on the amount of imbalance.

Based on these considerations, monophasic pulsing causes the greatest shift of the electrode potential during pulsing away from the equilibrium potential, thus causes the most accumulation of unrecoverable charge (corresponding to products of irreversible Faradaic reactions) of the three protocol types (monophasic, charge-balanced biphasic, charge-imbalanced biphasic). Furthermore, since during monophasic pulsing the electrode potential is not brought back toward the equilibrium potential by an anodic phase, there is accumulation of unrecoverable charge during the open-circuit interpulse interval.

2.5 Electrochemical reversal

The purpose of the reversal phase during biphasic stimulation is to reverse the direction of electrochemical processes that occurred during the stimulating phase, minimizing unrecoverable charge. A reversible process is one where the reactants are reformed from the products upon reversing the direction of current. Upon delivering current in the stimulation phase and then reversing the direction of current,

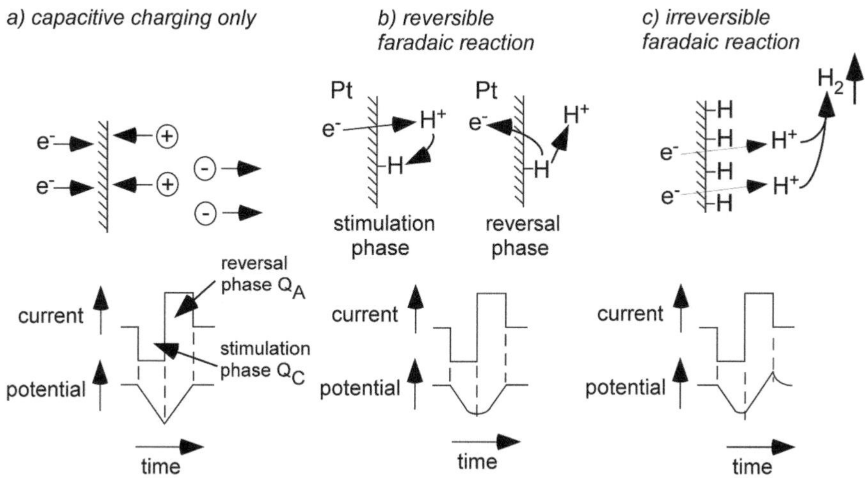

Fig. 11 Electrochemical Processes and Potential Waveforms During Charge-Balanced Stimulation (**a**) capacitive charging only (**b**) reversible hydrogen plating (**c**) irreversible hydrogen evolution

charge on the electrode capacitance will discharge, returning the electrode potential toward its prepulse value. If only double-layer charging occurred, then upon passing an amount of charge in the reversal phase equal to the charge delivered in the stimulation phase (a charge-balanced protocol), the electrode potential will return precisely to its prepulse potential by the end of the reversal phase and the potential curve will be a simple sawtooth as shown in Fig. 11(a) (corrected for solution resistance). If reversible Faradaic reactions occur during the stimulation phase, then charge in the reversal or secondary phase may go into reversing these reactions. Figure 11(b) illustrates an example reversible Faradaic process; in this case, charging of the pseudocapacitance (reduction of protons and plating of monatomic hydrogen onto the metal electrode surface) as may occur on platinum. During reversal, the plated hydrogen is oxidized back to protons. Because the electrochemical process occurring during the reversal phase is the exact opposite of that occurring during the stimulation phase, there is zero net accumulation of electrochemical species. Reversible Faradaic reactions include adsorption processes as in Fig. 11(b), as well as processes where the solution-phase product remains near the electrode due to mass-diffusion limitations. If irreversible Faradaic reactions occur, upon passing current in the reverse direction, reversal of electrochemical product does not occur as the product is no longer available for reversal (it has diffused away). An example shown in Fig. 11(c) is the formation of hydrogen gas after a monolayer of hydrogen atoms has been adsorbed onto the platinum surface. In the case where a Faradaic reaction has occurred during the stimulation phase, the potential waveform during the stimulation phase is not linear, but displays a slope inflection as Faradaic processes consume charge (this is charge that does *not* charge the capacitance, thus does not change the electrode potential). If irreversible Faradaic reactions occur, then when an equal amount of charge is passed in the reversal phase, the electrode potential goes positive of the prepulse potential. To return the electrode potential exactly to its prepulse value would require that the charge in the reversal phase be equal to only the amount of charge that went onto the capacitance during the stimulation phase (a charge-imbalanced waveform).

The use of biphasic stimulation (either charge balanced or charge imbalanced) moves the electrode potential out of the most negative ranges immediately after stimulation. In comparison (as shown in Fig. 10), the monophasic-stimulation protocol allows the electrode potential to remain relatively negative during the interpulse interval, and during this time Faradaic reduction reactions may continue. In the presence of oxygen, these reactions may include reduction of oxygen and formation of reactive oxygen species, which have been implicated in tissue damage [20, 21, 22, 23, 24]. The charge-imbalanced waveform has the added advantage that the electrode potential at the end of the anodic pulse is less positive than with charge-balanced biphasic pulsing, thus less charge goes into irreversible oxidation reactions such as corrosion when using the charge-imbalanced protocol. Charge-imbalanced biphasic waveforms provide a method to reduce unrecoverable charge in the cathodic direction (with respect to monophasic stimulation) and in the anodic direction (with respect to charge-balanced biphasic stimulation), thus are an attractive solution to minimizing damage to either the stimulated tissue or the metal electrode.

2.6 Charge delivery by a voltage source between the working electrode and counter electrode

An alternative form of charge injection involves the direct connection of a voltage source between the working and counter electrodes. Figure 12 compares the current, working electrode to reference electrode voltage ($V_{WE\text{-}RE}$), and working electrode to counter electrode voltage ($V_{WE\text{-}CE}$) waveforms during monophasic pulsing under current control versus $V_{WE\text{-}CE}$ control. The $V_{WE\text{-}RE}$ waveforms represent the working electrode interfacial potentials and do not imply that a reference electrode is required for either control scheme. Upon applying a voltage pulse with amplitude V_{app} between the working electrode and counter electrode in $V_{WE\text{-}CE}$ control, the current is at its maximum value at the beginning of the pulse as the double-layer capacitances of the two electrodes charge and the current is predominantly capacitive. Given a long duration pulse, the current will asymptotically approach a value where V_{app} maintains a steady-state Faradaic current, with current density given by equation (18). Figure 12 illustrates the steady-state waveforms when using an exhausting circuit [25, 26], where at the end of the monophasic voltage pulse the working and counter electrodes are shorted together, causing the charge on the working electrode capacitance to rapidly discharge, and the working electrode potential to attain the counter electrode potential. If the counter electrode is sufficiently large, its potential will not be notably perturbed away from its equilibrium potential during pulsing, and upon shorting the working electrode to the counter electrode, the

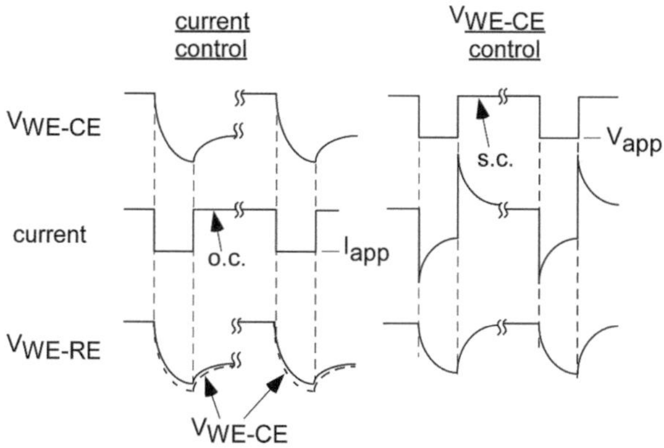

Fig. 12 Steady-State Voltage and Current Waveforms Using Current Control and $V_{WE\text{-}CE}$ Control Note the rapid discharge of the working electrode during the short-circuit interpulse interval of $V_{WE\text{-}CE}$ control relative to the open-circuit interpulse interval of current control. s.c.=short circuit, o.c.=open circuit, I_{app}=applied current, V_{app}=applied voltage

working electrode potential will be brought back to the counter electrode equilibrium potential. Donaldson [26] showed that during cathodic-monophasic pulsing of real electrodes with an exhausting scheme, the potentials of both the working electrode and counter electrode moved positive in response to a continuous train, increasing the risk of oxidizing reactions such as corrosion. The discharge of the working electrode is relatively rapid during $V_{WE\text{-}CE}$ control with an exhausting circuit, as the working electrode is directly shorted to the counter electrode. This is contrasted by the relatively slow discharge using monophasic current control, as shown in Fig. 12, with an open circuit during the interpulse interval. During the open-circuit period, the working-electrode capacitance discharges through Faradaic reactions at the working-electrode interface. This leads to a greater accumulation of unrecoverable charge during the open-circuit interpulse interval (current control) than with the short-circuit interpulse interval ($V_{WE\text{-}CE}$ control). However, in current control, appropriate biphasic pulsing waveforms (Fig. 9) can promote rapid electrode discharge.

Advantages of the $V_{WE\text{-}CE}$ control scheme over the current control scheme include (1) the circuitry is simpler (it may be a battery and an electronic switch); and (2) unrecoverable charge accumulation is lower during the interpulse interval than it would be with monophasic current control. Disadvantages of the $V_{WE\text{-}CE}$ control scheme include (1) maximum stimulation of excitable tissue occurs only at the beginning of the pulse when current is maximum, and stimulation efficiency decreases throughout the pulse as current decreases, whereas with current control the current is constant throughout the pulse; (2) an increase in resistance anywhere in the electrical conduction path will cause an additional voltage drop, decreasing the current and potentially causing it to be insufficient for stimulation, whereas with current control the current is constant (assuming the required voltage is within the range of the stimulator); and (3) neither the current driven nor the charge injected are under direct control using voltage control [27]. Because the level of neuronal membrane depolarization is related to the applied current, these factors result in a reduction in reproducibility between experiments, as well as between clinical implants, during $V_{WE\text{-}CE}$ control. Moreover, because tissue properties can change over time, stimulation efficacy may change when using $V_{WE\text{-}CE}$ control.

3 Materials Used as Electrodes for Charge Injection and Reversible Charge Storage Capacity

The ideal material for use as a stimulating electrode satisfies the following six requirements. (1) The passive (unstimulated) material must be biocompatible, so it should not induce a toxic or necrotic response in the adjacent tissue, nor an excessive foreign body or immune response. (2) The material must be mechanically acceptable for the application. It must maintain mechanical integrity given the intended tissue, surgical procedure, and duration of use. The material must not buckle if it is to pass through the meninges. If a device is to be used chronically, it must be flexible enough to withstand any small movement between the device and tissue following implantation. (3) The complete device must be efficacious. This requires that sufficient charge can be injected with the chosen material and electrode area to elicit action potentials. The required charge is quantified by the charge-duration curve, discussed in Section 4. (4) During electrical stimulation, Faradaic reactions should not occur at levels that are toxic to the surrounding tissue. The level of reaction product that is tolerated may be significantly higher for acute stimulation than chronic stimulation. (5) During electrical stimulation, Faradaic corrosion reactions should not occur at levels that will cause premature failure of the electrode. This again depends greatly on the intended duration of use. During acute stimulation, corrosion is rarely a concern, whereas a device that is intended for a 30-year implant must have a very low corrosion rate. (6) The material characteristics must be acceptably stable for the duration of the implant. For a chronic electrode, the device electrical impedance must be stable. The conducting and insulating properties of all materials must remain intact.

Dymond et al. [28] tested the toxicity of several metals implanted into the cat cerebral cortex for 2 months. Materials were deemed toxic if the reaction to the implanted metal was significantly greater than the reaction to a puncture made from the same metal that was immediately withdrawn (Table 1). Stensaas and Stensaas [29] reported on the biocompatibility of several materials implanted passively into the rabbit cerebral cortex (Table 1). Materials were classified into one of three categories depending upon changes occurring at the implant/cortex interface: (1) Nonreactive. For these materials, little or no gliosis occurred, and normal CNS tissue with synapses was observed within 5 μm of the interface. (2) Reactive. Multinucleate giant cells and a thin layer (10 μm) of connective tissue surrounded the implant. Outside of this was a zone of astrocytosis. Normal CNS tissue was observed within 50 μm of the implant. (3) Toxic. These materials are separated from the cortical tissue by a capsule of cellular connective tissue and a surrounding zone

Table 1 Classification of material biocompatibility

	Classification by Dymond	Classification by Stensaas and Stensaas	Other references
Conductors:			
Aluminum		Non-reactive	
Cobalt		Toxic	
Copper		Toxic	Toxic [34, 35, 36]
Gold	Nontoxic	Non-reactive	
Gold-nickel-chromium	Nontoxic		
Gold-palladium-rhodium	Nontoxic		
Iron		Toxic	
Molybdenum		Reactive	
Nickel-chromium (Nichrome)		Reactive	Nontoxic [34]
Nickel-chromium-molybdenum	Nontoxic		
Nickel-titanium (Nitinol)			Biocompatible [37, 38]
Platinum	Nontoxic	Non-reactive	Biocompatible [31, 32]
Platinum-iridium	Nontoxic		Biocompatible [33]
Platinum-nickel	Nontoxic		
Platinum-rhodium	Nontoxic		
Platinum-tungsten	Nontoxic		
Platinized platinum (Pt black)	Nontoxic		
Rhenium	Nontoxic		
Silver	Toxic	Toxic	Toxic [34, 35, 36]
Stainless steel	Nontoxic		Nontoxic [34]
Tantalum		Reactive	
Titanium			Biocompatible [32]
Tungsten		Non-reactove	
Insulators:			
Alumina ceramic		Non-reactive	Biocompatible [32]
Araldite (epoxy plastic resin)		Reactive	
Polyethylene		Non-reactive	
Polyimide			Biocompatible [154]
Polypropylene		Non-reactive	
Silastic RTV	Toxic		
Silicon dioxide (Pyrex)		Reactive	
Teflon TFE (high purity)		Non-reactive	
Teflon TFE (shrinkable)		Reactive	
Titanium dioxide		Reactive	
Semiconductors:			
Germanium		Toxic	
Silicon		Non-reactive	Biocompatible [40, 155, 156]
Assemblies			
Gold-silicon dioxide passivated microcircuit		Reactive	

of astrocytosis. Loeb [30] studied the histological response to materials used by the microelectronics industry implanted chronically in the subdural space of cats, and found reactions to be quite dependent on specific material formulations and surface preparations.

Platinum has been demonstrated as biocompatible for use in an epiretinal array [31] and in cochlear implants [32]. Both titanium and ceramic [32] and platinum-iridium wire [33] have been shown as biocompatible in cochlear implants. Babb and Kupfer [34] have shown stainless steel and nickel-chromium (Nichrome) to be nontoxic. Copper and silver are unacceptable as stimulating electrodes, as these metals cause tissue necrosis even in the absence of current [28, 29, 34, 35, 36]. Nickel-titanium shape memory alloys have good biocompatibility response [37], up to a nickel content of 50% [38].

The first intracortical electrodes consisted of single-site conductive microelectrodes made of material stiff enough to penetrate the meninges, as either an insulated metallic wire or a glass pipette filled with conductive electrolyte. Advances in materials science and microelectronics technology have allowed the development of multiple-site electrodes built onto a single substrate using planar photolithographic and silicon micromachining technologies. Such devices have been made from silicon [39, 40] and polyimide [41]. In further advancements, bioactive components have been added to the electrode to direct neurite growth toward the electrode, minimizing the distance between the electrode and stimulated tissue [42, 43, 44].

Chronic implantation of any device into the central nervous system, even those materials considered biocompatible, elicits a common response consisting of encapsulation by macrophages, microglia, astrocytes, fibroblasts, endothelia, and meningeal cells [45]. The early response to material implantation is inflammation [29, 45, 46]. The chronic response is characterized by a hypertrophy of the surrounding astrocytes [29], which display elevated expression of intermediate filament proteins such as GFAP and vimentin [47], an infiltration of microglia and foreign body giant cells [29], and a thickening of the surrounding tissue that forms a capsule around the device [40, 46].

The reversible charge-storage capacity (CSC) of an electrode, also known as the reversible charge injection limit [48], is the total amount of charge that may be stored reversibly, including storage in the double-layer capacitance, pseudocapacitance, or any reversible Faradaic reaction. In electrical stimulation of excitable tissue, it is desirable to have a large reversible charge-storage capacity so that a relatively large amount of charge may be injected (thus being efficacious for stimulation) prior to the onset of irreversible Faradaic reactions (which may be deleterious to the tissue being stimulated or to the electrode itself). The reversible charge-storage capacity depends upon the material used for the electrode, the size and shape of the electrode, the electrolyte composition, and parameters of the electrical stimulation waveform.

The slow cyclic voltammogram for a material is a graphic display of the current density into various electrochemical processes as a function of the electrode potential as the potential is slowly cycled. At any point in time, the amount of current going into a particular process is determined by the potential as well as by the reactant concentration, as given by equation (18). The water window is defined as the potential region between the oxidation of water to form oxygen and the reduction of water to form hydrogen. Because water does not become mass transport limited in an aqueous solution, once the electrode potential attains either of these two water-window boundaries, all further injected charge goes into the irreversible processes of water oxidation (anodically) or water reduction (cathodically). In many studies, the reversible charge-storage capacity has been defined as the maximum charge density that can be applied without the electrode potential exceeding the water window during pulsing. It should be noted that in fact irreversible processes might occur at potentials within the water window, including such reactions as irreversible oxygen reduction [49, 50] that may become mass transport limited.

The noble metals, including platinum Pt, gold Au, iridium Ir, palladium Pd, and rhodium Rh, have been commonly used for electrical stimulation, largely due to their relative resistance to corrosion [28, 51, 52]. These noble metals do exhibit some corrosion during electrical stimulation, as shown by dissolution [53, 54, 55, 56, 57] and the presence of metal in the neighboring tissue [58, 59]. In addition to corrosion of the electrode, there is evidence of long-term toxic effects on the tissue from dissolution [60, 61, 62].

Platinum and platinum-iridium alloys are common materials used for electrical stimulation of excitable tissue. Brummer and Turner [63, 64, 65, 66] have reported on the electrochemical processes of charge injection using a platinum electrode. They reported that three processes could store charge reversibly, including charging of the double-layer capacitance, hydrogen atom plating and oxidation (pseudocapacity, reaction 6), and reversible oxide formation and reduction on the electrode surface, and that 300–350 $\mu C/cm^2$ (real area) could theoretically be stored reversibly by these processes in artificial cerebrospinal fluid (equivalently 420–490 $\mu C/cm^2$ (geometric area)). This is a maximum reversible charge-storage capacity under optimum conditions, including relatively long pulse widths (>0.6 ms). Rose and Robblee [67] reported on the charge-injection limits for a platinum electrode using 200 μs charge-balanced biphasic pulses. The reversible charge-injection limit was defined as the maximum charge density that could be applied without the electrode potential exceeding the water window during pulsing. The authors determined the charge-injection limit to be 50–100 $\mu C/cm^2$ (geometric) using anodic first pulses, and 100–150 $\mu C/cm^2$ (geometric) using cathodic first pulses. These values are considerably lower than the theoretical values determined by Brummer and Turner [66], since the electrode potential at the beginning of a pulse begins somewhere intermediate to oxygen and hydrogen evolution and not all of the three reversible processes accommodate charge during the stimulating pulse. Dissolution of platinum in saline increases linearly with the injected charge during biphasic stimulation [55]. Anodic first pulses cause more dissolution than cathodic first pulses, as the electrode potential attains more positive values during the stimulating (first) phase. Robblee et al. [56] have shown that in the presence of protein such as serum albumin, the dissolution rate of platinum decreases by an order of magnitude.

The reversible charge-storage capacity is dependent upon the electrode real surface area and geometry. The geometric area of an electrode is usually easily calculated, but the real area is the value that determines the total charge capacity. Brummer and Turner [65] have reported on a method to experimentally determine the real area of a platinum electrode in vitro, however this may not be applicable to the in vivo situation. It should also be noted that the real area of an electrode may change during the course of stimulation. A nonuniform (nonspherical) electrode geometry will cause a nonuniform current density [68] with maximum current at the electrode edges, which may lead to localized electrode corrosion [69] or tissue burns [70] at the electrode edges.

Platinum is a relatively soft material and may not be mechanically acceptable for all stimulation applications. Platinum is often alloyed with iridium to increase the mechanical strength. Alloys of platinum with 10–30% iridium have similar charge-storage capacity to pure platinum [57]. Iridium is a much harder metal than platinum, with mechanical properties that make it suitable as an intracortical electrode. The reversible charge-storage capacities of bare iridium or rhodium are similar to that of platinum. However, when a surface oxide is present on either of these materials, they have greatly increased charge-storage capacity over platinum. These electrodes inject charge using valency changes between two oxide states, without a complete reduction of the oxide layer.

Iridium oxide is a popular material for stimulation and recording, using reversible conversion between Ir^{3+} and Ir^{4+} states within an oxide to achieve high reversible charge-storage capacity. Iridium oxide is commonly formed from iridium metal in aqueous electrolyte by electrochemical activation (known as anodic iridium oxide films on bulk iridium metal, or AIROF), which consists of repetitive potential cycling of iridium to produce a multilayered oxide [10, 48, 57, 71, 72]. Such activated iridium oxide films have been used for intracortical stimulation and recording using iridium wire [73, 74, 75, 76, 77] or with micromachined silicon electrodes using sputtered iridium on the electrode sites [78, 79]. The maximum charge density that can be applied without the electrode potential exceeding the water window was reported for activated iridium oxide using 200 μs charge-balanced pulses as +/−2 mC/cm^2 (geometric) for anodic first pulsing and −/+1 mC/cm^2 for cathodic first [80, 81]. By using an anodic bias, cathodic charge densities of 3.5 mC/cm^2 (geometric) have been demonstrated both in vitro [80, 81] and in vivo [82]. Iridium oxide films can also be formed by thermal decomposition of an iridium salt onto a metal substrate (known as thermally prepared iridium oxide films, or TIROF) [83], or by reactive sputtering of iridium onto a metal substrate (known as sputtered iridium oxide films, or SIROF) [84]. Meyer and Cogan [77] reported on a method to electrodeposit iridium oxide films onto substrates of gold, platinum, platinum-iridium, and 316LVM stainless steel achieving reversible charge-storage capacities of >25 mC/cm^2.

The stainless steels (types 303, 316, and 316LVM) as well as the cobalt-nickel-chromium-molybdenum alloy MP35N are protected from corrosion by a thin passivation layer that develops when exposed to atmospheric oxygen and which forms a barrier to further reaction. In the case of stainless steel, this layer consists of iron oxides, iron hydroxides, and chromium oxides. These metals inject charge by reversible oxidation and reduction of the passivation layers. A possible problem with these metals is that if the electrode potential becomes too positive (the transpassive region), breakdown of the passivation layer and irreversible metal dissolution may occur at an unacceptable rate [51, 85, 86], potentially leading to failure of the electrode. A cathodic charge imbalance has been shown to allow significantly increased charge injection without electrode corrosion [87, 88]. Titanium and cobalt-chromium alloys are also protected from corrosion by a surface oxide passivation layer and demonstrate better corrosion resistance than does stainless steel [89]. 316LVM stainless steel has good mechanical properties and has been used for intramuscular electrodes. The charge-storage capacity of 316LVM is only 40–50 $\mu C/cm^2$ (geometric), potentially necessitating large surface area electrodes.

Capacitor electrodes inject charge strictly by capacitive action, as a dielectric material separates the metal electrode from the electrolyte preventing Faradaic reactions at the interface [90, 91, 92]. The tantalum/tantalum pentoxide (Ta/Ta_2O_5) electrode has a high charge-storage capacity achieved by using sintered tantalum or electrolytically etched tantalum wire to increase the surface area [93]. Guyton and Hambrecht [90, 91] have demonstrated a sintered Ta/Ta_2O_5 electrode with a charge-storage capacity of 700 $\mu C/cm^2$ (geometric). The Ta/Ta_2O_5 electrodes have sufficient charge-storage capacity for electrodes in the range of 0.05 cm^2 and charge densities up to 200 $\mu C/cm^2$ (geometric); however, they may not be acceptable for microelectrode applications where the required charge densities may exceed 1 mC/cm^2 [92]. Tantalum capacitor electrodes must operate at a relatively positive potential to prevent electron transfer across the oxide. If pulsed cathodically, a positive bias must be used on the electrode.

Stimulation of muscle, peripheral nerve, or cortical surface requires relatively high charge per pulse (on the order of 0.2–5 μC), thus platinum or stainless steel electrodes must be of fairly large surface area to stay within the reversible charge-storage capacity. Intracortical stimulation requires much less total charge per pulse; however, in order to achieve selective stimulation, the electrode size must be very small, resulting in high charge-density requirements. With a geometric surface area of $20 \times 10^{-6} cm^2$, the charge per pulse may be on the order of 0.008–0.064 μC yielding a charge density of 400–3200 $\mu C/cm^2$ [82, 94]. Such high charge densities may be achieved using iridium oxide electrodes with anodic pulses, or cathodic pulses with an anodic bias.

Table 2 lists several parameters of interest for materials commonly used for stimulation.

4 Charge Injection for Extracellular Stimulation of Excitable Tissue

The goal of electrical stimulation of excitable tissue is often the triggering of action potentials in axons, which requires the artificial depolarization of some portion of the axon membrane to threshold. In the process of extracellular stimulation, the extracellular region is driven to relatively more negative potentials, equivalent to driving the intracellular compartment of a cell to relatively more positive potentials. Charge is transferred across the membrane due to both passive (capacitive and resistive) membrane properties as well as through active ion channels [95]. The process

Table 2 Reversible charge-storage capacity and other parameters in electrode material selection

	Reversible charge Storage capacity ($\mu C/cm^2$)	Reversible charge injection processes	Corrosion characteristics	Mechanical characteristics
Platinum				
AF, 200 μs:	300–350 r [66]	double layer charging, hydrogen atom plating, and oxide formation and reduction	relatively resistant; greatly increased resistance with protein	relatively soft
CF, 200 μs:	50–100 g [67] 100–150 g [67]			
Platinum/Iridium Alloys	Similar CSC to Pt			stronger than Pt
Iridium	Similar CSC to Pt			stronger than Pt
Iridium Oxide	AF: +/− 2200 g [80, 81] CF: −/+ 1200 g [80, 81] AB: −/+ 3500 g [80, 81, 82]	Oxide valency charges	highly resistant [57, 82]	
316LVM Stainless Steel	40–50 g	passive film formation and reduction	resistant in passive region; rapid breakdown in transpassive region	strong and flexible
Tantalum/ Tantalum Pentoxide	700 g [90, 91] 200 g [92]	capacitive only	corrosion resistant [93, 157, 158, 159]	

r = real area, g = geometric area, AF = anodic first, charge-balanced CF = cathodic first, charge-balanced AB = cathodic first, charge balanced, with anodic bias.

of physiological action-potential generation is well reviewed in the literature (in particular, see Principles of Neural Science by Kandel, Schwartz, and Jessell, 2000) [96]), and models have been proposed [97, 98] for mammalian myelinated axons in terms of the parameters “m” and “h” as defined by Hodgkin and Huxley [99, 100, 101, 102] in their studies of the squid giant axon.

The mechanisms underlying electrical excitation of nerve have been reviewed elsewhere [1, 103, 104, 105, 106]. In the simplest case of stimulation, a monopolar electrode (a single current carrying conductor) is placed in the vicinity of excitable tissue. Current passes from the electrode, through the extracellular fluid surrounding the tissue of interest, and ultimately to a distant counter electrode. For a current I (in amps) flowing through the monopolar electrode located a distance r away from a segment of excitable tissue, and uniform conductivity in the fluid of σ (S/m), the extracellular potential V_e at the tissue is

$$V_e = \frac{I}{4\pi\sigma r} \tag{35}$$

Bipolar and other electrode configurations have more complex voltage and current patterns and will not be discussed here. Durand [106] has reviewed solutions for electrical-potential profiles of various systems.

During current-controlled stimulation, the current is constant throughout the period of the pulse; thus the V_e at any point in space is constant during the pulse. During $V_{WE\text{-}CE}$ control, current is not constant throughout the period of the pulse (Fig. 12) and the V_e at any point decreases during the pulse.

The electric field generated by a monopolar electrode will interact with an axon membrane (these principles may be generalized to any excitable tissue). During cathodic stimulation, the negative charge of the working electrode causes a redistribution of charge on the axon membrane, with negative charge collecting on the outside of the membrane underneath the cathode (depolarizing the membrane). Associated with the depolarization of the membrane under the cathode is movement of positive charge intracellularly from the distant axon to the region under the electrode, and hyperpolarization of the membrane at a distance away from the electrode. If the electrode is instead driven as an anode (to more positive potentials), hyperpolarization occurs under the anode, and depolarization occurs at a distance away from the anode. During such anodic stimulation, action potentials may be initiated at the regions distant from the electrode where depolarization occurs, known as virtual cathodes. The depolarization that occurs with anodic stimulation is roughly 1/7–1/3 that of the depolarization with cathodic stimulation; thus cathodic stimulation requires less current to bring an axon to threshold. During cathodic stimulation, anodic surround block may occur at sufficiently high current levels where the hyperpolarized regions of the axon distant from the cathode may suppress an action potential that has been initiated near the electrode. This effect is observed at higher current levels than the threshold values required for initiation of action potentials with cathodic stimulation.

In a mammalian axon, hyperpolarizing with a pulse that is long compared with the time constant of the sodium inactivation gate will remove the normal partial inactivation. If the hyperpolarizing current is then abruptly terminated (as with a rectangular pulse), the sodium activation gate conductance increases back to the rest value relatively quickly, but the activity of the slower inactivation gate remains high for a period of milliseconds; thus the net sodium conductance is briefly higher than normal and an action potential may be initiated. This phenomenon, known as anodic break, may be observed with either cathodic or anodic stimulation, since both cause some region of hyperpolarization in the axon. Anodic break may be prevented by using stimulating waveforms with slowly decaying exponential phases instead of abrupt terminations [107, 108, 109, 110].

Prolonged subthreshold stimuli can produce the phenomenon of accommodation. A long-duration cathodic pulse to mammalian axon that produces subthreshold depolarization will increase sodium inactivation, reducing the number of axons that can be recruited and so increasing the threshold. This is not a problem with brief pulses that are shorter than the time constant of sodium channel inactivation, but can be with more prolonged pulses.

It is often desirable to have some degree of selectivity during electrical excitation of tissue. Selectivity is the ability to activate one population of neurons without activating a neighboring population. Spatial selectivity is the ability to activate a localized group of neurons, such as restricting activation to a certain fascicle or fascicles within a nerve trunk. Changes in the transmembrane potential due to electrical excitation are greatest in fibers closest to the stimulating electrode because the induced extracellular potential decreases in amplitude with distance from the stimulation electrode (equation 35), as does the second spatial derivative of the extracellular potential, which is responsible for excitation [111]. Thus, activation of neurons closest to the electrode requires the least current. As the distance between the electrode and desired population of neurons for activation increases, larger currents are required, which generally means neurons between the electrode and desired population are also activated. Fiber diameter selectivity is the ability to activate fibers within a certain range of diameters only. Fibers with greater internodal distance and larger diameter experience greater changes in the transmembrane potential due to electrical excitation [112]. Using conventional electrical stimulation waveforms with relatively narrow pulses, the largest diameter fibers are activated at the lowest stimulus amplitude. In motor nerves, activating large-diameter fibers first corresponds to activating the largest motor units first. This recruitment order is opposite of the physiological case where the smallest motor units are recruited first. Fang and Mortimer [110] have demonstrated a waveform that allows a propagated action potential in small-diameter fibers but not large-diameter fibers. Hyperpolarizing pulses have a greater effect on larger fibers than smaller, just as for depolarizing pulses. This means that sustained hyperpolarization can be used to block action-potential initiation selectively in the large fibers, so that the corresponding depolarizing stimuli can selectively activate small fibers. Electrical-stimulation protocols have also been developed [113] for triggering of action potentials in specific cell types (e.g., interneurons) and structures (e.g., nerve terminals).

The relationship between the strength (current) of an applied constant current pulse required to initiate an action potential and the duration of the pulse, known as the strength–duration curve, is shown in Fig. 13(a). The threshold current I_{th} decreases with increasing pulse width. At very long pulse widths, the current is

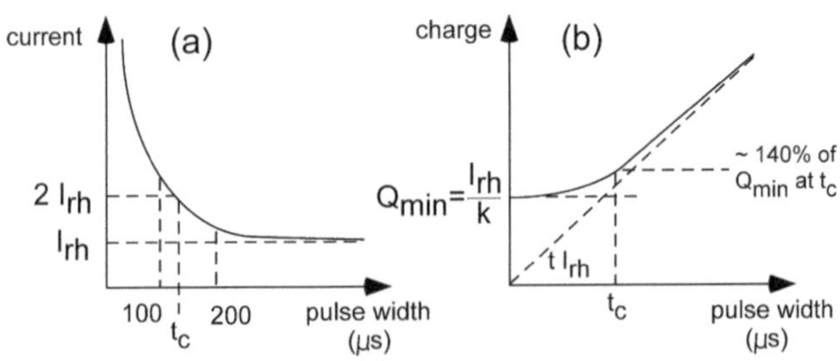

Fig. 13 Strength–Duration and Charge–Duration Curves for Initiation of an Action Potential Rheobase current I_{rh} is the current required when using an infinitely long pulse width. Chronaxie time t_c is the pulse width corresponding to two times the rheobase current

a minimum, called the rheobase current I_{rh}. The following relationship has been derived experimentally to quantify the strength–duration curve [114]:

$$I_{th} = \frac{I_{rh}}{1 - \exp\left(-W/\tau_m\right)} \tag{36}$$

where I_{th} is the current required to reach threshold, I_{rh} is the rheobase current, W is the pulse width, and τ_m is the membrane time constant. The qualitative nature of the strength–duration curve shown is representative of typical excitable tissue. The quantitative aspects, e.g., the rheobase current, depend upon factors such as the distance between the neuron population of interest and the electrode and are determined empirically. Figure 13(b) illustrates the charge–duration curve, which plots the threshold charge $Q_{th} = I_{th}W$ versus pulse width. At longer pulse widths, the required charge to elicit an action potential increases due to two phenomena. First, over a period of tens to hundreds of μs, charge is redistributed through the length of the axon and does not all participate in changing the transmembrane potential at the site of injection [115, 116]. Second, over a period of several ms, accommodation (increased sodium inactivation) occurs. The minimum charge Q_{min} occurs as the pulse width approaches zero. In practice, the Q_{th} is near Q_{min} when narrow pulses are used (tens of microseconds).

It is generally best to keep the pulse width narrow in order to minimize any electrochemical reactions occurring on the electrode surface. The narrowness of a pulse is often limited by the amount of current that can be delivered by a stimulator, especially if it is battery operated. Furthermore, some kinds of stimulation, such as selective activation of certain axons of a nerve, require pulses longer than tens of microseconds.

5 Mechanisms of Damage

An improperly designed electrical stimulation system may cause damage to the tissue being stimulated or damage to the electrode itself. Damage to an electrode can occur in the form of corrosion if the electrode is driven anodically such that the electrode potential exceeds a value where significant metal oxidation occurs. An example of such a reaction is the corrosion of platinum in a chloride-containing medium such as extracellular fluid, equation (10). Corrosion is an irreversible Faradaic process. It may be due to dissolution where the electrochemical product goes into solution or the product may form an outer solid layer on a passivation film that cannot be recovered. Charge-balanced waveforms (Fig. 9(b)) are more likely to reach potentials where corrosion may occur during the anodic-reversal phase and the open-circuit interpulse interval than are monophasic waveforms (Fig. 10). The charge-imbalanced waveform (Fig. 9(c)) has advantages both in preventing tissue damage due to sustained negative potentials during the interpulse interval, and in preventing corrosion by reducing the maximum positive potential during the anodic-reversal phase (Section 2).

The mechanisms for stimulation-induced tissue damage are not well understood. Two major classes of mechanisms have been proposed. The first is that tissue damage is caused by intrinsic biological processes as excitable tissue is overstimulated. This is called the mass-action theory and proposes that damage occurs from the induced hyperactivity of many neurons firing, or neurons firing for an extended period of time, thus changing the local environment. Proposed mass-action mechanisms include depletion of oxygen or glucose, or changes in ionic concentrations both intracellularly and extracellularly, e.g., an increase in extracellular potassium. In the CNS, excessive release of excitatory neurotransmitters such as glutamate may cause excitotoxicity. The second proposed mechanism for tissue damage is the creation of toxic electrochemical reaction products at the electrode surface during cathodic stimulation at a rate greater than that which can be tolerated by the physiological system.

McCreery et al. [117] have shown that both charge per phase and charge density are important factors in determining neuronal damage to cat cerebral cortex. In terms of the mass-action theory of damage, charge per phase determines the total volume within which neurons are excited, and the charge density determines the proportion of neurons close to an electrode that are excited; thus both factors determine the total change in the extracellular environment. The McCreery data show that as the charge per phase increases the charge density for safe stimulation decreases. When the total charge is small (as with a microelectrode) a relatively large charge density may safely be used. Tissue damage that has been attributed to mass-action effects may be alternatively explained by electrochemical means, as charge and charge density may influence the quantity of irreversible reaction products being generated at the electrode interface. Shannon [118] reprocessed the McCreery data and developed an expression for the maximum safe level for stimulation, given by

$$\log (Q/A) = k - \log (Q) \tag{37}$$

where Q is charge per phase (μC/phase), Q/A is charge density per phase (μC/cm^2/phase), and $2.0 > k > 1.5$, fit to the empirical data.

Figure 14 illustrates the charge vs. charge density relationship of equation (37) using k values of 1.7, 1.85, and 2.0, with histological data from the 1990 McCreery study using cat parietal cortex as well as data from Yuen et al. [119] on cat parietal cortex, Agnew et al. [120] on cat peroneal nerve, and Bhargava [121] on cat sacral anterior roots. Above the threshold for damage, experimental data demonstrate tissue damage, and below the threshold line, experimental data indicate no damage.

McCreery et al. [122] have reviewed damage from electrical stimulation of peripheral nerve. They concluded that damage may be from mechanical constriction of the nerve as well as neuronal hyperactivity and irreversible reactions at the electrode.

Supporting the concept that damage is due to electrochemical reaction products is the work by Lilly et al. [123], which demonstrated that loss of electrical excitability

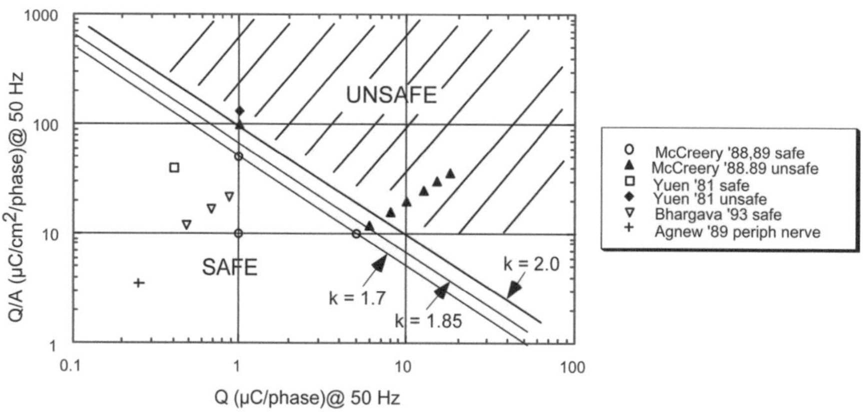

Fig. 14 Charge (Q) vs. Charge Density (Q/A) for Safe Stimulation A microelectrode with relatively small total charge per pulse might safely stimulate using a large charge density, whereas a large surface area electrode (with greater total charge per pulse) must use a lower charge density

and tissue damage occur when the cerebral cortex of monkey is stimulated using monophasic current pulses. Later, Lilly et al. [124] showed that biphasic stimulation caused no loss of excitability or tissue damage after 15 weeks of stimulation for 4–5 h per day. Lilly interpreted these results as due to movement of charged particles such as proteins out of physiological position. The concept that monophasic is a more damaging form of stimulation than charge-balanced biphasic was confirmed by Mortimer et al. [125], who reported that breakdown of the blood brain barrier during stimulation of the surface of cat cerebral cortex occurs when monophasic pulses were used at power densities greater than 0.003 W/in^2 (0.5 mW/cm^2), but does not occur with charge-balanced biphasic pulses until a power density of 0.05 W/in^2 (8 mW/cm^2) is exceeded. Pudenz et al. [126, 127] further showed that monophasic stimulation of the cat cerebral cortex causes vasoconstriction, thrombosis in venules and arterioles, and blood brain barrier breakdown within 30 s of stimulation when used at levels required for a sensorimotor response; however, charge-balanced biphasic stimulation could be used for up to 36 h continuously without tissue damage if the charge per phase was below 0.45 μC (4.5 μC/cm^2). Also supporting the hypothesis that damage is due to electrochemical products are observations of cat muscle that suggest some nonzero level of reaction product can be tolerated [88, 128]. These studies showed that monophasic stimulation causes significantly greater tissue damage than a nonstimulated implant at 1 μC/mm^2 per pulse, but not 0.2 μC/mm^2 per pulse, and that charge-balanced biphasic stimulation does not cause significant tissue damage at levels up to 2 μC/mm^2 per pulse. However, in order to prevent electrode corrosion, the charge-balanced waveform must not exceed 0.4 μC/mm^2 per pulse, otherwise the electrode potential is driven to damaging positive potentials during the anodic (reversal) phase and interpulse interval. Scheiner and Mortimer [88] studied the utility of charge-imbalanced biphasic stimulation demonstrating that this waveform allows greater cathodic charge densities than monophasic prior to the onset of tissue damage as reactions occurring during the cathodic phase are reversed by the anodic phase, and also that greater cathodic charge densities can be used than with the charge-balanced waveform prior to electrode corrosion since the anodic phase is no longer constrained to be equal to the cathodic phase, thus the electrode potential reaches less positive values during the anodic phase and interpulse interval. Scheiner found that cat muscle tissue was significantly damaged using monophasic stimulation at 0.4 $\mu C/mm^2$ per phase, and that when charge-imbalanced biphasic stimulation was used, tissue was damaged with 1.2 $\mu C/mm^2$ per phase cathodic and 0.2 $\mu C/mm^2$ per phase anodic, and could safely tolerate 1.2 $\mu C/mm^2$ per phase cathodic and 0.5 $\mu C/mm^2$ per phase anodic. No electrode corrosion was observed under any of the conditions studied.

In 1975, Brummer and Turner [63] gave an alternative explanation to Lilly's for why biphasic pulses were less damaging than monophasic. They proposed that two principles should be followed to achieve electrochemically safe conditions during tissue stimulation:

> "(1) Perfect symmetry of the electrochemical processes in the two half-waves of the pulses should be sought. This implies that we do not generate any electrolysis products in solution. One approach to achieve this would appear to involve the use of perfectly charge-balanced waveforms of controlled magnitude. (2) The aim should be to inject charge via non-Faradaic or surface-Faradaic processes, to avoid injecting any possibly toxic materials into the body."

Their model for safe stimulation interprets the charge-balanced waveform in electrochemical terms. Any process occurring during the first (stimulating) phase, whether it is charging of the electrode or a reversible Faradaic process, is reversed during the second (reversal) phase, with no net charge delivered. The observation that monophasic stimulation causes greater tissue damage than biphasic stimulation at the same amplitude, pulse width, and frequency is explained by the fact that during monophasic stimulation, all injected charge results in generation of electrochemical reaction products.

Reversible processes include charging and discharging of the double-layer capacitance, as well as surface-bound reversible Faradaic processes such as reactions (3), (4), (5), (6), (7), (8), and (13). Reversible reactions often involve the production or consumption of hydrogen or hydroxyl ions as the charge counterion. This causes a change in the pH of the solution immediately adjacent to the electrode surface. Ballestrasse et al. [129] gave a mathematical description of these pH changes and determined that the pH may range from 4 to 10 near a 1-μm diameter electrode during biphasic current pulses, but this change extended for only a few microns. Irreversible processes include Faradaic reactions where the product does not remain near the electrode surface, such as reactions (1) and (9), (10), (11), and (12).

Free radicals are known to cause damage to myelin, the lipid cell membrane, and DNA of cells. A likely candidate for a mechanism of neural tissue damage due to electrochemical products is peroxidation of the myelin by free radicals produced on the electrode surface. Several researchers [130, 131, 132, 133, 134, 135] have demonstrated the great susceptibility of myelin to free radical damage. Damage occurs as fatty acyl chains move apart and the myelin goes from a crystalline (ordered) state to a liquid (disordered) state.

Morton et al. [136] have shown that oxygen reduction occurs on a gold electrode in phosphate-buffered saline under typical neural stimulating conditions. Oxygen-reduction reactions that may occur during the cathodic-stimulating phase include reactions that generate free radicals such as superoxide and hydroxyl, and hydrogen peroxide, collectively known as reactive oxygen species. These species may have multiple deleterious effects on tissue [20, 21, 22, 23, 24]. As free radicals are produced, they may interfere with chemical signaling pathways that maintain proper perfusion of nervous tissue. Nitric oxide has been identified as the endothelium-derived relaxing factor, the primary vasodilator [137, 138, 139]. Nitric oxide is also known to prevent platelet aggregation and adhesion [140, 141, 142]. Beckman et al. [143] have shown that the superoxide radical reacts with nitric oxide to form the peroxynitrite radical. Oxygen-derived free radicals from the electrode may reduce the nitric oxide concentration and diminish its ability as the principal vasodilator and as an inhibitor of platelet aggregation. Superoxide depresses vascular smooth muscle relaxation by inactivating nitric oxide, as reviewed by Rubanyi [144].

An electrochemical product may accumulate to detrimental concentrations if the rate of Faradaic reaction, given by the current–overpotential relationship of equation (18), exceeds the rate for which the physiological system can tolerate the product. For most reaction products of interest there is some sufficiently low concentration near the electrode that can be tolerated over the long term. This level for a tolerable reaction may be determined by the capacity of an intrinsic buffering system. For example, changes in pH are buffered by several systems including the bicarbonate buffer system, the phosphate buffer system, and intracellular proteins. The superoxide radical, a product of the reduction of oxygen, is converted by superoxide dismutase and cytochrome c to hydrogen peroxide and oxygen. The diffusion rate of a toxic product must be considered, as it may be the case that high concentrations only exist very near the site of generation (the electrode surface).

6 Design Compromises for Efficacious and Safe Electrical Stimulation

A stimulating system must be both efficacious and safe. Efficacy of stimulation generally means the ability to elicit the desired physiological response, which can include initiation or suppression of action potentials. Safety has two primary aspects. First, the tissue being stimulated must not be damaged, and second, the stimulating electrode itself must not be damaged, as in corrosion. An electrode implanted into a human as a prosthesis may need to meet these requirements for decades. In animal experimentation, damage to the tissue or the electrode can seriously complicate or invalidate the interpretation of results.

Efficacy requires that the charge injected must exceed some threshold (Fig. 13). However, as the charge per pulse increases, the overpotential of the electrode increases, as does the fraction of the current going into Faradaic reactions (which may be damaging to tissue or the electrode if the reaction is irreversible). Judicious design of stimulation protocols involves acceptable compromises between stimulation efficacy, requiring a sufficiently high charge per pulse, and safety, requiring a sufficiently low charge per pulse, thus preventing the electrode from reaching potentials where deleterious Faradaic reactions occur at an intolerable rate. The overpotential an electrode reaches, and thus Faradaic reactions that can occur, depend on several factors in addition to the charge per pulse, including (1) waveform type (Fig. 10), (2) stimulation frequency, (3) electrode material (a high charge-storage capacity allows relatively large charge storage prior to reaching overpotentials where irreversible Faradaic reactions occur), (4) electrode geometric area and roughness (determining real area) and therefore total capacitance, and (5) train effects (Section 2). Increasing either the stimulus phase pulse width or the reversal phase pulse width of a charge-balanced stimulation protocol has the effect of increasing unrecoverable charge into irreversible reactions. Any factor which either drives the electrode potential into a range where irreversible reactions occur (such as a long stimulus phase pulse width) or fails to quickly reverse the electrode potential out of this range (such as a long reversal phase pulse width) will allow accumulation of unrecoverable charge.

The overpotential an electrode must be driven to before any given current will be achieved is highly dependent on the kinetics of the system, characterized by the exchange current density i_0. For a system with a large exchange current density (e.g., $i_0 = 10^{-3}$ A/cm^2), no significant overpotential may be achieved before a large Faradaic current ensues (equation (18)). When i_0 is many orders of magnitude smaller (e.g., $i_0 = 10^{-9}$ A/cm^2), a large overpotential must be applied before there is substantial Faradaic current. When i_0 is very low, a large total charge can be injected through the capacitive mechanism before significant Faradaic reactions commence. This is the generally desirable paradigm for a stimulating electrode, minimizing Faradaic reactions that lead to either electrode damage or tissue damage.

The fundamental design criteria for an electrochemically safe stimulation protocol can be stated: *the electrode potential must be kept within a potential window where irreversible Faradaic reactions do not occur at levels that are intolerable to the physiological system or the electrode*. If irreversible Faradaic reactions do occur, one must ensure that they can be tolerated (e.g., that physiological buffering systems can accommodate any toxic products) or that their detrimental effects are low in magnitude (e.g., that corrosion occurs at a very slow rate, and the electrode will last for longer than its design lifetime).

The charge–duration curve shown in Fig. 13 demonstrates that to minimize the total charge injected in an efficacious stimulation protocol, one should use short-duration pulses. In practice, pulses on the order of tens of microseconds approach the minimum charge and are often reasonable design solutions. During this relatively short duration, one may be able to avoid Faradaic reactions that would occur at higher levels of total charge with longer pulses. While it is desirable to use short-duration pulses on the order of tens of microseconds, there are applications for which biological constraints require longer-duration pulses. The time constants of several key ion channels in the membranes of excitable tissue are measured in hundreds of microseconds to milliseconds. By using stimulating pulses with comparable durations one can selectively manipulate the opening and closing of these ion channels to accomplish various specific behaviors. Certain waveforms have been developed that allow selectivity during electrical excitation of tissue (Section 4). Grill and Mortimer [145] have reviewed stimulus waveforms used for spatial and fiber diameter selective neural stimulation, illustrating the response of the neural membrane to different waveforms. Selective waveforms often require stimulation or reversal phases with long pulse widths relative to conventional stimulus waveforms; thus waveforms optimized for physiological responses may not be efficient for reversing electrochemical processes. Judicious design of electrical protocols has allowed the designers of neural prostheses to selectively inactivate the larger neurons in a nerve trunk [146], selectively inactivate the superficial fibers in a nerve by preconditioning [147], and prevent anodic break. Lastly, there are applications where tonic polarization mandates the use of very long (> 1 s) monophasic pulses; for example, tonic hyperpolarization of the soma to control epileptic activity [148, 149]. The use of these various waveforms with long pulse widths allows greater accumulation of any electrochemical product, thus requiring additional diligence by a neurophysiologist or prosthesis designer to prevent electrochemical damage.

In addition to biological constraints on the pulse durations, the required current for a short pulse width may also be a limitation. In order to inject the minimum charge required for effect, a large current is required (Fig. 13). This is not always possible, as may be the case with a battery-powered stimulator with limited current output.

Certain applications, such as clinical Deep Brain Stimulation [150, 151] and experimental long-term potentiation [152], require the use of high-frequency (> 50 Hz) pulsing. As discussed in Section 2, this can lead to a ratcheting of the electrode potential not achieved during single-pulse stimulation. Appropriate design of stimulation protocols can minimize damage by careful attention to the effects of high-frequency stimulation on the electrode potential.

Fig. 15 summarizes key features of various stimulation-waveform types. The cathodic-monophasic waveform illustrated in Fig. 15(a) consists of pulses of current passed in one direction, with an open-circuit condition during the interpulse interval. At no time does current pass in the opposite direction. Commonly the working electrode is pulsed cathodically for stimulation of tissue (as shown), although anodic stimulation may also be used (Section 4). Of the waveforms illustrated in Fig. 15, the monophasic is the most efficacious for stimulation. However, monophasic pulses are not used in long-term stimulation where tissue damage is to be avoided. Greater negative potentials are reached during monophasic pulsing than with biphasic pulsing (Fig. 10). Furthermore, the electrode potential during the interpulse interval of cathodic-monophasic pulsing remains relatively negative as the charged-electrode capacitance slowly discharges through Faradaic reactions, allowing reduction reactions which may be deleterious to tissue to proceed throughout the entire period of stimulation. Biphasic waveforms are illustrated in Fig. 15(b) to (f). The first (stimulating) phase elicits the desired physiological effect such as initiation of an

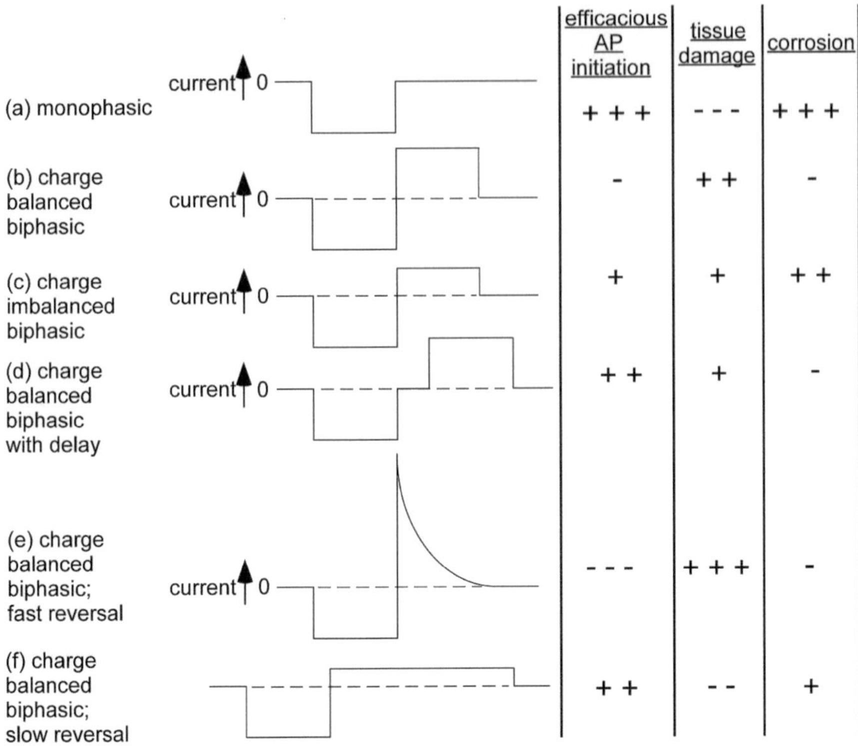

Fig. 15 Comparison of Stimulating Waveforms Six prototypical waveforms are rated for relative merit in efficacy and safety. "+++" = best (most efficacious, least damaging to tissue or the electrode), "–" = worst

action potential and the second (reversal) phase is used to reverse the direction of electrochemical processes occurring during the stimulating phase (Section 2). If all processes of charge injection during the stimulating phase are reversible, then the reversal phase will prevent net changes in the chemical environment of the electrode, as desired. The charge-balanced biphasic waveform (Fig. 15(b)) is widely used to prevent tissue damage. It should be noted that charge balance does not necessarily equate to electrochemical balance. As given by equations (32) and (33), during certain instances of stimulation there are irreversible Faradaic reactions during the cathodic phase (e.g., oxygen reduction), and then different irreversible reactions during the anodic phase (e.g., electrode corrosion) that are not the reverse of the cathodic Faradaic reactions. Such electrochemical imbalance leads to a potential waveform as illustrated in Fig. 10(b), where the potential at the end of the anodic phase is positive of the prepulse potential, allowing irreversible reactions such as electrode corrosion to occur. The charge-imbalanced waveform, illustrated in Fig. 15(c), may be used to reduce the most positive potentials during the anodic phase with respect to the charge-balanced waveform and prevent electrode corrosion [88]. Ideally, the charge in the reversal phase is equal to the charge going into reversible processes during the stimulation phase, in which case the electrode potential returns to its prepulse value at the end of the reversal phase.

In addition to electrode corrosion, a second concern with the charge-balanced biphasic waveform is that the reversal phase not only reverses electrochemical processes of the stimulation phase, but may also reverse some of the desired physiological effect of the stimulation phase, i.e., it may suppress an action potential that would otherwise be induced by a monophasic waveform. This effect causes an increased threshold for biphasic stimulation relative to monophasic. Gorman and Mortimer [153] have shown that by introducing an open-circuit interphase delay between the stimulating and reversal phases, the threshold for biphasic stimulation is similar to that for monophasic. This is illustrated in Fig. 15(d). Although the introduction of an interphase delay improves threshold, it also allows the electrode potential to remain relatively negative during the delay period. A delay of 100 μs is typically sufficient to prevent the suppressing effect of the reversal phase, and may be a short enough period that deleterious Faradaic reaction products do not accumulate to an unacceptable level.

As illustrated in Fig. 15(e) and (f), the more rapidly charge is injected during the anodic-reversal phase, the more quickly the electrode potential is brought out of the most negative range, and thus the less likely that tissue damage will occur. A high current reversal phase however means more of a suppressing effect on action-potential initiation, and also means the electrode potential will move positive during the reversal phase, thus risking electrode corrosion.

When evaluating the electrochemistry of a stimulating electrode system, both the working electrode and counter electrode should be considered. If the area, and thus total capacitance, of a counter electrode is relatively large, there is a small potential change for a given amount of injected charge. Such an electrode will not be perturbed away from its resting potential as readily as a small electrode, and all charge injection across this large counter electrode is assumed to be by capacitive charging, not Faradaic processes. If the working electrode is driven cathodically first in a biphasic waveform (and thus the counter electrode anodically), then during the reversal phase the working electrode is driven anodically and the counter electrode cathodically. In such a system, the working electrode is often referred to simply as the cathode. Strictly speaking, the working electrode is the cathode during the stimulus phase, and during the reversal phase the roles are reversed so that the working electrode is the anode and the counter electrode is the cathode.

References

1. Merrill DR, Bikson M, Jefferys JGR (2005) Electrical stimulation of excitable tissue: design of efficacious and safe protocols. J Neurosci Methods 141(2):171–198
2. Helmholtz von HLF (1853) Ueber einige gesetze der vertheilung elektrischer strome in korperlichen leitern mit anwendung auf die thierisch-elektrischen versuche. Ann Physik 89:211–233
3. Guoy G (1910) Constitution of the electric charge at the surface of an electrolyte. J Physique 9:457–467

4. Chapman DL (1913) A contribution to the theory of electrocapillarity. Philos Mag 25: 475–481
5. Stern O (1924) Zur theorie der elektrolytischen doppelschicht. Z Elektrochem 30:508–516
6. Grahame DC (1947) The electrical double layer and the theory of electrocapillarity. Chem Rev 41:441–501
7. Randles JEB (1947) Rapid electrode reactions. Disc Faraday Soc 1:11–19
8. Gileadi E, Kirowa-Eisner E, Penciner J (1975) Interfacial Electrochemistry: An Experimental Approach. Addison-Wesley, Reading, MA, Section II
9. Bard AJ, Faulkner LR (1980) Electrochemical Methods. John Wiley and Sons, New York
10. Rand DAJ, Woods R (1974) Cyclic voltammetric studies on iridium electrodes in sulfuric acid solutions. Nature of oxygen layer and metal dissolution. J Electroanal Chem Interfacial Electrochem 55:375–381
11. Frazer EJ, Woods R (1979) The oxygen evolution reaction on cycled iridium electrodes. J Electroanal Chem 102:127–130
12. Gottesfeld S (1980) The anodic rhodium oxide film:a two-color electrochromic system. J Electrochem Soc 127:272–277
13. Dautremont-Smith WC (1982) Transition metal oxide electrochromic materials and displays: a review. Part 2. Oxides with anodic coloration. Displays 3(3):67–80
14. Delahay P (1965) Double Layer and Electrode Kinetics. Interscience Publishers, New York
15. Pletcher D, Walsh FC (1990) Industrial Electrochemistry, second edition. Chapman and Hall, London
16. Silbey RJ, Alberty RA (2001) Physical Chemistry, third edition. John Wiley and Sons, New York
17. Ives DJG, Janz GJ (1961) Reference Electrodes: Theory and Practice. Academic Press, New York
18. Rand DAJ, Woods R (1971) The nature of adsorbed oxygen on rhodium, palladium, and gold electrodes. J Electroanal Chem Interfacial Electrochem 31:29–38
19. Michael DJ, Wightman RM (1999) Electrochemical monitoring of biogenic amine neurotransmission in real time. J Pharm Biomed Anal 19:33–46
20. Halliwell B (1992) Reactive oxygen species and the central nervous system. J Neurochem 59(5):1609–1623
21. Stohs SJ (1995) The role of free radicals in toxicity and disease. J Basic Clin Physiol Pharmacol 6(3–4):205–228
22. Hemnani T, Parihar MS (1998) Reactive oxygen species and oxidative DNA damage. Indian J Physiol Pharmacol 42(4):440–452
23. Imlay JA (2003) Pathways of oxidative damage. Annu Rev Microbiol 57:395–418
24. Bergamini CM, Gambetti S, Dondi A, Cervellati C (2004) Oxygen, reactive oxygen species and tissue damage. Curr Pharm Des 10(14):1611–1626
25. Donaldson NdN, Donaldson PEK (1986) When are actively balanced biphasic ('Lilly') stimulating pulses necessary in a neural prosthesis? I.Historical background, Pt resting potential, dQ studies. Med and Biol Eng and Comput 24:41–49
26. Donaldson NdN, Donaldson PEK (1986) When are actively balanced biphasic ('Lilly') stimulating pulses necessary in a neural prosthesis? II.pH changes, noxious products, electrode corrosion, discussion. Med and Biol Eng and Comput 24:50–56
27. Weinman J, Mahler J (1964) An analysis of electrical properties of metal electrodes. Med Electron Biol Eng 2:229–310
28. Dymond AM, Kaechele LE, Jurist JM, Crandall PH (1970) Brain tissue reaction to some chronically implanted metals. J Neurosurg 33:574–580
29. Stensaas SS, Stensaas LJ (1978) Histopathological evaluation of materials implanted in the cerebral cortex. Acta Neuropathol 41:145–155
30. Loeb GE, Walker AE, Vematsu S, Konigsmark BW (1977) Histological reaction to various conductive and dielectric films chronically implanted in the subdural space. J Biomed Mater Res 11(2):195–210

31. Majji AB, Humayun MS, Weiland JD, Suzuki S, D'Anna SA, deJuan E Jr (1999) Long-term histological and electrophysiological results of an inactive epiretinal electrode array implantation in dogs. Invest Opthalmol Vis Sci 40(9):2073–2081
32. Chouard CH, Pialoux P (1995) Biocompatibility of cochlear implants. Bull Acad Natl Med 179(3):549–555
33. Niparko JK, Altschuler RA, Xue XL, Wiler JA, Anderson DJ (1989) Surgical implantation and biocompatibility of central nervous system auditory prostheses. Ann Otol Rhinol Laryngol 98(12 Pt 1):965–970
34. Babb TL, Kupfer W (1984) Phagocytic and metabolic reactions to chronically implanted metal brain electrodes. Exp Neurol 86(2):171–182
35. Fisher G, Sayre GP, Bickford RC (1961) Histological changes in the cat's brain after introduction of metallic and plastic-coated wire. In: Sheer DE (ed) Electrical Stimulation of the Brain. Univ. of Texas Press, Austin, 55–59
36. Sawyer PN, Srinivasan S (1974) In Ray CD (ed) Medical Engineering. Chicago 1099–1110.
37. Ryhanen J, Kallioinen M, Tuukkanen J, Junila J, Niemela E, Sandvik P, Serlo W (1998) In vivo biocompatibility evaluation of nickel-titanium shape memory alloy: muscle and perineural tissue responses and encapsule membrane thickness. J Biomed Mater Res 41(3):481–488
38. Bogdanski D, Koller M, Muller D, Muhr G, Bram M, Buchkremer HP, Stover D, Choi J, Epple M (2002) Easy assessment of the biocompatibility of Ni-Ti alloys by in vitro cell culture experiments on a functionally graded Ni-NiTi-Ti material. Biomaterials 22(23):4549–4555
39. Jones KE, Campbell PK, Normann RA (1992) A glass/silicon composite intracortical electrode array. Ann Biomed Eng 20(4):423–437
40. Hoogerwerf AC, Wise KD (1994) A three-dimensional microelectrode array for chronic neural recording. IEEE Trans Biomed Eng 41:1136–1146
41. Rousche PJ, Pellinen DS, Pivin DP, Williams JC, Vetter RJ, Kipke DR (2001) Flexible polyimide-based intracortical electrode arrays with bioactive capability. IEEE Trans Biomed Eng 48(1):361–370
42. Kennedy PR (1989) The cone electrode:A long-term electrode that records from neurites grown onto its recording surface. J Neurosci Methods 29(3):181–193
43. Kennedy PR, Bakay RA (1998) Restoration of neural output from a paralyzed patient by a direct brain connection. Neuroreport 9(8):1707–1711
44. Kennedy PR, Bakay RA, Moore M, Adams K, Montgomery G (1999) Neural activity during acquisition of cursor control in a locked-in patient. Soc Neurosci Abstr 25(1):894
45. Rudge JS, Smith GM, Silver J (1989) An in vitro model of wound healing in the central nervous system: analysis of cell reaction and interaction at different times. Exp Neurol 103:1–16
46. Turner JN, Shain W, Szarowski DH, Andersen M, Martins S, Isaacson M, Craighead H (1999) Cerebral astrocyte response to micromachined silicon implants. Exp Neurol 156: 33–49
47. Bignami A, Dahl D (1976) The astroglial response to stabbing: immunoflourescence studies with antibodies to astrocyte-specific protein (GFA) in mammalian and sub-mammalian vertebrate. Neuropathol Appl Neurobiol 251:23–43
48. Robblee LS, Rose TL (1990) Electrochemical guidelines for selection of protocols and electrode materials for neural stimulation. In: Agnew WF, McCreery DB (ed) Neural Prostheses: Fundamental Studies. Prentice-Hall, Englewood Cliffs 25–66.
49. Merrill DR (2002) Electrochemical processes occurring on gold in sulfuric acid under neural stimulation conditions, Ph.D. thesis, Case Western Reserve University, Dept. of Biomedical Engineering, Cleveland,OH
50. Merrill DR, Stefan IC, Scherson DA, Mortimer JT (2005) Electrochemistry of gold in aqueous sulfuric acid solutions under neural stimulation conditions. J Electrochem Soc 152(7):E212–E221

51. White RL, Gross TJ (1974) An evaluation of the resistance to electrolysis of metals for use in biostimulation probes. IEEE Trans Biomed Eng BME-21:487–490
52. Johnson PF, Hench LL (1977) An in vitro analysis of metal electrodes for use in the neural environment. Brain Behav Evol 14:23–45
53. Brummer SB, McHardy J, Turner MJ (1977) Electrical stimulation with Pt electrodes: Trace analysis for dissolved platinum and other dissolved electrochemical products. Brain Behav Evol 14:10–22
54. Black RD, Hannaker P (1979) Dissolution of smooth platinum electrodes in biological fluids. Appl Neurophysiol 42:366–374
55. McHardy J, Robblee RS, Marsten M, Brummer SB (1980) Electrical stimulation with platinum electrodes. IV. Factors influencing platinum dissolution in inorganic saline. Biomater B1:129–134
56. Robblee RS, McHardy J, Marsten M, Brummer SB (1980) Electrical stimulation with platinum electrodes. V. The effects of protein on platinum dissolution. Biomater B1:135–139
57. Robblee RS, Lefko JL, Brummer SB (1983) Activated iridium: An electrode suitable for reversible charge injection in saline solution. J Electrochem Soc 130:731–733
58. Robblee RS, McHardy J, Agnew WF, Bullara LA (1983) Electrical stimulation with Pt electrodes. VII. Dissolution of Pt electrodes during electrical stimulation of the cat cerebral cortex. J Neurosci Methods 9:301–308
59. Tivol WF, Agnew WF, Alvarez RB, Yuen TGH (1987) Characterization of electrode dissolution products on the high voltage electrode microscope. J Neurosci Methods 19:323–337
60. Rosenberg B, VanCamp L, Krigas T (1965) Inhibition of cell division in Escherichia coli by electrolysis products from a platinum electrode. Nature 205:698–699
61. Rosenberg B (1971) Some biological effects of platinum compounds: New agents for the control of tumours. Platin Met Rev 15:42–51
62. Macquet JP, Theophanides T (1976) DNA-Platinum interactions. Characterization of solid DNA-$K_2[PtCl_4]$ complexes. Inorg Chim Acta 18:189–194
63. Brummer SB, Turner MJ (1975) Electrical stimulation of the nervous system: the principle of safe charge injection with noble metal electrodes. Bioelectrochem Bioenerg 2:13–25
64. Brummer SB, Turner MJ (1977) Electrochemical considerations for safe electrical stimulation of the nervous system with platinum electrodes. IEEE Trans Biomed Eng BME-24: 59–63
65. Brummer SB, Turner MJ (1977) Electrical stimulation with Pt electrodes. I. A method for determination of 'real' electrode areas. IEEE Trans Biomed Eng BME-24:436–439
66. Brummer SB, Turner MJ (1977) Electrical stimulation with Pt electrodes. II. Estimation of maximum surface redox (theoretical non-gassing) limits. IEEE Trans Biomed Eng BME-24:440–443
67. Rose TL, Robblee LS (1990) Electrical stimulation with Pt electrodes. VIII. Electrochemically safe charge injection limits with 0.2 ms pulses. IEEE Trans Biomed Eng 37(11):1118–1120
68. Bruckenstein S, Miller B (1970) An experimental study of non-uniform current distribution at rotating disk electrodes. J Electrochem Soc 117:1044–1048
69. Shepherd RK, Murray MT, Houghton ME, Clark GM (1985) Scanning electron microscopy of chronically stimulated platinum intracochlear electrodes. Biomater 6:237–242
70. Wiley JD, Webster JJ (1982) Analysis and control of the current distribution under circular dispersive electrodes. IEEE Trans Biomed Eng BME-29:381–385
71. Zerbino JO, Tacconi NR, Arvia AJ (1978) The activation and deactivation of iridium in acid electrolytes. J Electrochem Soc 125:1266–1276
72. Mozota J, Conway BE (1983) Surface and bulk processes at oxidized iridium electrodes-I: monolayer stage and transition to reversible multilayer oxide film behavior. Electrochimica Acta 28:1–8
73. Bak MK, Girvin JP, Hambrecht FT, Kufta CV, Loeb GE, Schmidt EM (1990) Visual sensations produced by intracortical microstimulation of the human occipital cortex. Med Biol Eng Comput 28:257–259

74. McCreery DB, Yuen TGH, Agnew WF, Bullara LA (1992) Stimulation with chronically implanted microelectrodes in the cochlear nucleus of the cat: histologic and physiologic effects. Hear Res 62:42–56
75. Loeb GE, Peck RA, Martyniuk J (1995) Toward the ultimate metal microelectrode. J Neurosci Methods 63:175–183
76. Liu X, McCreery DB, Carter RR, Bullara LA, Yuen TGH, Agnew WF (1999) Stability of the interface between neural tissue and chronically implanted intracortical microelectrodes. IEEE Trans Rehabil Eng 7(3):315–326
77. Meyer RD, Cogan SF (2001) Electrodeposited iridium oxide for neural stimulation and recording electrodes. IEEE Trans Neural Syst Rehabil Eng 9(1):2–10
78. Anderson DJ, Najafi K, Tanghe SJ, Evans DA, Levy KL, Hetke JF, Xue X, Zappia JJ, Wise KD (1989) Batch-fabricated thin-film electrodes for stimulation of the central auditory system. IEEE Trans Biomed Eng 36:693–704
79. Weiland JD, Anderson DJ (2000) Chronic neural stimulation with thin-film, iridium oxide electrodes. IEEE Trans Biomed Eng 47(7):911–918
80. Beebe X, Rose TL (1988) Charge injection limits of activated iridium oxide electrodes with 0.2 ms pulses in bicarbonate buffered saline. IEEE Trans Biomed Eng BME-35:494–495
81. Kelliher EM, Rose TL (1989) Evaluation of charge injection properties of thin film redox materials for use as neural stimulation electrodes. Mater Res Soc Symp Proc 110:23–27
82. Agnew WF, Yuen TGH, McCreery DB, Bullara LA (1986) Histopathologic evaluation of prolonged intracortical electrical stimulation. Exp Neurol 92:162–185
83. Robblee LS, Mangaudis MM, Lasinsky ED, Kimball AG, Brummer SB (1986) Charge injection properties of thermally-prepared iridium oxide films. Mater Res Soc Symp Proc 55:303–310
84. Klein JD, Clauson SL, Cogan SF (1989) Morphology and charge capacity of sputtered iridium oxide films. J Vac Sci Technol A7:3043–3047
85. Loucks RB, Weinberg H, Smith M (1959) The erosion of electrodes by small currents. Electroenceph Clin Neurophysiol 11:823–826
86. Greatbatch W, Chardack WM (1968) Myocardial and endocardiac electrodes for chronic implantation. Ann N.Y Acad Sci 148:234–251
87. McHardy J, Geller D, Brummer SB (1977) An approach to corrosion control during electrical stimulation. Ann Biomed Eng 5:144–149
88. Scheiner A, Mortimer JT (1990) Imbalanced biphasic electrical stimulation: muscle tissue damage. Ann Biomed Eng 18:407–425
89. Gotman I (1997) Characteristics of metals used in implants. J Endourol 11(6):383–389
90. Guyton DL, Hambrecht FT (1973) Capacitor electrode stimulates nerve or muscle without oxidation-reduction reactions. Science 181:74–76
91. Guyton DL, Hambrecht FT (1974) Theory and design of capacitor electrodes for chronic stimulation. Med and Biol Eng 7:613–620
92. Rose TL, Kelliher EM, Robblee LS (1985) Assessment of capacitor electrodes for intracortical neural stimulation. J Neurosci Methods 12(3):181–193
93. Johnson PF, Bernstein JJ, Hunter G, Dawson WW, Hench LL (1977) An in vitro and in vivo analysis of anodized tantalum capacitive electrodes: corrosion response, physiology and histology. J Biomed Mater Res 11:637–656
94. McCreery DB, Bullara LA, Agnew WF (1986) Neuronal activity evoked by chronically implanted intracortical microelectrodes. Exp Neurol 92:147–161
95. Hille B (1984) Ionic Channels of Excitable Membranes. Sinauer Associates, Sunderland, MA
96. Kandel ER, Schwartz JH, Jessell TM (2000) Principles of Neural Science, fourth ed. McGraw Hill, New York
97. Chiu SY, Ritchie JM, Rogart RB, Stagg D (1979) A quantitative description of membrane current in rabbit myelinated nerve. J Physiol 292:149–166
98. Sweeney JD, Durand D, Mortimer JT (1987) Modeling of mammalian myelinated nerve for functional neuromuscular stimulation. Proc. 9th Intl. Conf. IEEE-EMBS, 1577–1578

99. Hodgkin AL, Huxley AF (1952) Currents carried by sodium and potassium ions through the membrane of the giant axon of Loligo. J Physiol 116:449–472
100. Hodgkin AL, Huxley AF (1952) The Components of membrane conductance in the giant axon of logigo. J Physiol 116:473–496
101. Hodgkin AL, Huxley AF (1952) The dual effect of membrane potential on sodium conductance in the giant axon of logigo. J Physiol 116:497–506
102. Hodgkin AL, Huxley AF (1952) A quantitative description of the membrane current and its application to conduction and excitation in nerve. J Physiol 117:500–544
103. McNeal DR (1976) Analysis of a model for excitation of myelinated nerve. IEEE Trans Biomed Eng 23(4):329–337
104. Ranck JB (1981) In: Patterson MM, Kesner RP (ed) Electrical Stimulation Research Techniques. Academic Press, New York, chapter 1
105. Mortimer JT (1990) In: Agnew WF, McCreery DB (ed) Neural Prostheses: Fundamental Studies. Prentice-Hall. Englewood Cliffs, chapter 3.
106. Durand D (1995) In: Bronzino JD (ed) Biomedical Engineering Handbook. CRC Press, Boca Raton, chapter 17
107. van den Honert C, Mortimer JT (1979) Generation of unidirectionally propagated action potentials in a peripheral nerve by brief stimuli. Science 206(4424):1311–1312
108. van den Honert C, Mortimer JT (1981) A technique for collision block of peripheral nerve: single stimulus analysis. IEEE Trans Biomed Eng 28:373–378
109. van den Honert C, Mortimer JT (1981) A technique for collision block of peripheral nerve: frequency dependence. IEEE Trans Biomed Eng 28:379–382
110. Fang Z, Mortimer JT (1991) Selective activation of small motor axons by quasitrapezoidal current pulses. IEEE Trans Biomed Eng 38(2):168–174
111. Rall W (1977) In: Handbook of Physiology-The Nervous System I, vol.1/Part 1. American Physiological Society, chapter 3.
112. Rattay F (1989) Analysis of models for extracellular fiber stimulation. IEEE Trans Biomed Eng 36:676–682
113. McIntyre CC, Grill WM (2002) Extracellular stimulation of central neurons: influence of stimulus waveform and frequency on neuronal output. J Neurophysiol 88(4):1592–1604
114. Lapicque L (1907) Recherches quantitatives sur l'excitation electrique des nerfs traites comme une polarization. J Physiol (Paris) 9:622–635
115. Warman EN, Grill WM, Durand D (1992) Modeling the effects of electric fields on nerve fibers: determination of excitation thresholds. IEEE Trans Biomed Eng 39(12):1244–1254
116. Plonsey R, Barr RC (1988) Bioelectricity: A Quantitative Approach. Plenum press, New York
117. McCreery DB, Agnew WF, Yuen TGH, Bullara LA (1990) Charge density and charge per phase as cofactors in neural injury induced by electrical stimulation. IEEE Trans Biomed Eng 37(10):996–1001
118. Shannon RV (1992) A model of safe levels for electrical stimulation. IEEE Trans Biomed Eng 39(4):424–426
119. Yuen TGH, Agnew WF, Bullara LA, Jacques S, McCreery DB (1981) Histological evaluation of neural damage from electrical stimulation: considerations for the selection of parameters for clinical application. Neurosurg 9(3):292–299
120. Agnew WF, McCreery DB, Yuen TGH, Bullara LA (1989) Histologic and physiologic evaluation of electrically stimulated peripheral nerve: considerations for the selection of parameters. Ann Biomed Eng 17:39–60
121. Bhargava A (1993) Long-term effects of quasi-trapezoidal pulses on the structure and function of sacral anterior roots. M.S. thesis, Case Western Reserve University, Dept. of Biomedical Engineering, Cleveland, OH
122. McCreery DB, Agnew WF, Yuen TGH, Bullara LA (1992) Damage in peripheral nerve from continuous electrical stimulation: comparison of two stimulus waveforms. Med Biol Eng Comput 30(1):109–114

123. Lilly JC, Austin GM, Chambers WW (1952) Threshold movements produced by excitation of cerebral cortex and efferent fibers with some parametric regions of rectangular current pulses (cats and monkeys). J Neurophysiol 15:319–341
124. Lilly JC, Hughes JR, Alvord EC, Garkin TW (1955) Brief noninjurious electric waveforms for stimulation of the brain. Science 121:468–469
125. Mortimer JT, Shealy CN, Wheeler C (1970) Experimental nondestructive electrical stimulation of the brain and spinal cord. J Neurosurg 32(5):553–559
126. Pudenz RH, Bullara LA, Dru D, Talalla A (1975) Electrical stimulation of the brain II: effects on the blood-brain barrier. Surg Neurol 4:265–270
127. Pudenz RH, Bullara LA, Jacques P, Hambrecht FT (1975) Electrical stimulation of the brain III: the neural damage model. Surg Neurol 4:389–400
128. Mortimer JT, Kaufman D, Roessmann U (1980) Intramuscular electrical stimulation: tissue damage. Ann Biomed Eng 8:235–244
129. Ballestrasse CL, Ruggeri RT, Beck TR (1985) Calculations of the pH changes produced in body tissue by a spherical stimulation electrode. Ann Biomed Eng 13:405–424
130. Chan PH, Yurko M, Fishman R (1982) Phospholipid degradation and cellular edema induced by free radicals in brain slice cortical slices. J Neurochem 38:525–531
131. Chia LS, Thompson JE, Moscarello MA (1983) Disorder in human myelin induced by superoxide radical: an in vitro investigation. Biochem and Biophys Res Commun 117(1):141–146
132. Konat G, Wiggins RC (1985) Effect of reactive oxygen species on myelin membrane proteins. J Neurochem 45:1113–1118
133. Sevanian A (1988) In: Lipid Peroxidation, Membrane Damage, and Phospholipase A2 Action. CRC Reviews, Cellular Antioxidant Defense Mechanisms, Vol.II, 77–95
134. Griot C, Vandevelde RA, Peterhans E, Stocker R (1990) Selective Degeneration of oligiodendrocytes mediated by reactive oxygen species. Free Radic Res Commun 11(4,5):181–193
135. Buettner GR (1993) The pecking order of free radicals and antioxidants: lipid peroxidation, alpha-tocopherol, and ascorbate. Arch Biochem Biophys 300(2):535–543
136. Morton SL, Daroux ML, Mortimer JT (1994) The role of oxygen reduction in electrical stimulation of neural tissue. J Electrochem Soc 141:122–130
137. Furchgott RF (1988) Studies on relaxation of rabbit aorta by sodium nitrite: the basis for the proposal that the acid-activatable inhibitory factor from retractor penis is inorganic nitrite and the endothelium-derived relaxing factor is nitric oxide, vasodilatation. In: Vascular Smooth Muscle, Peptides, Autonomic Nerves and Endothelium. Raven Press, New York, 401–414
138. Ignarro LJ, Byrns RE, Wood KS (1988) Biochemical and pharmacological properties of endothelium-derived relaxing factor and its similarity to nitric oxide radical. In: Vascular Smooth Muscle, Peptides, Autonomic Nerves and Endothelium. Raven Press, New York, 427–436
139. Umans J, Levi R (1995) Nitric oxide in the regulation of blood flow and arterial pressure. Ann Rev Physiol 57:771–790
140. Azuma H, Ishikawa M, Sekizaki S (1986) Endothelium-dependent Inhibition of Platelet Aggregation. British J Pharm 88:411–415
141. Radomski MW, Palmer RMJ, Moncada S (1987) Endogenous nitric oxide inhibits human platelet adhesion to vascular endothelium. Lancet 2:1057–1058
142. Moncada S, Palmer RMJ, Higgs EA (1991) Nitric oxide: Physiology, Pathology and Pharmacology. Pharmacol Rev 43(2):109–142
143. Beckman JS, Beckman TW, Chen J, Marshall PA, Freeman BA (1990) Apparent hydroxyl radical production by peroxynitrite: implications for endothelial injury from nitric oxide and superoxide. Proc Natl Acad Sci USA 87:1620–1624
144. Rubanyi GM (1988) Vascular Effects of Oxygen Derived Free Radicals. Free Radic Biol and Med 4:107–120
145. Grill WM, Mortimer JT (1995) Stimulus waveforms for selective neural stimulation. IEEE Eng Med Biol 14:375–385

146. Fang Z, Mortimer JT (1991) A method to effect physiological recruitment order in electrically activated muscle. IEEE Trans Biomed Eng 38(2):175–179
147. Grill WM, Mortimer JT (1997) Inversion of the current-distance relationship by transient depolarization. IEEE Trans Biomed Eng 44 (1):001–009
148. Gluckman BJ, Neel EJ, Netoff TI, Ditto WL, Spano ML, Schiff SJ (1996) Electric field suppression of epileptiform activity in hippocampal slices. J Neurophysiol 76(6):4202–4205
149. Ghai RS, Bikson M, Durand DM (2000) Effects of applied electric fields on low-calcium epileptiform activity in the CA1 region of rat hippocampal slices. J Neurophysiol 84(1): 274–280
150. McIntyre CC, Thakor NV (2002) Uncovering the mechanisms of deep brain stimulation for Parkinson's disease though functional imaging, neural recording, and neural modeling. Crit Rev Biomed Eng 30(4–6):249–281
151. O'Suilleabhain PE, Frawley W, Giller C, Dewey RB (2003) Tremor response to polarity, voltage, pulsewidth and frequency of thalamic stimulation. Neurology 60(5):786–790
152. Bliss TV, Lomo T (1973) Long-lasting potentiation of synaptic transmission in the dentate area of the anaesthetized rabbit following stimulation of the perforant path. J Physiol 232(2):331–356
153. Gorman PH, Mortimer JT (1983) The effect of stimulus parameters on the recruitment characteristics of direct nerve stimulation. IEEE Trans Biomed Eng BME-30:407–414
154. Stieglitz T, Meyer JU (1999) Implantable microsystems. Polyimide-based neuroprostheses for interfacing nerves. Med Dev Technol 10(6):28–30
155. Schmidt S, Horch K, Normann R (1993) Biocompatibility of silicon-based electrode arrays implanted into feline cortical tissue. J Biomed Mater Res 27(11):1393–1399
156. Kristensen BW, Noraberg J, Thiebaud P, Koudelka-Hep M, Zimmer J (2001) Biocompatibility of silicon-based arrays of electrodes coupled to organotypic hippocampal brain slice cultures. Brain Res 896:1–17
157. Bernstein JJ, Hench LL, Johnson PF, Dawson WW, Hunter G (1977) Electrical stimulation of the cortex with Ta_2O_5 capacitive electrodes. In: Hambrecht FT, Reswick JB (ed) Functional Electrical Stimulation. Marcel-Dekker, New York, 465–477
158. Donaldson PEK (1974) The stability of tantalum-pentoxide films in vivo. Med Biol Eng 12:131–135
159. Lagow CH, Sladek KJ, Richardson PC (1971) Anodic insulated tantalum oxide electrocardiograph electrodes. IEEE Trans Biomed Eng 18:162–164

In Situ Characterization of Stimulating Microelectrode Arrays: Study of an Idealized Structure Based on Argus II Retinal implants

Vincent Kandagor, Carlos J. Cela, Charlene A. Sanders, Elias Greenbaum, Gianluca Lazzi, David D. Zhou, Richard Castro, Sanjay Gaikwad, and Jim Little

Abstract The development of a retinal prosthesis for artificial sight includes a study of the factors affecting the structural and functional stability of chronically implanted microelectrode arrays. Although neuron depolarization and propagation of electrical signals have been studied for nearly a century, the use of multielectrode stimulation as a proposed therapy to treat blindness is a frontier area of modern ophthalmology research. Mapping and characterizing the topographic information contained in the electric field potentials and understanding how this information is transmitted and interpreted in the visual cortex is still very much a work in progress. In order to characterize the electrical field patterns generated by the device, an in vitro prototype that mimics several of the physical and chemical parameters of the in vivo visual implant device was fabricated. We carried out multiple electrical measurements in a model "eye," beginning with a single electrode, followed by a 9-electrode array structure, both idealized components based on the Argus II retinal implants. Correlating the information contained in the topographic features of the electric fields with psychophysical testing in patients may help reduce the time required for patients to convert the electrical patterns into graphic signals.

Contents

E. Greenbaum (✉)
Molecular Bioscience and Biotechnology, Chemical Sciences Division, Oak Ridge National Laboratory, Oak Ridge, TN, USA
e-mail: greenbaum@ornl.gov

D.D. Zhou, E. Greenbaum (eds.), *Implantable Neural Prostheses 2*,
Biological and Medical Physics, Biomedical Engineering, DOI 10.1007/978-0-387-98120-8_5,

1 Introduction

Key hypotheses in the field of multielectrode retinal prosthetic implants are that visual information can be encoded in the spatial patterns of the electric potentials that are generated by energized electrodes and that the information can be transmitted to the visual cortex by stimulation of the retinal neurons. Recent progress in the science, engineering, and clinical testing of retinal implants supports these hypotheses. For example, advances in the design, fabrication, testing, and surgical implantation of 16- and 60-electrode stimulating devices have been reported [1]. Second Sight's first-generation 16-electrode devices, implanted in patients in the United States, have contributed to the proof-of-principle for functional restoration of partial sight. There is, however, the need for a device with more electrodes. It has been calculated that for visual acuity of 20/80, an optoelectronic device with a stimulating density of up to 2500 pix/mm^2 [2, 3] is desired. Although significant progress in research and surgical techniques has been achieved, challenges remain. These include power dissipation in the ocular cavity, cross talk between neighboring electrodes, and collateral electrochemical reactions electrode tissue interface. The focus of this chapter is on cross talk between neighboring electrodes and their respective track lines and how this additional information might affect the visual content of the field maps. We report fabrication of an in vitro prototype that mimics several of the physical and chemical parameters of the in vivo device. We also report topographic electric potential maps of an idealized multielectrode structure that mimics the size and spacing of the individual Argus II retinal implant electrodes and avoids the additional field information of the track lines that bring electric charge to the electrodes. Studies of 200+ and the 1000+ electrode devices are planned.

2 Physical Analysis of Argus II Electrode Array and Representative Analogs

2.1 The Argus II Electrode Array

The artificial retina system is composed of multiple components (Fig. 1) [3]. It contains a light-sensing device for scanning the surrounding environment and capturing image data, a visual-processing information module that converts the image into

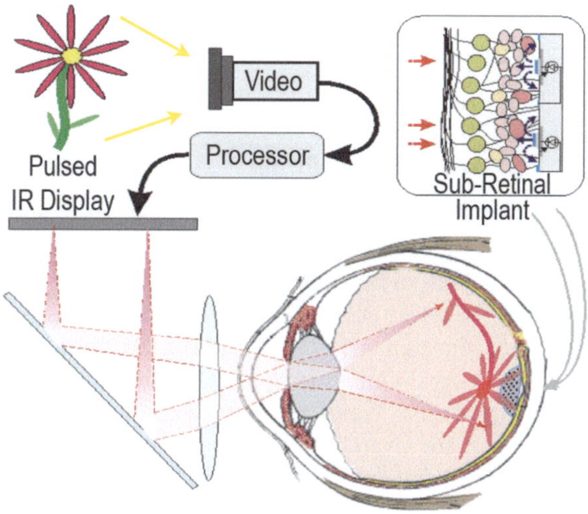

Fig. 1 Simplified diagram of a photovoltaic retinal prosthetic system developed by the Palanker lab at Stanford University (Reproduced from [3] with permission from Springer)

an electrode stimulation pattern, and a multielectrode array that delivers the patterned stimulus to the retina [4]. Several technologies have proven successful in fabricating the light-sensing device. However, the implanted device, in particular the electrode array that delivers the stimuli into the retina remains a scientific and engineering challenge. The characteristics of the interface between retinal tissue and device require a combination of materials that deliver the stimulation while maintaining a stable relationship between the array and tissue. The Argus II prosthesis is biocompatible, conforms to retinal tissue, and is mechanically robust. Also, other components of the system, such as the electronic chip set that is encased in a metal can (and serves as the return electrode), have a significant influence on the electrical characteristics of the system. In order to characterize the field patterns generated by the device, we carried out multiple electrical measurements, beginning with a single electrode, followed by a 9-electrode array structure, both idealized components based on the Argus II electrode size and array geometry.

2.2 Electrical Properties of the Vitreous

The microelectrode array is implanted 10–20 μm from the retinal surface in a viscous fluid, the vitreous humor [5]. The vitreous is a viscoelastic, gel-like substance that fills the ocular cavity [6]. The experimental setup for characterization of the retinal prosthesis components requires a medium that has the same chemical formulation as the vitreous. Since the early 1900s, different materials have been used as vitreous replacements. These substitutes serve two main purposes. First, they are used to replace a dysfunctional vitreous humor in cases where clouding or physical collapse and liquefaction of the vitreous have occurred. The second is temporary or permanent application during retinal surgery. Since arrays are intended for implantation, in vitro characterization should be carried out in chemically similar media [7].

2.2.1 Stimulation Waveforms

Vitreous humor is primarily a saline solution consisting of approximately 98% water. It is an electrolytically conductive and corrosive medium. The retinal prosthesis involves stimulation of neural tissue with metal electrodes. The interface between the metal and delicate retinal tissue places constraints on the magnitude of charge injection and time profiles of the waveforms that can be applied. Failure to observe these limitations will result in tissue damage, gas evolution, and electrode degradation. Waveforms are chosen to ensure that the net charge injection is zero. There are several methods that can be used to achieve this: a stimulator can supply charge-balanced cathodic and anodic current pulses, or an in-line capacitor can slowly discharge after monophasic stimulation. In cases where charge balance is not ensured, the resulting effects might be damaging to retinal tissue. Over time, net charge accumulation can increase the potential at the electrode/tissue interface resulting in electrolysis and accumulation of oxygen and hydrogen [8, 9].

2.2.2 AC and DC Impedance

It might be desirable to perform real-time impedance measurements while monitoring the patient. It is a challenging goal, but one that may not be achievable due to practical limitations. However, the in situ characterization of the electrode array can be done under conditions similar to those in the eye. An electric current from an AC or DC source is typically injected through the electrodes and the resulting voltage measured, from which the impedance can be calculated using Ohm's law.

2.2.3 Retinal Tissues

A key reason for the in situ characterization of the prosthetic model is to study the characteristics of the device under real-world operating conditions including upper-limits for safe charge injection. Even in a retinal environment with overall charge balance, it is possible to momentarily exceed the established charge limits [10]. Although previous experiments [9] have established electrochemical safe limits, charge-injection limits may vary from patient to patient. Humayun et al. [11] reported that with an electrode of about 200-μm diameter, the safe charge-density limit that can be used before neural damage occurs is 1 mC/cm^2. For a given current, the smaller the electrode diameter, the higher the charge density, and the development of 200+ and 1000 electrode arrays should therefore lead to smaller threshold currents per electrode and higher overall resolution compared to the 16- and the 60-electrode array.

2.3 *Electrical Stimulation, Electrodes, and Systems*

2.3.1 Electrode Configuration

As noted above, a key priority in electrode configuration and design is to provide safe communication between a specific location in the retina and the externally controlled electrical signals. Safe stimulation can be achieved with close proximity of the electrodes to the retinal tissue. This helps prevent irreversible reactions associated with high-stimulus charge density at the electrode/tissue interface.

2.3.2 Electrode Size and Spacing

The size of the region of the retina that is the target for stimulation limits the overall dimensions of the electrode array, the number of electrodes, and the diameter per electrode. The target region in the retina is approximately 5 mm × 5 mm. The initial 16-electrode implantable array was a first-generation device. An Argus 16-electrode array in a human subject is shown in Fig. 2 (*Left*). Patients could sense light and darkness, motion, and recognize large objects [12]. The Argus II device consists of 60 electrodes in the same 5 mm × 5 mm area (see Fig. 2 (*right*)). Correspondingly, the diameters of the electrodes were reduced to 200 μm with the pitch decreasing to accommodate 60 electrodes in the same area. The operational stability for chronic implants is an important clinical parameter. The track record is good. The first Argus 16 array was implanted in February 2002. The longest Argus 16 currently in use was implanted in June 2004. The first Argus II electrode was implanted in September 2006. A retinal array that is capable of higher resolution image recognition will probably require at least 1000 electrodes [13].

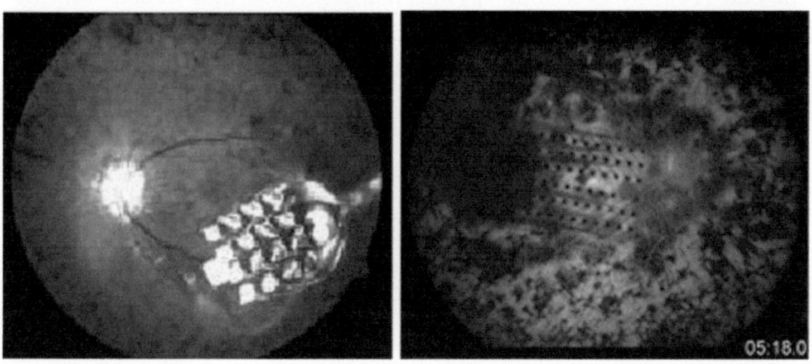

Fig. 2 (*Left*) An Argus 16-Pt/silicone electrode array in a human subject; and (*right*) an Argus II thin-film 60-electrode array in a human subject

2.3.3 Materials and Stability

The main objective of prosthetic devices is to restore functionality in a manner that will be safe, reliable, and physiologically acceptable. The choice of materials for achieving these goals is determined by the specific function of the prosthesis. The complexity and delicate nature of the retinal tissue limits the choice of the materials for implants. Important factors are biocompatibility and stability [14].

Gold and platinum are suitable electrode materials for other prosthetic implants, but current electrode implants have been fabricated by depositing Pt electrodes and their corresponding microtrack-lines onto a polycarbonate substrate [15] followed by a thin layer of the polymer to cover the Pt track-lines so that only the surfaces of the stimulating electrodes are exposed. The 4×4 Pt array with an electrode diameter of 520 μm that was implanted in patients has shown stability similar to that achieved with Pt cochlear implant electrodes.

2.4 Return Electrode

2.4.1 Functions

The multielectrode array needs a return electrode in a mono-polar stimulation configuration. Although a minimum charge density is needed to trigger an action potential in neural cells, no such limitation is imposed on the charge density of the return electrode. The return electrode can be larger and conveniently chosen to integrate with the overall circuit design of the system [9]. For the work reported here, a metal can, the housing for the electronic circuitry of the device, was provided by Second Sight.

2.4.2 Design

The metal can (Fig. 3) is a small cylindrical structure that houses the integrated circuitry of the electrical infrastructure. It is equipped with a small wire that is a connection point between the electrical settings inside and the outside.

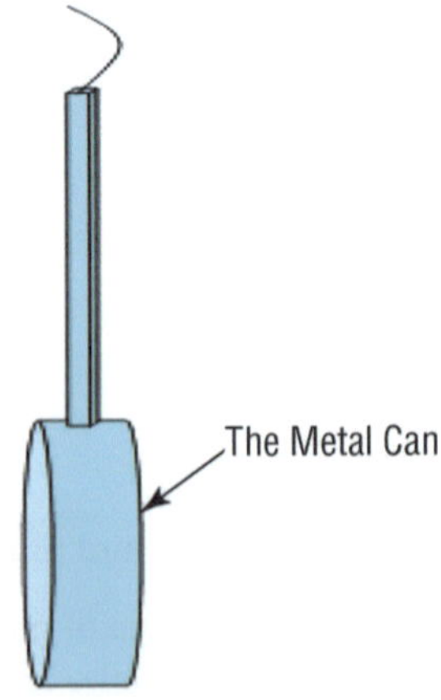

Fig. 3 Schematic of the metal can (the return path)

3 Characterization of Simplified Argus II Analogs

Although a number of studies have been carried out on the Argus II electrode array, less has been done on individual electrode interactions and the contribution of variable density microtrack lines to the overall electrical environment. As part of the characterization of the Argus II electrode array structure, we designed and fabricated idealized analog electrode arrays based on the Argus II configurations.

3.1 Single Electrode

3.1.1 Design and Material

The idealized array structure contains an electrode configuration similar to the electrodes in the Argus II set. It was fabricated by fusing a 200-μm diameter Pt wire in a Pyrex® glass capillary tube (see Fig. 4). One end was polished flush with the tube while the Pt wire extended from the other end for making electrical contact. Depending on preparation, Pt can have a high effective surface area, a desirable feature because it increases interfacial capacitance and raises the threshold for the onset of Faradaic redox chemistry. Other forms of Pt may have higher and more desired surface area characteristics [8, 16].

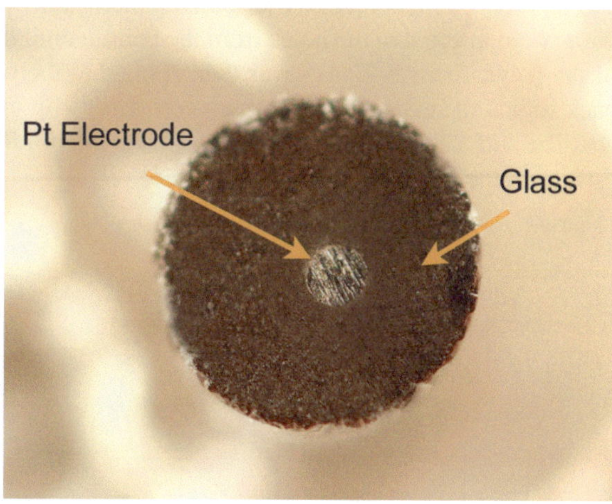

Fig. 4 A picture of the polished surface of the single-electrode array structure

3.2 9-Electrode Array Structure

In order to further characterize the Argus II electrode array, we fabricated a 9-electrode array structure that mimics any subset of 9 (3 × 3) electrodes in the Argus II pattern (Fig. 5). However, unlike the Argus II which contains closely packed microelectrode traces that are in the plane of the array, the traces of the

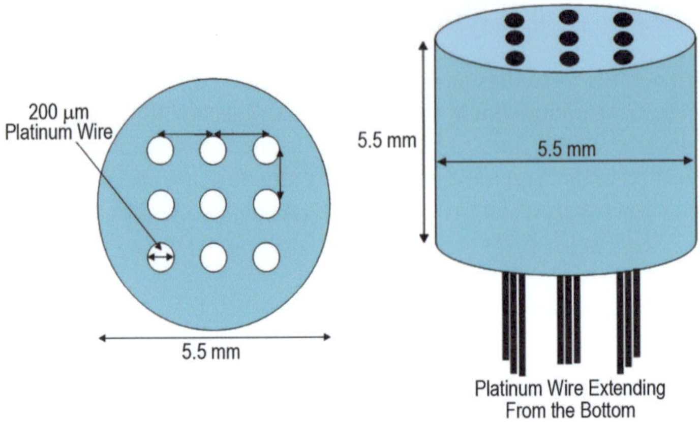

Fig. 5 *Top* and isometric views of the 9-electrode array. The 200-μm diameter wires are embedded in a polycarbonate block

model 9-electrode array are perpendicular to the array plane. This simplification allows separation of the effects of the traces' electric field potential from those of the electrode disks per se. The 9-electrode array can also be used for simultaneous stimulation of multiple electrodes.

The structure was fabricated by drilling 9 holes of 200-μm diameter in a polycarbonate cylindrical block. The spatial pattern was the same as in the Argus II. Pt wires were inserted in the holes from the bottom until they extended through the top of the polycarbonate. Epoxy was then applied to the wires which were then pulled back so that the ends of the Pt wires were flush with the surface of the polycarbonate. The surface was then polished so that a smooth Pt disk of each electrode was flush with the polycarbonate plane (Fig. 6).

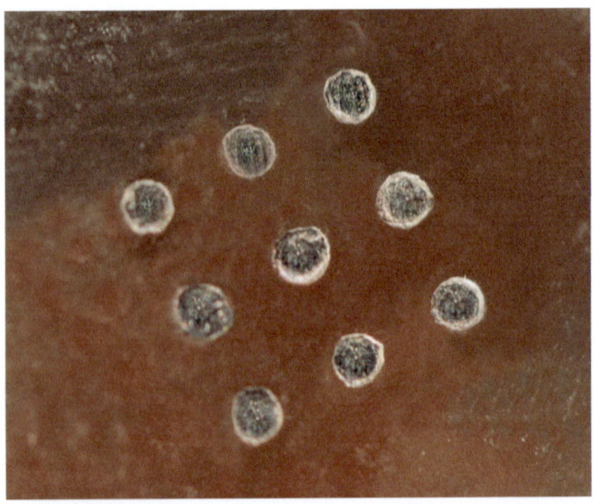

Fig. 6 Photograph of the polished surface of the 9-electrode array structure

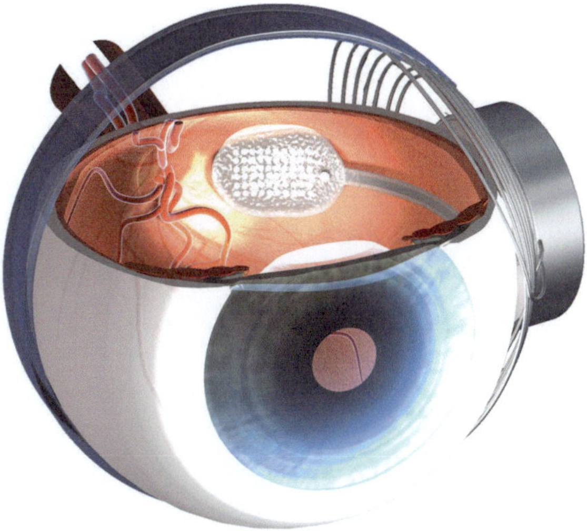

Fig. 7 The position of the metal can with respect to the Argus II electrode array in the retina

The test apparatus for characterizing the single- and 9-electrode array was based on an engineering diagram provide by Second Sight Medical Products containing the position of the metal can and electrode array in relation to the retina. We constructed a model "eye" to perform the electrical characterization, which was based on the positions shown in Fig. 7.

The electrodes were introduced into the bath from the bottom (Figs. 8 and 9). This ensured that only the surface area of the Pt was exposed to the electrolytic

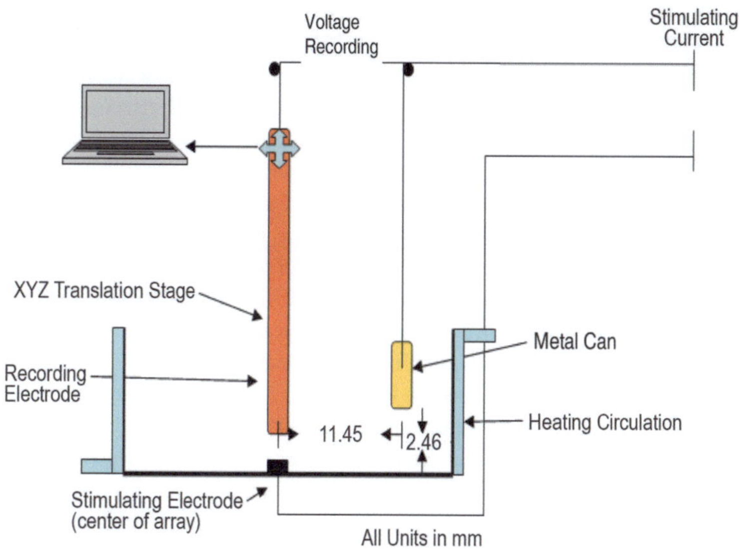

Fig. 8 Schematic illustration of the experimental set up translated from the Second Sight diagram

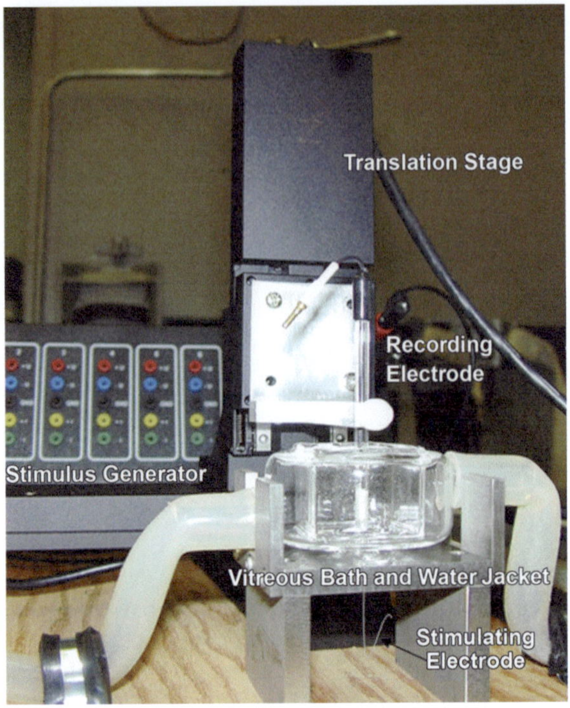

Fig. 9 Apparatus for automated field potential measurements. Positioning was achieved with a computerized Thorlabs XYZ translation stage. The recording electrode was attached to the stage hardware and controlled by the computer. Stage movement was programmable so that any position could be reached with mouse a click away from the experimental set up to minimize human intervention

bath. The recording electrode was initially centered 20 μm above the stimulating electrode by movement in the x-y plane until the measured voltage in both directions was maximized. Twenty μm is approximately the closest position that the Argus II electrode can be positioned near the retina. Since the retinal surface is not flat and consists of several layers, placing the electrode array too close to the retina may result in uneven and excessive pressure. A larger distance between the electrode array and the retina can result in inefficient stimulation and exaggeration of electrode cross talk [17].

The 200-μm Pt single electrode was stimulated with a charge-balanced biphasic, cathode-first, 38-μA current for 1 msec duration, giving a charge density of 100 $\mu C/cm^2$ with an interpulse delay of 16 mSec. The recording electrode, mounted to the programmable XYZ translation stage, was moved incrementally vertically and horizontally over the stimulating electrode. The metal can, which is the return electrode, was affixed on the inside wall of the bath in contact with the synthetic vitreous medium during the measurements. A high-speed digital oscilloscope was used to record the compliance voltage relative to the metal can. A 10× buffer amplifier was used to amplify and condition the signal. The recording electrode was then moved horizontally to "neighboring" positions that correspond to the immediate adjacent electrodes in the Argus II electrode array pattern as illustrated in Figs. 10 and 11.

The position of the stimulating electrode is indicated by the black spot at the center location in Fig. 10. The recording electrode was then moved to the other

Fig. 10 Position of the metal can with respect to the single electrode

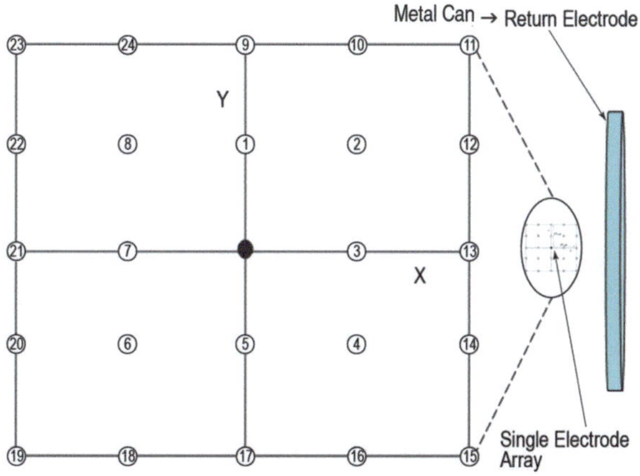

Fig. 11 Topographic potential map for a single electrode in Ames' media at 37°C. Z=20 μm

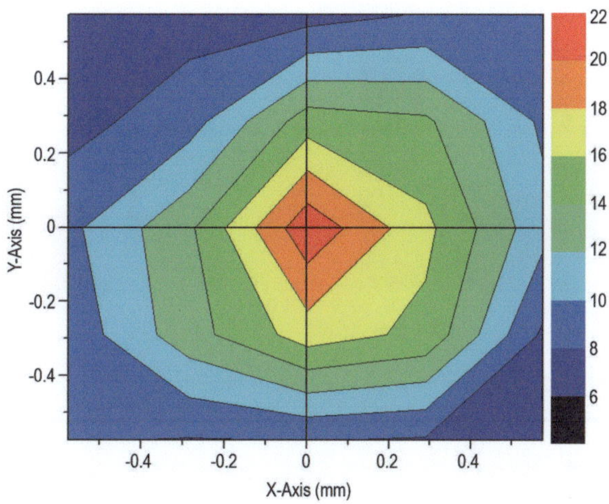

positions and the voltage was measured and recorded at each position, which corresponds to half and full distances between one electrode and another in the Argus II electrode array configuration. The experiment was repeated with the recording electrode positioned at a vertical distance (z) of 30 and 40 μm.

The same experimental procedure was repeated with the 9-electrode array configuration. It was also introduced into the synthetic media bath from the bottom. In this case, stimulation was done on each of the 9 electrodes individually, followed by simultaneous stimulation of multiple electrodes in predetermined patterns. For a set of 9 electrodes, analysis of all possible combinations of single and multiple electrode stimulation reveals that the maximum number of patterns is 511. In these experiments, not all 511 patterns were stimulated. A few selected combinations were chosen that were representations of the complex patterns. The characterization of the 9-electrode analog was performed 20 μm above the surfaces of the electrode.

The results obtained characterizing the single-electrode array revealed defining features of the system components. The potential maps indicate a bias toward the metal can irrespective of its vertical position. This result is consistent with previous modeling studies with a different return electrode configuration [18]. As

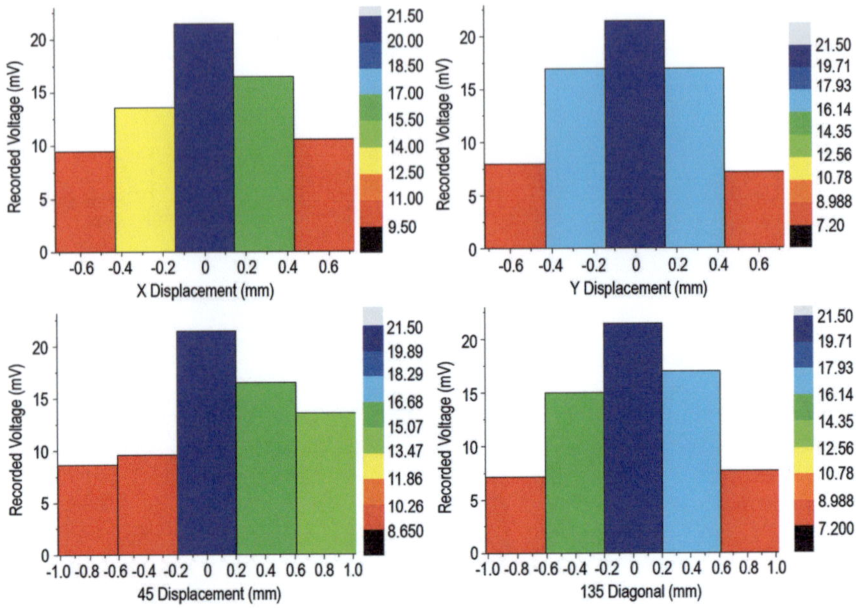

Fig. 12 Potential change in different directions for a single electrode in Ames' media at 37°C. Z=20 μm

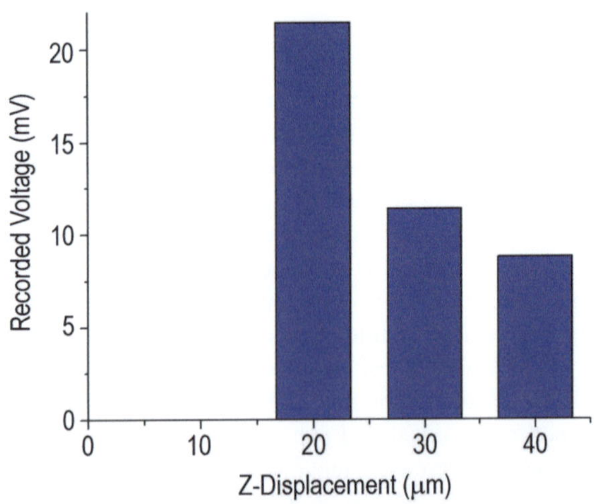

Fig. 13 Bar chart of potentials as a function of the vertical distance from the surface of the stimulating electrode

indicated in Figs. 12 and 13, the recorded voltage decreased symmetrically in both the positive and negative Y directions when the recording electrode was moved parallel to the metal can. When the recording electrode was moved in the direction toward or away from the return electrode, in the X direction, the voltage decreased.

However, the decrease in potential in the positive x direction (toward the metal can) was less than in the negative direction (away from the can), indicating that the presence of the metal can (the return electrode) has a significant effect on the field pattern. As expected, there was a decrease in the recorded voltage as the recording electrode was moved away from the stimulating electrode whether in the horizontal or vertical direction. The relationship between change in potential with change in x, y, or z distance between the electrode array and the recording electrode is summarized in Figs 12 and 13. All experiments were carried out at 37°C.

As expected, simultaneous stimulation of more than one electrode in different combinatorial patterns yields correspondingly different patterns. As noted, stimulation of a single electrode gives a decrease in potential with distance. The decrease in potential (in the case where three electrodes in a line pattern are stimulated) occurs only when the recording electrode is moved beyond the region of the stimulation as shown in Fig. 14. The same experiment was repeated with only the two outer electrodes stimulated. A different feature is revealed as shown in Fig. 15 below, where there is a drop in potential at the center where the electrode was not stimulated whereas the potential at the center of the stimulating electrodes remained high and dropped again when the recording electrode was scanned away from the end electrodes in each side.

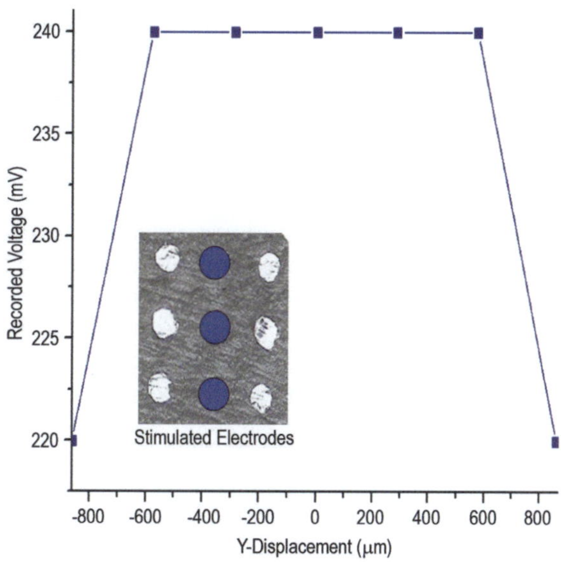

Fig. 14 Field potential mapping for simultaneous stimulation of three electrodes in Ames' medium at 37°C, Z=20 μm

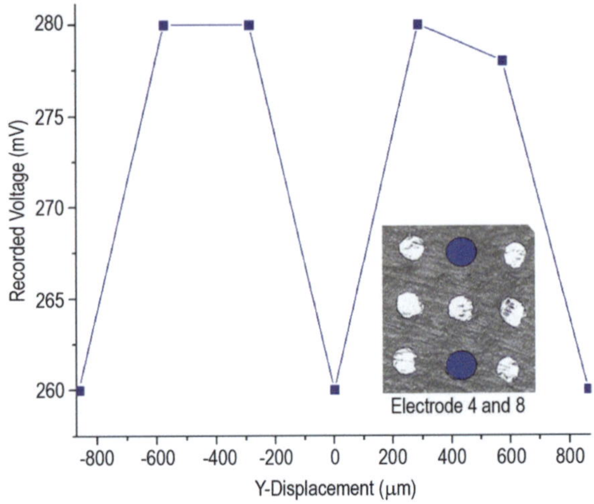

Fig. 15 Field potential mapping for simultaneous stimulation of two electrodes in the 9-electrode array in Ames' medium at 37°C, Z=20 μm

4 Numerical Simulation of Argus II Simplified Model

4.1 Numerical Simulation

In addition to characterizing aspects of the implantable arrays using analog and experimental measurements, electromagnetic numerical modeling and simulation was used to determine the distribution of electric potentials resulting from the charge injection of the stimulating electrodes. Models took into account the conductivity of the different materials involved and the configuration of the single electrode analog and a simplified 60-electrode array. Numerical simulations were performed using a three-dimensional variant of the Admittance Method [19].

4.2 Single Electrode

A first set of simulations was executed to determine the variation of the electric potential with the distance perpendicular to the tip of a stimulating electrode in a single-electrode setup. A three-dimensional simulation model, as shown in Fig. 16, was developed from the diagram of the experimental configuration. Simulation results were later contrasted with experimental measurements from the single-electrode analog setup to verify the model and determine if the presence of the measurement electrode affected the electric potential at the probe points.

This particular configuration used a wire electrode for current return instead of the electronics metal can that was used in some of the experimental setups. Both electrodes were immersed in a medium that had the same dielectric properties as

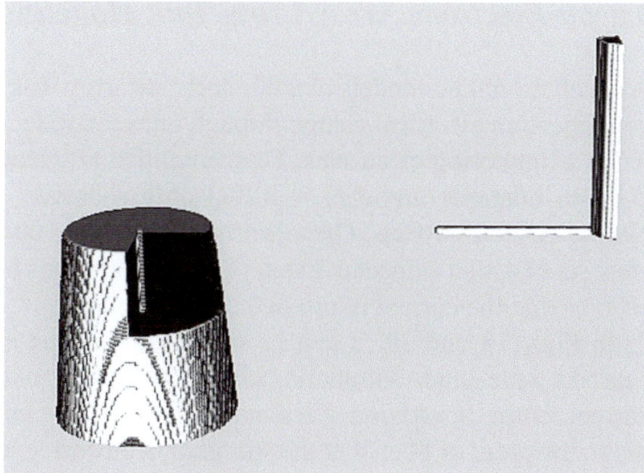

Fig. 16 Three-dimensional model of the experimental single-electrode analog setup without a probe electrode. Stimulating electrode (*left*) has a Pt conductor surrounded by glass for insulation. The counter electrode (*right*) includes a horizontal conducting wire and an insulation support. A section of the glass cladding of the stimulating electrode has been removed to show the inner Pt wire

vitreous humor. The stimulating electrode had a diameter of 150 μm and the stimulation waveform was a 1-ms pulse of 177 μA. The model had a spatial resolution of 35 μm and the resulting fields were interpolated to 20 μm for reporting. The comparison between measured and simulated potentials is shown in Fig. 17.

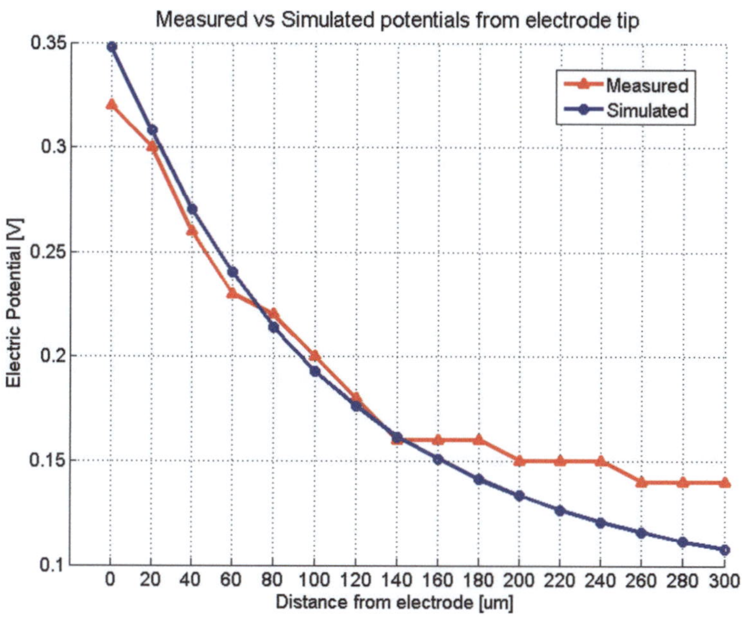

Fig. 17 Measured versus simulated potentials. Distances are from the tip of the stimulating electrode, in a line that passes through the center of the electrode, perpendicular to the electrode surface

4.3 60-Electrode Array Cross-Talk Modeling

A set of simplified models of a 60-electrode array was developed to help determine how injecting electrical charge through one electrode modifies the electrical potential at neighboring electrodes. This simplified 60-electrode array had electrodes of 200-μm diameter arranged in a dielectric substrate, spanning a total volume of 600 × 433 × 29 voxels. One electrode injected a 100-μA current pulse; the others were set to a high-impedance state. Several variations of the model were considered, each having the current return in a different place.

In Figs. 18 and 19, it can be observed how the electric potential distribution caused by the single stimulating electrode changes with the relative position of the current return. In addition, it was noted that a charge injection that produces a potential in the order of 80 mV at the stimulating electrode would cause a variation in the order of 30 mV on the immediate neighboring electrodes when they are left in a high-impedance state (floating).

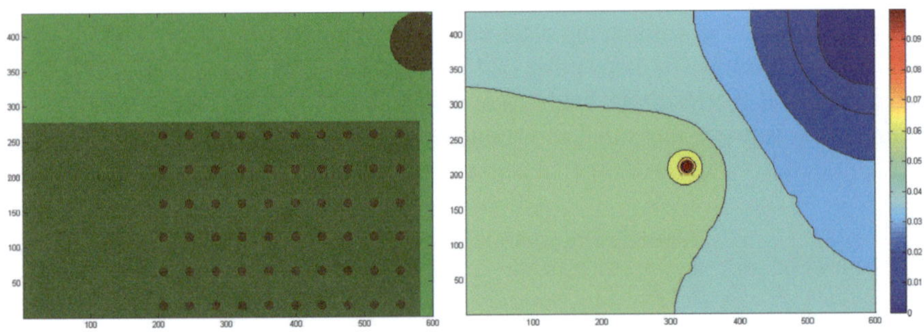

Fig. 18 Simplified 60-electrode model with current return in the upper right (*left*), and corresponding equipotential boundaries at the surface of the electrode array (*right*). The indents in the otherwise smooth equipotential lines are caused by the highly conductive surfaces of the neighboring electrodes

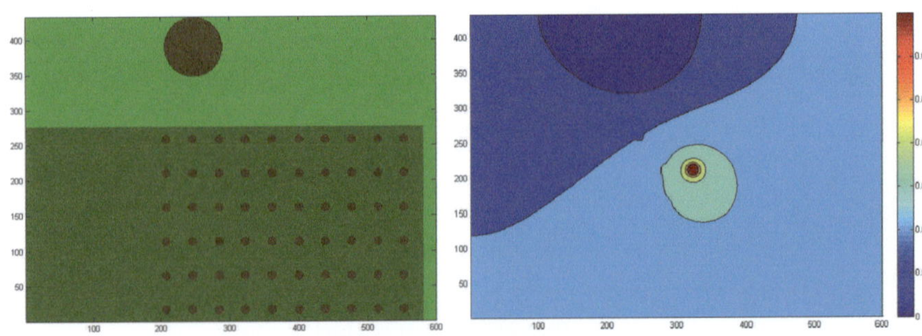

Fig. 19 Simplified 60-electrode model with current return in the upper center (*left*) and corresponding equipotential boundaries at the surface of the electrode array (*right*)

5 Conclusions

Although neuron depolarization and propagation of electrical signals have been studied for nearly a century, the use of multielectrode stimulation as a proposed therapy to treat blindness is a frontier area of modern ophthalmology research. Mapping and characterizing the topographic information contained in the electric field potentials and understanding how this information is transmitted and interpreted in the visual cortex is still very much a work in progress. Correlating the information contained in the topographic features of the electric fields with psychophysical testing in patients may help reduce the time required for patients to convert the electrical patterns into graphic signals.

Acknowledgements This work was supported by the Office of Biological and Environmental Research, U.S. Department of Energy. Oak Ridge National Laboratory is managed by UT-Battelle, LLC for the U.S. Department of Energy under Contract DE-AC05-00OR22725.

References

1. Humayun MS, Weiland JD, Fujii GY et al (2003) Visual perception in a blind subject with a chronic microelectronic retinal prosthesis. Vision Research 43: 2573–2581.
2. Santos A, Humayun MS, de Juan E, Jr. et al. (1997) Preservation of the inner retina in retinitis pigmentosa. A morphometric analysis. Archives of Ophthalmology 115:511–555.
3. Palanker D, Vankov A, Huie P et al. (2007) High-resolution opto-electronic retinal prosthesis: physical limitations and design. In Artificial Sight: Basic Research, Biomedical Engineering, and Clinical Advances, Biological and Medical Physics-Biomedical Engineering, MS Humayun, JD Weiland, G Chader, E Greenbaum eds., New York: Springer, pp. 255–278.
4. Shah S, Hines A, Zhou D et al. (2007) Electrical properties of retinal-electrode interface. Journal of Neural Engineering 4:S24–S29.
5. Colodetti L, Weiland JD, Colodetti S et al. (2007) Pathology of damaging electrical stimulation in the retina. Experimental Eye Research 85:23–33.
6. Kummer MP, Abbott JJ, Dinser S et al. (2007) Artificial vitreous humor for in vitro experiments. Conf Proc IEEE Eng Med Biol Soc, pp. 6407–6410.
7. Chirila TV, Hong Y, Dalton PD et al. (1998) The use of hydrophilic polymers as artificial vitreous. Progress in Polymer Science 23:475–508.
8. Sanders CA, Nagler EJ, Zhou D et al. (2007) Dynamic Interactions of retinal prosthesis electrodes with neural tissue and materials science in electrode design. In Artificial Sight:Basic Research, Biomedical Engineering, and Clinical Advances, Biological and Medical Physics-Biomedical Engineering, MS Humayun, JD Weiland, G Chader, E Greenbaum eds., New York: Springer, pp. 209–226.
9. Weiland JD, Humayun MS, Liu W et al. (2002) Stimulating neural activity. In Handbook of Neuroprosthetic Methods, WE Finn, PG LoPresti eds., Boca Raton: CRC Press, pp. 75–94.
10. Suesserman MF, Spelman FA, Rubinstein JT (1991) In vitro measurement and characterization of current-density profiles produced by nonrecessed, simple recessed, and radially varying recessed stimulating electrodes. IEEE Transactions on Biomedical Engineering 38:401–408.
11. Ameri H, Weiland JD, Humayun MS (2007) Biological considerations for an intraocular retinal prosthesis. In Artificial Sight:Basic Research, Biomedical Engineering, and Clinical Advances, Biological and Medical Physics-Biomedical Engineering, MS Humayun, JD Weiland, G Chader, E Greenbaum eds., New York: Springer, pp. 1–30.

12. Chu A, Morris K, Agazaryan A et al. (2007) In vitro determination of stimulus-induced pH changes in visual prostheses. In Artificial Sight: Basic Research, Biomedical Engineering, and Clinical Advances, Biological and Medical Physics-Biomedical Engineering, MS Humayun, JD Weiland, G Chader, E Greenbaum eds., New York: Springer, pp. 227–242.
13. Hench LL (1975) Prosthetic implant materials. Annual Review of Materials Science 5: 279–300.
14. Brummer SB, Turner MJ (1975) Electrical-stimulation of nervous-system – principle of safe charge injection with noble-metal electrodes. Bioelectrochemistry and Bioenergetics 2: 13–25.
15. Humayun MS, Scribner D, Justus B (2001) Intraocular retinal prosthesis test device. Proc. 23rd annual EMBS international conference, Istanbul, Turkey, pp. 3430–3435.
16. Zhou DM (2005) Platinum electrode and method for manufacturing the same. US patent 6,974,533.
17. Troelstra A, Garcia CA (1975) Electrical response of human eye to sinusoidal light stimulation. IEEE Transactions on Biomedical Engineering 22:369–378.
18. Lazzi G et al (2008) unpublished data.
19. Schmidt S, Cela CJ, Singh V, Weiland J, Humayun MS, Lazzi G (2007) Computational Modeling of Electromagnetic and Thermal Effects for a Dual-Unit Retinal Prosthesis: Inductive Telemetry, Temperature Increase, and Current Densities in the Retina, In Artificial Sight:Basic Research, Biomedical Engineering,and Clinical Advances, Biological and Medical Physics-Biomedical Engineering, M. S. Humayun, J. D. Weiland, G. Chader and E. Greenbaum eds., New York: Springer, 2007, pp. 279–306.

Thin-Film Microelectrode Arrays for Biomedical Applications

Karen C. Cheung

Microfabrication offers many advantages for the batch manufacture of reliable, microscale electrode arrays. Such arrays have been used for highly localized recording and stimulation of neural tissue. This chapter gives a survey of the most commonly used materials and methods in the fabrication of microelectrodes, including planar silicon-based electrodes, three-dimensional silicon-based electrodes, sieve electrodes, and polymer-based structures. Several techniques for electrode modification with nanostructures are described, including carbon nanotube and conductive polymer nanotube coatings. Biocompatibility is described in the context of central nervous system response to chronically implanted devices, which leads to the eventual development of a glial scar.

Contents

K.C. Cheung (✉)
Department of Electrical & Computer Engineering, University of British Columbia, Vancouver, BC, Canada
e-mail: kcheung@ece.ubc.ca

D.D. Zhou, E. Greenbaum (eds.), *Implantable Neural Prostheses 2*,
Biological and Medical Physics, Biomedical Engineering, DOI 10.1007/978-0-387-98120-8_6,

1 Introduction

One of the most successful commercially available implantable microelectrode arrays is the deep-brain stimulator. Deep-brain stimulation (DBS) uses chronically implanted electrodes to treat neurological conditions such as movement disorders. Electrodes deliver high-frequency electrical stimulation to targeted regions of the brain to treat symptoms of Parkinson's disease. Implantable visual prostheses promise to restore vision by providing stimulation at the retina, optic nerve, or visual cortex. Cochlear implants include an electrode array implanted in the inner ear to stimulate the auditory nerve. Other commercially available implantable neural interfaces include neurostimulation systems for treatment of chronic pain or urinary control.

Recent work has shown that cortical recordings from implanted intracranial multielectrode arrays can provide movement-related signals [1]. Such arrays record neuronal action potentials or local field potentials from within the brain. Recordings from the motor cortex of awake, behaving monkeys have been used to control the movement of computer cursors [2] as well as real and virtual arm movements in a three-dimensional virtual environment [3]. Such work shows promise for the development of brain-machine interfaces which could restore function to patients of stroke, neurodegenerative disease, or injury. Although noninvasive recordings such as those using electroencephalographic (EEG) activity or the electrocorticogram (ECoG) have also been used to control computer activity, these methods lack spatial resolution. Only single-cell recordings can give information about neural coding and tuning. At a slightly larger scale, ensemble recordings from groups of neurons show how individual neurons interact as coherent assemblies.

For all of these applications, microelectrodes should be able to record neuronal action potentials or local field potentials or provide highly localized stimulation to select populations of neurons. Given that typical neuronal cell bodies are on the order of 10 to 30 μm in diameter, and electrode selectivity scales inversely with electrode size, microfabrication offers many advantages for the batch manufacture of reliable, microscale electrode arrays:

- Electrodes must remain stable for long periods – a wide variety of metals, insulators, and polymers (not to mention semiconductors) are machinable using microfabrication processes.
- The ideal implant would have a very small cross section for minimal tissue damage – microfabrication by definition aims to produce devices with small lengthscales.

- The ideal array would have a large number of electrode sites to be able to record from or stimulate separate neurons – again, microfabrication by definition aims to mass produce devices with a high functional density.

Microfabrication processes, as described in the following section, are based on thin-film deposition, patterning, and etching and are the basis for integrated-circuit manufacture as well as an expanding universe of microsystems.

2 Microfabrication Methods and Materials

The development of integrated-circuit (IC) fabrication processes, including photolithographic techniques and silicon etching technology, enabled the later development of micro-electro-mechanical systems (MEMS). The invention of the transistor in 1947 at Bell Telephone Laboratories was followed by the integrated circuit in 1958 made using germanium devices, and then the planar silicon IC. The availability of pure, single-crystal silicon made silicon an attractive semiconductor material. Planar batch-fabrication processes developed in the 1960s reduced the cost of semiconductor devices and also improved their reliability [4]. Although early fabrication processes had a minimum feature size of 20 μm, today that dimension has shrunk to submicron levels, and it continues to decrease.

The first sensor made of silicon was the silicon strain gauge, which relies on the piezoresistive effect; silicon strain gauges were marketed commercially in the late 1950s. Silicon pressure sensors were also developed in the 1960s. The first batch-fabricated MEMS device was the resonant gate transistor, developed in the 1960s [5]. Howe and coworkers developed a MEMS resonant bridge vapor sensor, the vibration of which was detected capacitively with an integrated NMOS circuit. The use of silicon fabrication technology thus allowed the monolithic integration of mechanical structures and electronic circuitry.

Up until 1988, MEMS devices featured silicon diaphragms, beams, and other suspended structures which had limited motion. The work of Fan et al. was the first to present freely rotating gears, springs, and sliders [6]. MEMS devices have since expanded tremendously to include micromotors, out-of-plane hinges, switches, microfluidic pumps, valves, and countless other sensors and actuators.

2.1 Micromachining

The term "micromachining" refers to the mechanical aspect of fabrication processes. MEMS microfabrication techniques, while based on conventional IC fabrication technology, also include more specialized and refined processes which permit the formation of mechanical structures. The key for both MEMS and IC fabrication is *photolithography*, which permits high volume, batch production of devices with microscale dimensions. In photolithography, a thin, photosensitive polymer film ("photoresist") is selectively exposed to UV light using a photomask (Fig. 1).

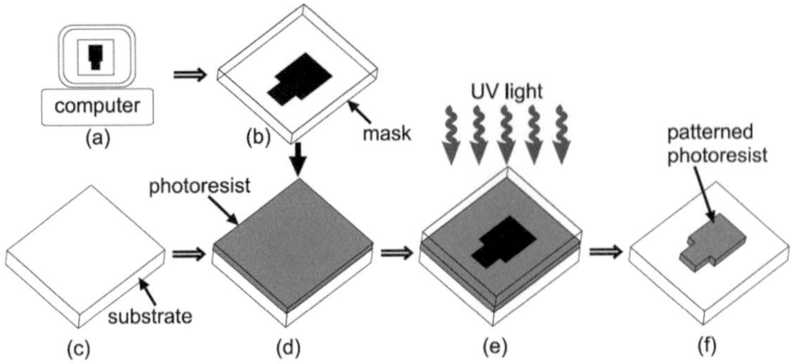

Fig. 1 Photolithographic process [7]. (**a**) A feature is designed and transferred to a mask (**b**), often using a direct laser writer. The mask is typically a glass plate that has chrome, which is opaque to UV light. (**c**) The substrate is spin coated with photoresist (**d**). The pattern is transferred from the mask to the substrate using UV light, which is blocked where the chrome defines the feature (**e**). The photoresist is submerged into the developer solution and the feature is revealed (**f**). The photoresist can be used as a masking layer for further *etching* steps, in which material is removed, or in *deposition* steps, in which material is added

The features on the photomask are transferred to the photoresist, and their size depends on the resolution of the photomask. The features are revealed after immersion in the developer solution. The photoresist can be used as a masking layer for further *etching* steps, in which material is removed, or in *deposition* steps, in which material is added.

"Surface micromachining" refers to the use of the silicon substrate as a mechanical support, while the actual MEMS structures comprise the relatively thin features at the surface, typically from 1 to 100 μm in thickness. Each layer in the structure is created through thin-film deposition, photolithographic patterning, and etching. This sequence of steps is repeated for each successive layer. Surface micromachining uses these standard IC processes while adding new materials, and the use of sacrificial layers. Bulk micromachining refers to the use of the silicon substrate as a functional component, such as in inkjet nozzles, pressure sensors, and molding structures. Bulk micromachining techniques include deep reactive ion etching and silicon molding. MEMS now includes non-silicon substrates, such as polymers and glass used in biomedical applications (bioMEMS). For a detailed overview of microfabrication technology, see *Fundamentals of Microfabrication: The Science of Miniaturization* by M. Madou [8].

2.2 Microfabricated Microelectrodes

Over the past several decades, microelectrodes for extracellular neural recording and stimulation have evolved from single wires to microfabricated arrays [9]. In the development of a breakthrough technique, Strumwasser used single 80-μm diameter stainless steel microwires to record from unanesthetized, freely moving animals in 1958 [10]. Today, wire electrodes can be made of platinum, tungsten, stainless steel, indium, Pt/W, or Pt/Ir. Commercially available stainless steel and tungsten microwires are still used today. They are extremely robust and have demonstrated recording ability for more than 18 months [11] (Fig. 2). One commonly used configuration is the tetrode, which comprises a bundle of four wires [12]. The tetrode is used to record simultaneously from multiple neurons, and these recordings can then be used to isolate single neurons. Although microwires have demonstrated robustness, one drawback is that wires may bend during implantation, so that the distance from one recording site to another is not constant. Also, with only one active site per wire, the volume of displaced tissue increases linearly with the number of electrodes. A tetrode wire has also been commercially introduced. The tetrode wire is made of a quartz glass fiber containing multiple platinum/tungsten alloy cores. The wire cores are arranged with one at the center and three surrounding it radially [13].

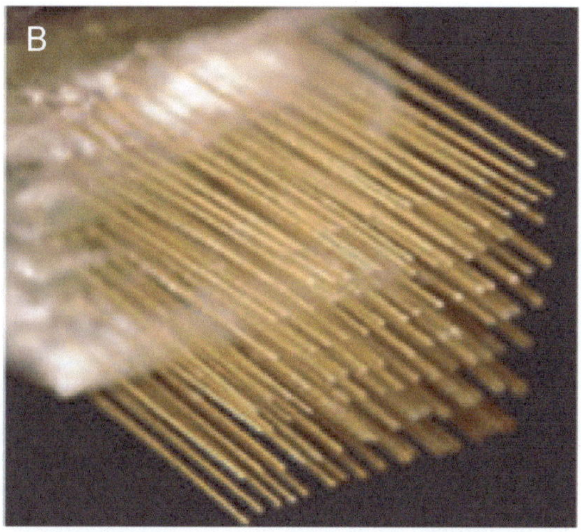

Fig. 2 Multiple two-dimensional microwire arrays are stacked to create a three-dimensional array with up to 128 microwires [11]. *Reproduced with permission*

2.3 Silicon-Based Thin-Film Electrodes

Given the historical evolution of MEMS from IC fabrication, it was natural that the first microfabricated, thin-film microelectrode arrays were made in silicon. Photolithography gives the designer freedom to create any two-dimensional arrangement of electrode sites to suit the specific application. In contrast to microwires, which have one recording or stimulating site at the end of each shank, planar microfabricated probes can have multiple electrode sites along each probe shank, with the distance between each defined precisely in the photolithographic process. In addition, the monolithic integration of elements such as preamplifier circuits could improve the performance of the sensors by reducing noise pickup.

2.3.1 Planar Silicon-Based Electrodes

The first planar silicon-based multielectrode arrays were developed in the 1970s by Wise and Angell [14, 15]. These electrodes were fabricated from 50-μm thick oxidized silicon wafers. Backside patterning of the oxide was used to define the probe shape, while topside patterning defined the insulating base for the gold metallization. The finished probes were separated from the wafer, bonded to insulated output wires, and attached to a carrier handle. In comparison to microwire bundles, these photolithographically defined electrode arrays have several advantages: the interelectrode spacing is controlled and can be as small as the photolithographic limit; the interelectrode spacing is fixed and will not change after implantation; batch fabrication can produce many identical devices with low marginal cost.

The "Michigan array" evolved from these early structures to become widely used internationally. In this thin-film array, several microelectrode sites are patterned on each shank of the structure (Fig. 3). The Center for Neural Communication Technology with funding from the National Institutes of Health (NIH) made standard arrays as well as custom designs available to the research community. They are currently available commercially through NeuroNexus Technologies, a University of Michigan spinoff company.

The microfabrication process begins with high-temperature diffusion of boron ions into the silicon substrate creating a highly doped p^{++} region defined using an oxide mask [16]. The shape of the boron-diffusion area defines the overall probe shape. This is followed by deposition of several lower dielectric layers, which are balanced to minimize residual stress and prevent warpage of the final structure. This oxide/nitride/oxide base dielectric layer serves to insulate the polysilicon conductors from the substrate. Next, the conducting layer is deposited and patterned. After deposition of the upper dielectric layers, the electrode sites are defined by a lift-off process. The probes are released by dissolving the wafer. During this step, the boron-diffused areas act as an etch stop since the etchant does not attack silicon with high boron concentration ($>5 \times 10^{19}/\text{cm}^3$). The released structure is then mounted and packaged to leads.

These thin-film structures provide reproducibility, high spatial resolution, and high density of recording sites. Active probes with monolithically integrated CMOS-signal buffers [17] and preamplifiers [18] have been used to record single-unit activity and local field potentials. High-density recording arrays with more than 50 electrode sites have been fabricated [19].

In addition to the deep boron-diffusion process to define silicon probe shape used in Michigan probes, silicon-on-insulator (SOI) technology has also been used to fabricate planar silicon-based microelectrode arrays [22]. The top single-crystal silicon device layer of an SOI wafer is separated from the silicon substrate by a dielectric layer. This dielectric is typically silicon dioxide from 1 to 2 μm in thickness. In this fabrication process, the probe thickness is defined by the device layer of the

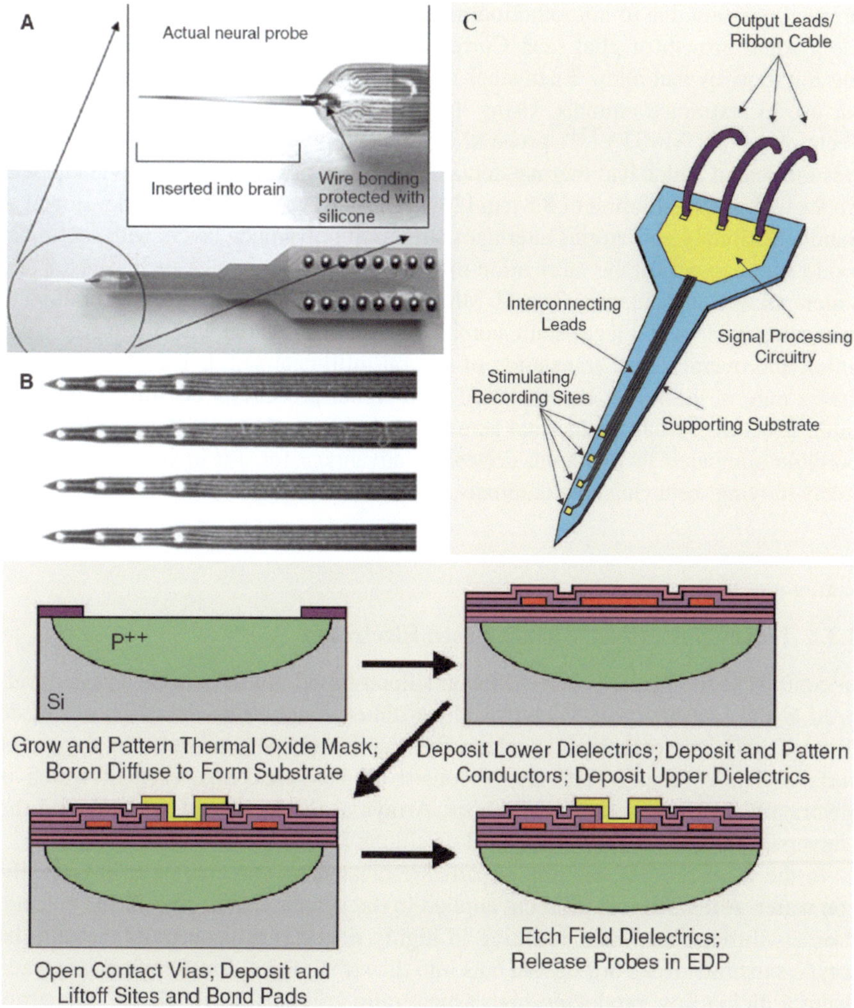

Fig. 3 (**A, B**) Silicon thin-film microelectrodes developed at the University of Michigan [20]. (**C**) Schematic of a planar, silicon-based thin-film microelectrode array. The dissolved wafer process used to fabricate the Michigan implantable probes (not to scale) [21] Reproduced with permission

SOI wafer; these are commercially available in a range of thickness combinations. Features are patterned on both sides of the wafer using aligned front/backside photolithography. The buried oxide acts as an etch stop during a backside deep reactive ion etch of the silicon wafer [23, 24]. Using direct write laser (DWL) lithography, SOI probes with semi-customized designs were fabricated by defining which electrodes out of a standardized array would be active [25].

The ability to move individual electrodes within an array would allow the user to reposition each one during chronic recordings if the electrode drifts from its original position due to micromotion, relative motion in freely moving animals, or the gradual growth of glial scar. Currently available commercial systems, featuring microdrives that allow 1 μm steps on individual microelectrodes, are designed for use in restrained animals. Using the Sandia's Ultraplanar Multi-level MEMS Technology (SUMMiTV™) process, a 5-layer polysilicon microelectrode array was fabricated which had thermal actuators to allow positioning of individual electrodes with step resolution of 8.8 μm [26]. The SUMMiTV™ process developed at Sandia National Laboratories alternates structural polysilicon layers with sacrificial oxide layers to permit the fabrication of freely moving structures such as gears and sliders as well as actuators. The SUMMiTV™ process was also used to fabricate microelectrodes with electrostatic comb-drive microactuators with 1 μm step resolution and overall linear translation of several millimeters [27]. These electrostatic drives may in the future be combined with signal processing circuitry for monolithic structures. Both the thermal actuator and electrostatic actuators are light and portable compared to hydraulic drives, an advantage for use in studies involving freely moving, behaving small animals.

2.3.2 Three-Dimensional Silicon-Based Electrodes

In contrast to the planar boron-diffusion silicon-based multielectrode arrays developed at the University of Michigan, three-dimensional electrode arrays are made through bulk micromachining, where the shanks are perpendicular to the wafer surface. These "needle-bed" arrays consist of sharpened tips, each of which is electrically isolated from its neighbors. Arrays at the University of Utah and the University of Twente were both based on silicon [28, 29].

In the Utah process, aluminum pads are created on one side of an n-type silicon wafer. A temperature gradient applied to the silicon wafer drives Si-Al eutectic droplets through the wafer resulting in highly doped p^+ silicon trails through the n-type substrate. Deep orthogonal cuts into the remaining n-doped regions are made using a dicing saw, producing high aspect ratio columns of p^+ silicon. The array is then placed into a stirred acid-etchant bath, which shapes and sharpens the silicon columns. Finally, the needle tips are pushed through a metal foil, and the tips are coated with platinum to form electrodes [30]. Insulated gold output wires are connected on the back side of the array. The Utah array needle shafts can be up to 1.5 mm in length with 400 μm spacing. The Bionic array produced commercially by Cyberkinetics, Inc. is based on the Utah design. The Twente array has needles of 250, 425, or 600 μm in length and 120 μm spacing (Fig. 4).

Utah arrays have been used to stimulate the cat auditory and visual cortex [33]. This array was used to record chronically from the monkey parietal cortex, with some electrodes still viable after 18 months [34]. The Utah Slant Array features electrodes with different lengths, compared with the original Utah array where all the electrodes are of the same length. Thus the slanted array, in which the microelectrode length varied from 0.5 to 1.5 mm, can stimulate nerve fibers at different

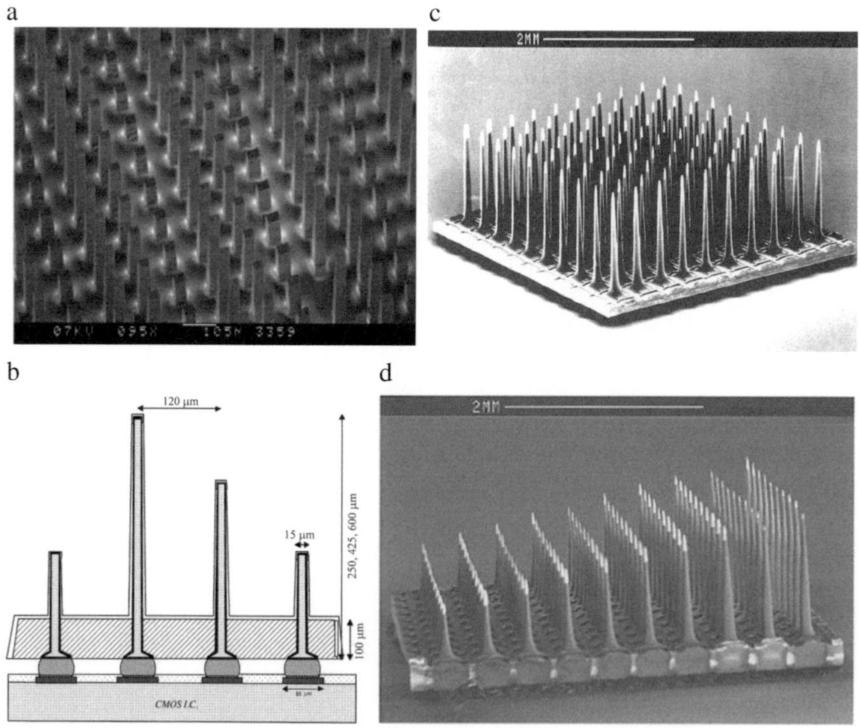

Fig. 4 Both the Twente and Utah processes start by sawing deep grooves into a silicon wafer. (**a, b**) The University of Twente array features 128 electrodes of three different lengths embedded in a glass matrix. The array is mounted on a CMOS-mixed mode chip. [31] (**c, d**) The Utah array features 100 silicon-based microelectrodes, each with a platinum-coated tip. The slanted array has been used to stimulate fibers at different depths in the cochlear nerve [32] Reproduced with permission

depths when implanted in the cochlear nerve [32]. The varying depth of focal stimulation can also be used in peripheral nerves for graded recruitment of muscles, by accessing individual fibers in each fascicle [35, 36].

Preliminary work implanting the Utah array into human subjects used a pneumatic insertion device [37]. This short-term study showed variability in the tissue reaction to individual shanks in the array, and further work is required to correlate the histological findings with electrode function.

The Utah Array is used in the BrainGate system, developed by the Cyberkinetics company. A current clinical trial has such implants in several disabled patients.

2.3.3 Sieve Electrodes

Sieve microelectrodes have been developed for regeneration and functional reinnervation of sensory and motor nerve fibers in the peripheral nervous system. In this approach, a device is aligned between the severed ends of a peripheral nerve, the proximal and distal nerve stumps. The goal is to guide regrowth of the nerves through the holes of the sieve. The holes of the sieve are integrated with electrodes so that action potentials can be recorded from individual axons, or stimulation can be applied to small fascicles.

Sieve electrodes were first made in silicon, and they were used to record electrical signals from regenerated nerve fibers [38–41]. In the case of amputation, it is envisioned that the proximal nerves will grow into the sieve, and the signals may be used to control a prosthetic limb (Fig. 5).

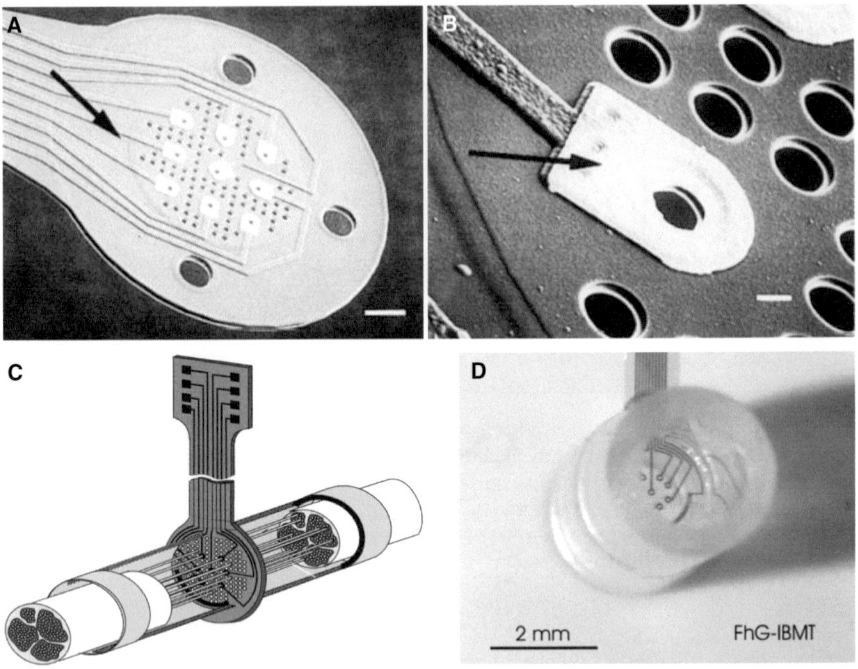

Fig. 5 (**A**) Silicon sieve electrode and (**B**) close-up view of the iridium-lined site [40]. *Reproduced with permission.* (**C**) Regenerative sieve electrode: severed nerve fibers grow through the holes of the sieve structure [20]. (**D**) Polyimide sieve electrode assembled with guidance channel [42]

2.4 Metal-Based Thin-Film Electrodes

Deep-brain stimulation (DBS) for the treatment of Parkinson's disease, in contrast to cortical recording, requires longer probes to reach the nuclei responsible for motor control. The most common target for DBS is the subthalamic nucleus, requiring probe lengths much greater than 10 mm in human patients. In order to batch-fabricate DBS electrodes for chronic use in an animal model, an electroplating process was developed to create long, rigid, tapered structures [43].

In this process, a patterned seed layer for electrodeposition was used to control the profile of the probe and create a needle-like shape with rounded edges. Platinum electrodes were patterned face down on a silicon substrate on top of a sacrificial silicon dioxide layer. After depositing an insulating silicon nitride layer, the seed layer of nickel was patterned as a series of isolated islands. Nickel was then electroplated onto the wafer. As the plated metal layer grew, the metal eventually connected to previously isolated regions of the seed layer producing a smooth, sloped shape. The nickel was then coated with a layer of gold for biocompatibility, and the probes were released from the substrate by etching the sacrificial oxide (Fig. 6a).

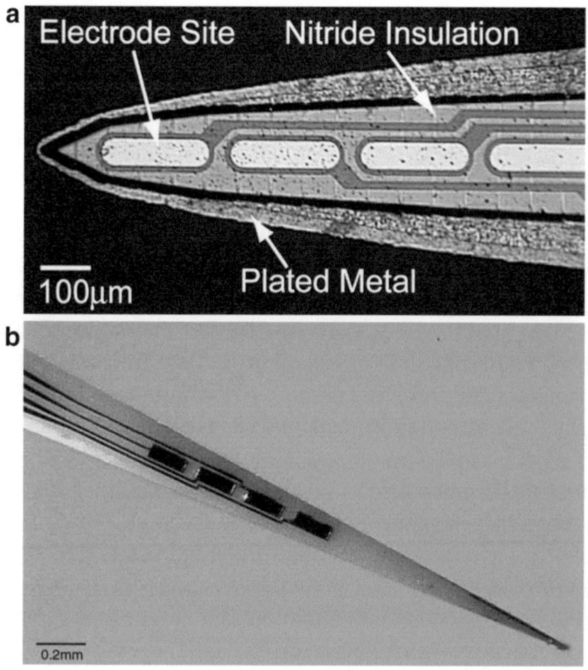

Fig. 6 (**a**) Metal-based electrode fabricated through an electroplating process [43]. Reproduced with permission (**b**) Ceramic-based electrode [44]

2.5 Ceramic-Based Thin-Film Electrodes

In silicon-based probes, high shunt capacitance and cross talk between the recording sites can give high background current and poor signal-to-noise ratio. Ceramic-based implantable electrodes were developed to take advantage of the insulating and higher-strength characteristics [45]. The Ti/Pt electrode sites were patterned using photolithography and a lift-off process, while the ceramic substrate was structured using a diamond saw. The probes were insulated using a polyimide coating. After packaging to a printed circuit board, the electrodes were dipped into Nafion, an anionic polymer which enhances selectivity for cationic neurotransmitters. These planar electrodes were designed to resemble the Michigan probes in shape. The electrodes were used in vivo for electrochemical measurements of dopamine and glutamate (Fig. 6b) [46, 47].

2.6 Polymer-Based Thin-Film Electrodes

In many acute applications, microelectrodes are fixed to stationary electronics for use in anesthetized animals. For chronic use, Michigan probes have been connected to flexible polyimide interconnect cables through thermocompression gold ball bonding [48]. In this case, the polyimide cable links the microfabricated silicon probe and an external connector for long-term recording.

In contrast, polymer-based thin-film electrodes offer the possibility of creating a monolithic device, in which the microelectrodes and the flexible interconnect are fabricated as one structure.

Continual, small relative movements between an implanted device and biological tissue contribute to tissue reaction at the implantation site [49]. The mechanical mismatch between rigid microelectrodes such as microwires or silicon-based structures and soft biological tissue exacerbates the damage due to relative movement between the two. The brain and surrounding fluid are in constant movement, due to changes in intracranial pressure caused by respiration and heartbeat. Using the rat model, surface micromotion in the somatosensory cortex was found to be 10–30 μm pulsations due to respiration and 2–4 μm due to heartbeat [50]; amplitudes have been found to be much larger, on the order of hundreds of microns, in the cat brainstem. For implants which are tethered to the skull, "micromotion" can also result from relative movement between the skull and brain. The Young's modulus of silicon is $\sim$170 GPa, while that of brain tissue is more than 6 orders of magnitude lower at $\sim$3 kPa. Shear-induced inflammation at the implant site may contribute to the foreign body reaction, which decreases the signal-to-noise ratio. Thus, flexibility in the long *interconnect* between the microelectrodes and the external world is vital for permitting the implant to float relative to the skull. In addition, a flexible *microelectrode* could reduce the micromotion, and thus improve the lifetime of the implanted device.

2.6.1 Polyimide

As with silicon, polyimide has also been used in IC fabrication processes. Polyimide is used as a dielectric and encapsulant because it provides good planarization, electrical insulation, and resistance to solvents. Polyimide thin films are patterned using standard microfabrication processes such as photolithography and reactive ion etching; both photosensitive and nonphotosensitive formulations are available.

Polyimide as a structural material for implantable microelectrodes was first investigated as an alternative to multicomponent devices featuring a microfabricated silicon probe that was subsequently bonded to larger electrical contacts. Using polyimide as the substrate instead of silicon, integrated, flexible interconnect between the microelectrodes and electrical contacts could be easily fabricated. Polyimide biocompatibility has been demonstrated both in vitro [51] and in vivo [52].

Thin-film polyimide-based implants use the polymer as both the structural and insulation material. They have been micromachined with multilayer metallization [54] for use as acute and chronic extracellular recording electrodes, and sieve and cuff regeneration electrodes. Thin-film metal layers (approximately 200–300 nm thickness) are sandwiched between the polymer layers. Reactive ion etching in oxygen opens contacts to the electrodes. This dry etch also defines the probe shape (Fig. 7).

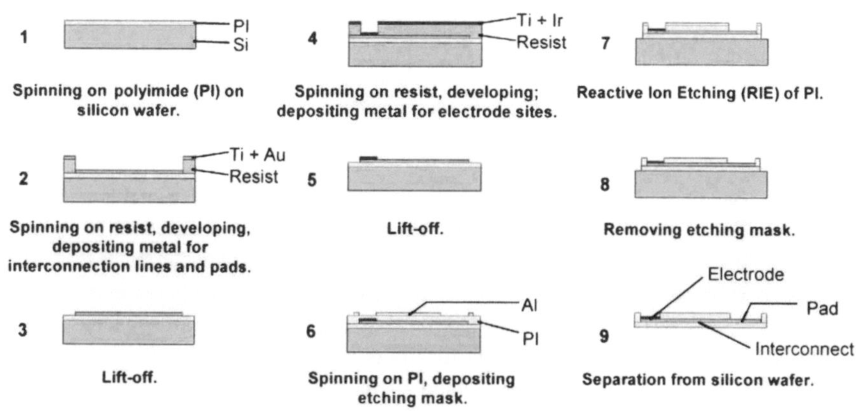

Fig. 7 Fabrication process for polyimide-based thin-film microelectrodes. The silicon substrate acts as a mechanical support during fabrication, and in the last process step the probes are removed from the wafer [53]

Polyimide sieve electrodes were designed to improve the regeneration of axons since the flexible, lightweight polyimide structures can reduce the mechanical stress transferred to the surrounding tissue compared to brittle silicon structures [55]. These flexible electrodes were bonded onto silicon integrated circuits using a "Microflex interconnection" technique [56]. This hybrid integration process is based on conventional wire-bonding techniques, in which the interconnect wire is ultrasonically bonded to the contact pads. Here, flexible ribbon cables are connected to electronic dies. The flexible ribbon structure has via holes in the contact pads. These via holes are aligned over the bond pads on the electronic substrate. The flexible cable and the substrate are bonded using gold balls created using a wire bonder. Since the gold ball studs are slightly larger than the via holes, this bond creates both electrical and mechanical interconnect between the flexible cable and the silicon die (Fig. 8). This Microflex technique was used to integrate a polyimide sieve electrode featuring a large number of electrodes with multiplexer ICs [57]. The multiplexers were used to reduce the number of input connections.

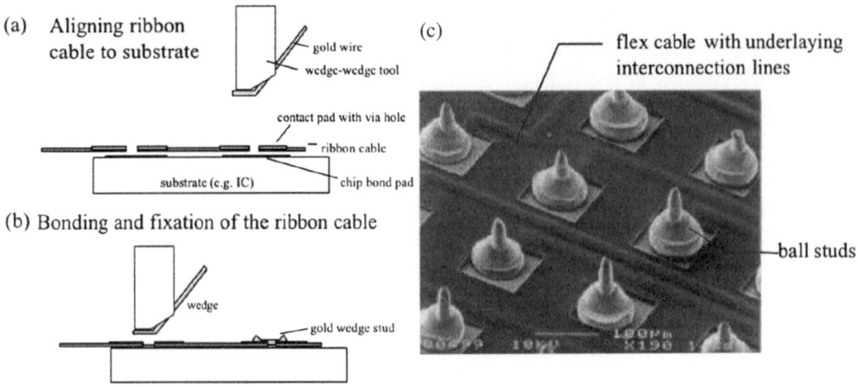

Fig. 8 (**a, b**) The Microflex interconnect technique is based on conventional wire bonding, in which a gold wire is bonded to contact pads. In this case, the goal is to place gold ball studs through via holes in the top flexible ribbon cable onto the IC substrate below. (**c**) The gold ball studs extend over the via holes, providing electrical and mechanical interconnect between the flex cable and the electronic die underneath [56]

Hybrid thin-film polyimide electrodes integrated in silicone guidance channels have been fabricated as cuffs for electrical stimulation of peripheral nerves [58]. Cuff electrodes wrap around a nerve trunk. They are used to record sensory signals from peripheral nerves or stimulate motor fibers to activate muscles [59].

A lamination technique has been used to fabricate polyimide microelectrodes with integrated microfluidic channels [60]. Open channels are patterned using two layers of spin-on, photosensitive polyimide. A second substrate is coated with another uniform, unpatterned spin-on layer of polyimide. This second uncured cover layer of polyimide was laminated onto the first patterned substrate to form closed channels. The final structures were defined using a reactive ion etch and then released from the silicon substrate in an anodic metal dissolution process [61]. Alternatively, embedded microfluidic channels and sealed cavities can be created using a heat-depolymerizable sacrificial polymer [62]. The sacrificial material is a polycarbonate which can be spun onto the substrate as a thin film. Both polyethylene carbonate and polypropylene carbonate have been used. The polycarbonate defines the channel geometry and is patterned on top of a polyimide substrate using reactive ion etching. A second cover layer of polyimide was spun on over the polycarbonate. During the second high-temperature polyimide curing step, the polycarbonate decomposes into a mixture of CO, CO_2, and H_2O. These volatile components diffuse through the polyimide, leaving behind a microfluidic channel or cavity.

In contrast to microfluidic channels, which flow straight from the reservoir to the delivery site, a U-shaped channel has an outlet close to the electrode sites at the apex of the "U." Continuous flow within the U-shaped channel aspirates interstitial fluid from the tissue into the channel. Small aliquots of fluid in the picoliter range can be delivered into the tissue by blocking the flow for a short time so that fluid is ejected from the delivery outlet (Fig. 9) [63].

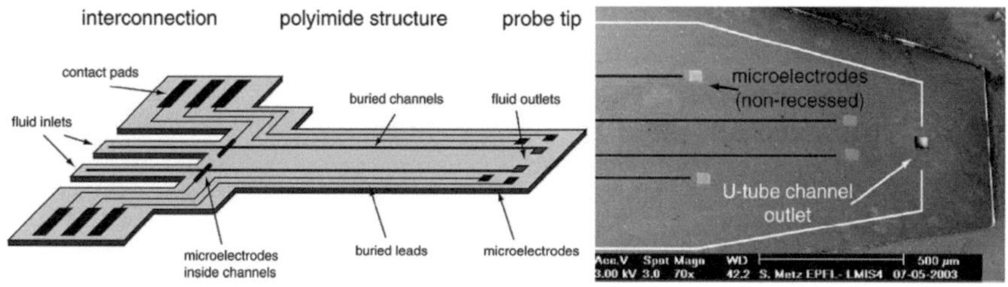

Fig. 9 Polyimide-based microelectrode with integrated microfluidic channel. The U-shaped channel has a fluid delivery outlet at the tip of the probe [63]

For applications involving filtration or dialysis, polyimide probes with integrated nanoporous filtration areas feature a membrane with submicron-sized pores [64]. The pores are created by irradiation with uranium or xenon ions to create ion tracks. These tracks are chemically etched in sodium hypochlorite and chlorine to produce a porous membrane.

2.6.2 Other Polymer-Based Microelectrodes

Parylene is an inert, biocompatible polymer that has been used to insulate microwire electrodes since the 1980s. Parylene is deposited in the vapor phase. Solid dimer particles are vaporized under vacuum at 150°C. The dimer molecules are cracked at 680 °C to form free radical monomers. The monomers condense and polymerize as a conformal coating on the component. Since the deposition is done at room temperature, the coating has low residual stress. Parylene films are pinhole-free and have low moisture permeability.

Parylene has been used as an insulating material for microwire electrodes used in chronic recording [65]. Laser ablation can be used to define openings at the tip of the microwires (Fig. 10) [66]. Microfabricated probes made of parylene have also been developed [67]. These probes featured an embedded microfluidic channel for localized drug delivery. The channel was created in the probe using photoresist as a sacrificial material. The photoresist was dissolved in a solvent, leaving behind an open channel. This structure is quite flexible. The channel was then filled with polyethylene glycol (PEG) to increase the stiffness of the probe during insertion. After insertion, the PEG dissolved into the aqueous environment, leaving behind a flexible polymer structure (Fig. 11).

Benzocyclobutene (BCB) has also been used as a packaging material in the microelectronics industry. Like polyimide, it has low dielectric constant and good chemical resistance. Its properties also include low moisture absorption and low cure temperature. BCB is a negatively photosensitive material. Since there are no outgassing products during curing and low volume shrinkage during curing, BCB is

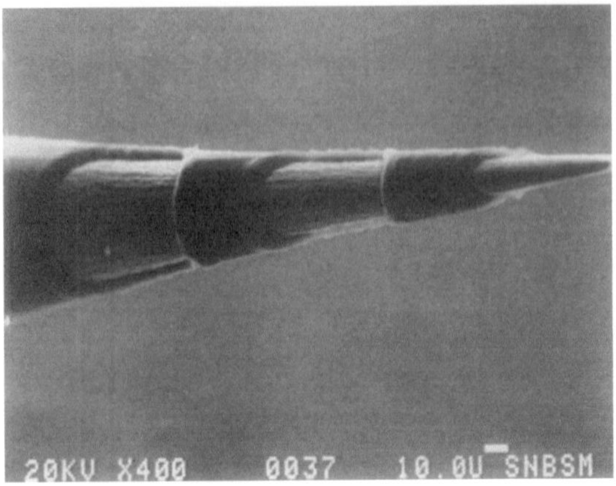

Fig. 10 Parylene-coated iridium microelectrode. Three sites have been opened using laser ablation [66]

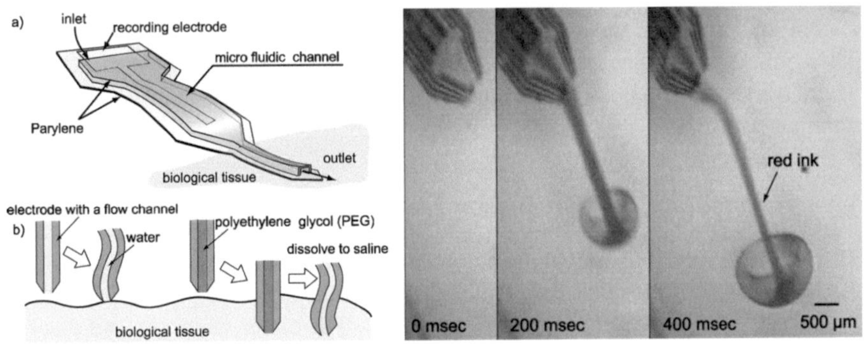

Fig. 11 Thin-film parylene-based microelectrode with embedded microfluidic channel. The parylene is structured in a reactive-ion etch. The photoresist serves as a sacrificial material, and it defines the channel shape. As with the polyimide fabrication process, the silicon substrate serves as a mechanical support during fabrication, and the parylene probe is removed in the last step. In this injection test, ink is passed through the microfluidic channel [67] – *Reproduced by permission of The Royal Society of Chemistry*

also a good candidate for adhesive bonding applications in MEMS and microfluidic devices [68, 69].

A hybrid BCB-silicon neural implant with embedded microfluidic channels has been fabricated and tested in acute recordings [70]. A thin layer of silicon was used to add mechanical stiffness to the implant. The fabrication process is based on SOI technology, where the device layer of the wafer was the 2-, 5-, or 10-µm silicon backbone of the BCB structure. The microfluidic channels were made with a sacrificial photoresist layer. Cytotoxicity tests of BCB have demonstrated its biocompatibility in glial and fibroblast cell culture [71] and using brain slice culture [72]. A summary of several microfabricated thin-film electrodes is presented in Table 1.

Table 1 Comparison of several microfabricated thin-film microelectrode technologies

	Microwires	Michigan array	Utah array	Polymer arrays	SOI	Metal	Ceramic	Sieve electrodes
Bulk material	Stainless steel [10]; tungsten; platinum/ iridium	Silicon [21]	Silicon [28]	Polyimide [63, 73]; parylene [67]; benzocyclobutene [70]	Silicon [24, 25]	Electroplated metal [43]	Ceramic [45]	(1) Silicon [38, 74]; (2) Polyimide [55]
Dielectric	Polyimide; teflon; parylene [65]	Silicon nitride; SiO_2	Polyimide	Self-insulation	Silicon nitride; SiO_2	Silicon nitride	Polyimide	(1) Silicon nitride; SiO_2 (2) Self-insulation
Layout	Three-dimensional array	Planar array (can be assembled into three-dimensional array [75, 76])	Three-dimensional array	Planar array	Planar array	Planar array	Planar array	Planar array
Typical shank thickness	~ 30–100 μm	~ 15 μm	~ 90 μm (base) • 2 μm (tip)	~ 15 μm	Varies according to SOI wafer (constant along shaft)	Varies along length of shaft	~ 125 μm	Assembled with guidance channel
Multiple electrode sites on each shank	No	Yes	No	Yes	Yes	Yes	Yes	Yes
Integrated electronics	–	[17, 77]	–	–	–	–	–	(1) – (2) [57]
Embedded microfluidic channel	No	Yes [78]	No	✓ Polyimide ✓ Parylene ✓ Benzocyclobutene	Yes [24]	No	No	N/A
Reported chronic recording duration	> 18 months [79]	> 4 months [80]	> 18 months [34]	> 8 weeks [73]	–	–	–	(1) > 60 days [40] (2) > 6 months [81]

2.7 Nanostructured Electrodes

Microelectrodes often exhibit high impedance due to their small geometric area, leading to high noise levels. By increasing the effective area of planar microelectrodes, the electrode impedance can be reduced while maintaining a small geometric footprint [82]. The geometric footprint of an electrode is one factor in determining the recording selectivity. Selectivity refers to the ability of an electrode to target a specific number of neurons, neuronal processes, nerves, or fascicles. In addition, high impedance brings higher thermal noise. Thermal noise of resistive origin is given by $V_{th,rms} = \sqrt{4kTBR_{tot}}$ where k is the Boltzmann constant, T is the temperature, B is the frequency bandwidth, and R_{tot} is the real part of the total electrode impedance.

Historically, microelectrode surface area has been increased by electroplating platinum powder, also known as platinum black, thus increasing the roughness of smooth electrodes [83]. This method of increasing surface area sometimes results in unstable impedance because the platinum powder adheres weakly to the electrode. For gold electrodes, a chemical etch using KI_3 preferentially attacks certain crystal axes in the amorphously deposited gold and roughens the surface. However, platinide electrodes etch very slowly and only in extremely harsh solutions such as aqua regia, which will also destroy the substrate material. Other methods of roughening platinide electrodes include ion milling and plasma etching. For iridium electrodes, the most common method of decreasing impedance is through chemical modification. Oxidized, or activated, iridium forms a hydrated and highly conductive layer and has high roughness [84–86]. Titanium nitride has also been used to increase the effective electrode surface. Titanium nitride is deposited on electrode sites by sputtering titanium in a nitrogen and argon atmosphere [87]. Depending on the process parameters, the titanium nitride can form a porous, columnar structure. This surface modification can lower microelectrode impedance from several MOhms to a few hundreds of kOhms [31].

Another effect of increasing the electrode roughness may be the improvement of the neuron-electrode contact. The neuron-electrode gap can be modeled as an ionic fluid. A tight seal between the neuron and electrode gives less leakage current through the extracellular medium. Recently, nanostructures such as nanofibers or carbon nanotubes have been integrated onto multielectrode arrays for improved electrical characteristics, cell interaction, or drug delivery. In neural tissue, laminin is the key extracellular matrix component. Laminin is a 50 nm $\times$ 70 nm cross-shaped protein whose main function is to mediate cell binding to collagen IV in the extracellular matrix [88]. Nanoscale topology on neural implants promises to improve tissue interaction, since cells naturally encounter nanoscale roughness in their environment. Nanostructures at the scale of tens of nanometers and above have been shown to attract attachment of primary cortical astrocytes while repelling transformed astrocytes [89]. The nanostructures offer topographical cues that can also guide neurite outgrowth [90]. Growth cones at the tip of the neurite can grasp onto carbon nanotubes that have been functionalized with neuron growth factors. Thus nanotubes can serve as scaffolds for neurons, and patterned areas of nanotubes can be used to direct neural network formation on planar electrode arrays. Carbon nanotubes also exhibit high mechanical strength and electrical conductivity.

2.7.1 Carbon Nanotube Coatings

Carbon nanotube coatings can be grown through chemical vapor deposition on a catalyst layer. Typically, nickel is evaporated onto the electrode sites to serve as the seed material. The nickel thickness determines the surface packing of the carbon nanotubes. Carbon nanotubes are formed by chemical vapor deposition using hydrogen and ethylene gases. Electrochemical testing has shown that the carbon nanotube-coated electrodes have high specific DC capacitance of 10 mF/cm^2 as well as low frequency-dependent AC impedance [91]. The nanotubes grown by Gabay et al. were intertwined, matted coatings on Ni/TiNi pads (Fig. 12a,b). Cell-electrode coupling to the rough surface is strong, with high signal-to-noise ratio for cells cultured on the nanotube-coated electrodes. The small amount of nickel catalyst was encapsulated inside the carbon nanotube coating, so that toxicity was reduced as shown through two weeks of cell culture on the nanotubes.

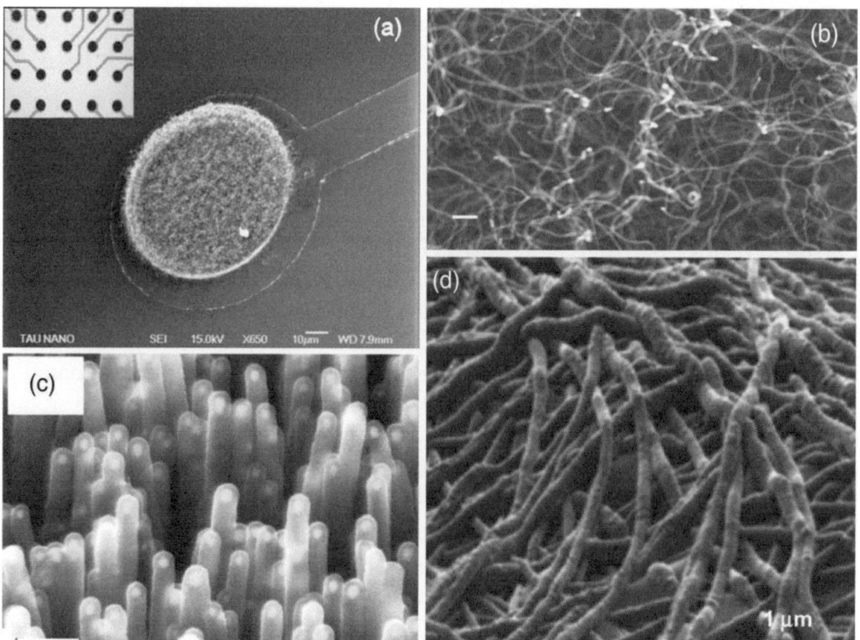

Fig. 12 (**a, b**) An 80 μm carbon nanotube island grown on a 100 μm conductive titanium nitride pad; scale bar is 10 μm [91]. These entwined nanotubes give a matted surface; scale bar is 100 μm. Reproduced with permission (**c**) Vertically aligned multiwalled carbon nanofiber array after electrodepositing a 40-nm-thick conformal polypyrrole film; scale bar is 500 nm [92]. *Reproduced with permission.* (**d**) PEDOT nanotubes formed using PLLA templates [93]

High aspect ratio, vertically aligned carbon nanotube coatings have also been produced (Fig. 12c). The as-grown arrays of nanotubes are hydrophobic, and they are not stable as capillary forces from water droplets during the drying process will tend to pull the carbon nanotubes together. Functionalization of the nanotubes can create a hydrophilic surface. Heating the nanotubes in air forms hydrophilic oxides at the ends and defect sites of the carbon nanotubes. Electrochemical deposition of a conformal film such as polypyrrole, an electrically conductive polymer, can increase the mechanical strength of the nanotubes, permitting the nanotubes to remain vertical and freestanding both when submerged in solution and in air. The application of extracellular matrix proteins such as collagen to the polypyrrole-coated carbon nanofibers improves the attachment of PC12 cells cultured on the nanofiber arrays [92].

The hydrophilicity of vertically aligned arrays of multiwalled carbon nanotubes can be modified through adsorption of molecules such as polyethylene glycol-lipid conjugates, or they can be incubated with cell culture medium. The charge-injection limit of microelectrodes with such modified carbon nanotubes is in the range of 1–1.6 mC/cm^2, compared to 0.1–0.3 mC/cm^2 for bare platinum and 2–3 mC/cm^2 for iridium oxide electrodes [94]. Tissue damage can occur due to prolonged stimulation; both charge density and charge per phase are important in determining the threshold for tissue damage [95]. The primary charge injection mechanism through carbon nanotubes is capacitive, thus avoiding electrochemical reactions at the electrode surface.

2.7.2 Conductive Polymer Nanotubes

Conductive polymer nanotubes have also been deposited onto microelectrodes. Polymers such as polypyrrole or poly(3,4-ethylenedioxythiophene) (PEDOT) feature a conjugated backbone with a high degree of π-orbital overlap. They can undergo oxidation or reduction and can contain dopants such as drug molecules. Abidian et al. have created polypyrrole and PEDOT nanotubes using a biodegradable poly(L:-lactic acid) (PLLA) template. The PLLA is first electrospun onto the neural probe. The polypyrrole or PEDOT is then electrochemically deposited around the electrospun PLLA fibers. Although the PLLA fibers coat the entire device, the conductive polymers nucleate at the electrode sites to give limited, local deposition. The core PLLA material is removed by soaking the devices in dichloromethane, leaving behind nanotubular conductive polymer structures [93] (Fig. 12d). The nanotube diameters range from 100 to 600 nm, while the wall thicknesses vary from 50 to 100 nm. The impedance of electrodes modified with polypyrrole nanotubes ($\sim$80 kΩ) and PEDOT nanotubes ($\sim$4 kΩ) is a much lower impedance than unmodified bare gold electrodes ($\sim$800 kΩ at 1 kHz). PEDOT nanotube-coated microelectrodes have been shown to have lower impedance and high signal-to-noise ratio than unmodified electrodes during in vivo tests for 7 weeks [96].

Conducting polymer nanotubes can also be used for localized drug delivery. The biodegradable polymer containing the drug is electrospun onto the structure. The conducting polymer is electrochemically deposited onto the drug-loaded core material. The drug is released from the nanotube through electrical stimulation of the nanotubes. Conducting polymers can undergo a change in volume upon oxidation or reduction, because they expand when counterions from solution enter the polymer and contract when ions leave the polymer. Abidian et al. have used PLLA and poly(lactic-co-glycolic acid) (PLGA) as nanofiber template materials since they can biodegrade, leaving behind conducting polymer nanotubes [97]. The PEDOT coating slows the release of the incorporated drug, dexamethasone, likely because it cannot diffuse easily through the PEDOT nanotube walls. During electrochemical actuation of the PEDOT nanotubes, the PEDOT contraction squeezes the PLGA degradation products and the dexamethasone out of the nanotube ends or through cracks in the surface of the nanotubes. This method gives controlled delivery of the drug (Fig. 13).

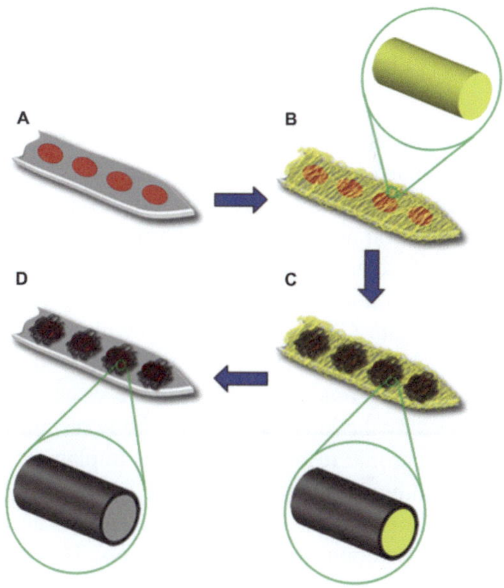

Fig. 13 Fabrication process for hollow conductive polymer nanotubes on a neural probe. The neural probe (**A**) is coated with biodegradable polymer nanofibers (PLLA) through an electrospinning process (**B**). Conductive polymers are electrochemically deposited around the nanofibers (**C**). Since they nucleate at the electrode sites, hollow conducting polymer nanotubes remain on the electrodes (**D**) after the core material is removed [93]

3 Central Nervous System Response to Implanted Devices

Although regeneration after injury to peripheral nervous system (PNS) is possible in adult mammals, injury to the central nervous system (CNS) is rarely followed by regeneration and return of lost function. In the PNS, the proximal stump of transected axons can bridge the gap and achieve reinnervation [98]. Nerve guidance channels are used to give a guide to the elongating neurites and maintain the local concentration of expressed growth factors.

In contrast, wound healing in the central nervous system (CNS) involves an initial period of cell damage and cell death followed by reactive gliosis, leading ultimately to the formation of a glial scar. This dense scar tissue is composed mainly of enlarged and entangled astrocytes which have increased expression of filament proteins. One function of this scar is to isolate the injury site from surrounding healthy tissue, protecting the CNS from further damage [99]. In studies involving transgenic mice in which reactive astrocytes are ablated, no glial scar forms and the inflammation is widespread with infiltration of leukocytes [100]. Reactive astrocytes are also involved in blood–brain barrier repair and regulation of blood vessel diameter. However, the glial scar creates a mechanical barrier to axon regeneration as well as a chemical environment that inhibits regeneration.

Although microfabricated electrodes have demonstrated effective recording and stimulation for acute applications, stability in chronic applications has not been achieved. Even using microwire arrays, which have demonstrated the greatest robustness, Nicolelis et al. reported a steady decline in the number of functional electrodes over 18 months [11]. The failure of chronically implanted microelectrodes is attributed to the CNS response.

3.1 Reactive Gliosis

Glial cells of the CNS include oligodendrocytes, astrocytes, and microglia [101]. Oligodendrocytes form myelin in the CNS. Astrocytes, which have a finely branched morphology, are involved in supportive functions such as neurotransmitter regulation and blood-brain barrier maintenance. Injury to the CNS activates the astrocytes, causing increased migration, proliferation, and hypertrophy. Microglia are resident macrophages in neural tissue and are activated during injury as well.

The acute CNS response starts within hours of injury and is characterized by migration of astrocytes and microglia to the injury site [102, 103]. The acute inflammatory reaction peaks within a week after injury [104] and is then gradually replaced by the long-term foreign body response.

The chronic response to devices implanted in the CNS comprises the glial scar surrounding the device, extending from tens to hundreds of microns away from the implant. In addition to astrocytes, the glial scar may also contain extracellular matrix (ECM) and connective tissue elements. This may occur if the CNS injury involves the meninges, which are the membranes that envelop the CNS. Concomitant damage to these membranes causes fibroblasts as well as macrophages to migrate to the lesion site [105]. The ECM components are inhibitory to axon regeneration. In particular, proteoglycans expressed by activated astrocytes, oligodendrocyte precursor cells, and meningeal cells have been found to be the most inhibitory. Chondroitin sulfate proteoglycans (CSPGs) are involved in cell migration and axon guidance during development [106]. Reactive astrocytes upregulate CSPGs in a gradient with the highest concentration at the center of the injury site. Here, CSPGs affect growth cones on the axons to form dystrophic endbulbs (Fig. 14).

3.2 Histology

The glial scar has been extensively characterized using immunohistological techniques. Activated astrocytes upregulate glial fibrillary acidic protein (GFAP), and

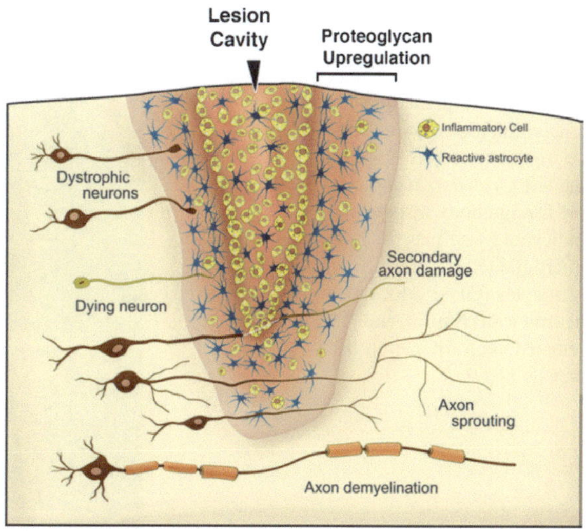

Fig. 14 Tissue damage in the central nervous system leads to reactive gliosis. The region of the glial scar has a gradient in extracellular matrix molecules such as proteoglycans, which inhibit the regeneration of axons [99]

immunostaining for GFAP is commonly used for their identification. Activated microglia can be identified using the ED-1 antibody, which recognizes CD68, a membrane protein [107]. Neurons can be identified with the NeuN stain. Reactive gliosis has been found in a variety of animal models using silicon-based implants, microwires, and plastic structures. Gliosis has also been seen in nonhuman primates using platinum microwires implanted for 3 years [108].

Previous work examining tissue reaction to implanted microelectrodes has shown that the long-term response varies widely. Early work by Stensaas et al. using chronically implanted plastic microneedles found that tissue response varied even around a single implant [110]. Edell et al. [111] reported more gliosis at the tips of implanted silicon arrays, while Schmidt et al. [112] found relatively constant gliosis along the length of each 1.5 mm shank in a Utah array. Both Turner et al. [103] and Szarowski et al. [102] have presented histological analyses of gliosis around micromachined silicon devices of different size and cross-section, showing a sheath of astrocytes around the insertion site that is independent of implant size, geometry, and surface roughness. Biran et al. [109] found significant neuron loss and reduction in neurofilament density at the microelectrode interface (Fig. 15).

3.3 Glial Scar and Tissue Impedance

The glial scar increases the distance between the electrode and neuron cell bodies. The amplitude of extracellular signals decreases rapidly with increasing distance from the cell body. Extracellular signal amplitude is often in the neighborhood of 50 μV, although strong spikes can be around 250 μV. Experimental data have shown that extracellular action potentials cannot be differentiated from noise for electrodes further than 140 μm from the cell body [113].

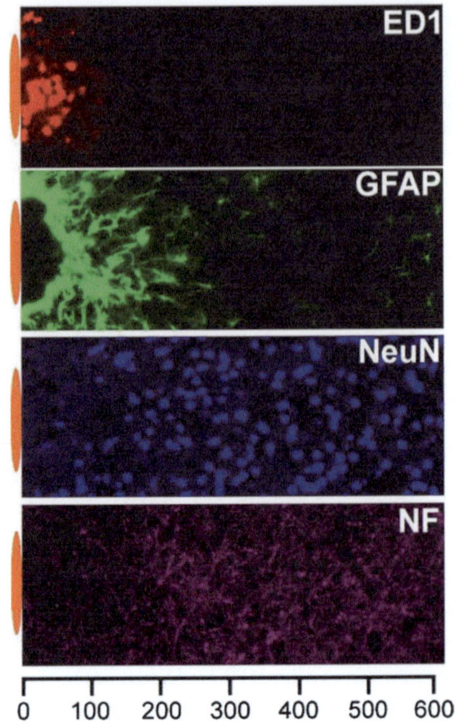

Fig. 15 Central nervous system reaction at the implant/tissue interface four weeks after implantation of a silicon microelectrode. The microelectrode was placed at the left (*drawn to scale*) and the foreign body response is indicated by activated microglia (ED-1 stain), astrocytes (GFAP), local neuron loss (NeuN), and reduced neurofilament (NF) density [109]

In addition to pushing neurons further from implanted electrodes, the glial scar may also insulate electrodes from the tissue and increase the effective impedance [114]. Electrical impedance can be used to monitor changes in extracellular/intracellular volume ratio or ionic composition [115]. Impedance monitoring of chronically implanted electrodes has shown an increase in impedance over time [116]. Changes in impedance spectra are influenced by changes in tissue surrounding the implant, and this impedance change peaks at 7 days after implantation, mirroring the transition from acute to chronic tissue response. Previous work has also used impedance spectroscopy to characterize the cell-electrode interface in cells cultured on planar multielectrode arrays [117–119].

4 Implant Biocompatibility

In the context of implantable thin-film microelectrodes for long-term use, the term "biocompatibility" includes both the bulk material as well as the surface characteristics. Finite element modeling (FEM) of the implant/tissue interface has shown that flexible polymeric materials such as polyimide, which has a Young's modulus of approximately 3 GPa, can reduce strain at the implant/tissue interface when compared to a rigid, silicon-based probe [120]. The lower strain may reduce micromotion, leading to lower chronic irritation. FEM has also shown that improved attachment between probe and tissue decreases both the micromotion-induced strain as well as the relative slip between the implant and the tissue [121].

Several strategies to modulate reactive gliosis surrounding implants have been proposed.

4.1 Microelectrode Structure

The cross-sectional area, size, and shape of the implant may have an effect on the tissue response. Using silicon implants, Szarowski et al. found that the initial response (<1 week) was proportional to implant cross-sectional area, but the sustained response (>4 weeks) was independent of electrode size [102]. In contrast, Stice et al. have found that decreased cross-sectional area leads to statistically significant reduction in glial scarring using biodegradable polymer-coated microwires [122].

4.2 Pharmacology

The CNS response to implanted microelectrodes may be modulated pharmacologically. Drug delivery has been effected either through a coating on the implant or through embedded microfluidic channels in the implant [123].

Historically, dexamethasone, an anti-inflammatory glucorticosteroid, has been used to reduce inflammation at the electrode-tissue interface of cardiac pacemakers and control fibrous capsule formation [124, 125]. Glucocorticoids have been found to reduce reactive gliosis when applied topically to the cortex [126]. Systemic doses of dexamethasone have been shown to reduce astroglia response around a silicon device implanted in the brain [127]. Localized dexamethasone elution from an implanted poly (ethyl-vinyl) acetate structure also showed reduction in the reactive response [128]. Dexamethasone incorporated into nitrocellulose coatings on implanted Michigan probes showed reduced reactive astrocyte and microglia numbers compared to uncoated probes [129]. Electrochemical release of dexamethasone using polypyrrole, a conducting polymer, has also been investigated [130].

Flexible polyimide electrodes were seeded with hydrogels that released neural growth factor (NGF) [131, 132]. NGF can encourage neurite growth toward the electrode for improved signal-to-noise ratio. Using micropipettes, the hydrogels were seeded into wells on the probes to provide time-controlled drug release.

Other candidates for attenuating astrogliosis include the anti-inflammatory cytokine IL-10 and antibodies to transforming growth factor (TGF)-β1. Chondroitinase ABC (chABC) selectively cleaves the glycosaminoglycan side chains from the CSPG protein core, and treatment with chABC has been shown to reduce the inhibitory effects and promote axon regeneration [106, 126]. At the same time, the implant can also incorporate treatments to protect tissue from damage associated with prolonged electrical stimulation [133].

4.3 Hybrid Structures

Hybrid structures which house living cells inside a microfabricated implant may give improved chronic recording ability. A silicon-based in vitro array featuring cages which house living, cultured neurons has been demonstrated [134]. The aim was to maintain a robust connection between electrodes and neuron cell bodies trapped in proximity to the electrodes. These neurowells have also been incorporated on implantable probes for in vivo use [135]. In this work, embryonic neurons were inserted into the wells and allowed to mature. They formed a seal with the gold electrode at the bottom of the well, and extended neurites out of the well into the surrounding tissue (Fig. 16).

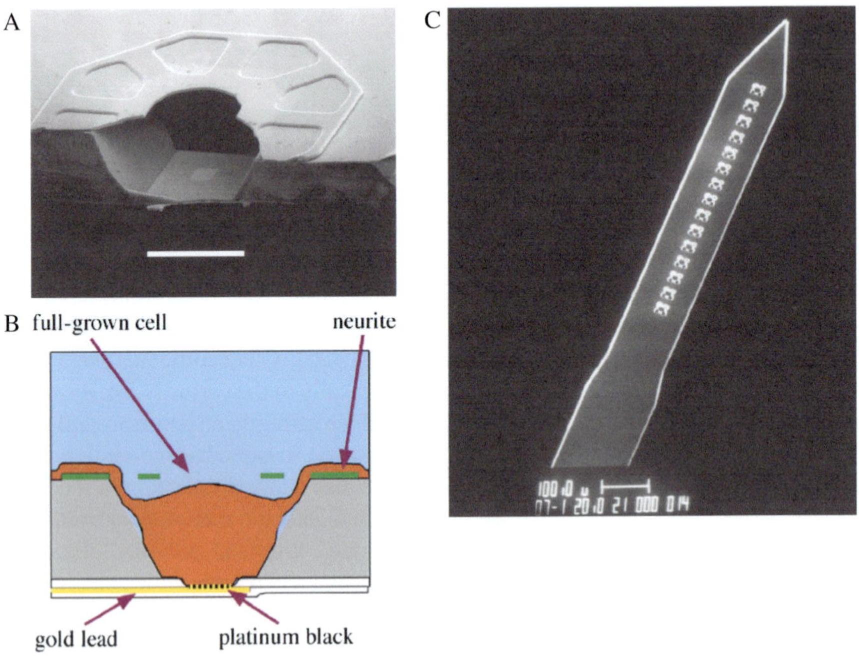

Fig. 16 (**A**) Silicon-based well, which will hold a cultured neuron. The canopy contains tunnels through which neural processes can extend. The bottom of the well is a gold electrode. *Scale bar*: 20 μm. [134] (**B**) Schematic of a neuron loaded into the well [136]. (**C**) Neuron wells on an implantable probe [135] Reproduced with permission

In addition to forming electrical connections, cell transplantation has been used for cell-mediated repair and sustained local delivery of therapeutic molecules. For example, PC-12 cells, which release dopamine, were transplanted using a semipermeable hollow fiber membrane device for application in treating Parkinson's disease. This device physically isolated the transplanted cells from the surrounding tissue [137]. In further work, genetically engineered fibroblasts were used as a model to study how the transplanted cells would affect local inflammation and glial scar formation [138].

In spinal cord injury models, cell transplantation has been reported to give functional recovery. Olfactory ensheathing cells (OECs) are glial cells that can enhance axon regeneration and remyelinate axons [139]. OECs express several molecules which promote axon adhesion, nerve growth factor, as well as extracellular matrix components such as laminin [140]. Separate studies using OEC transplantation into injured spinal cord models have shown restored locomotor behavior, spinal reflex function, as well as impulse conduction in the spinal cord. Factors which may affect degree of regeneration include delay between injury and cell transplantation, culture methods, and age of transplanted cells (time in culture). Cell transplantation thus has shown promise in the field of CNS repair, and cell-hybrid devices may also have application as implanted thin-film microelectrode arrays.

5 Conclusion

Electrode arrays have been microfabricated using a variety of methods and materials, including silicon, metal, ceramic, and polymers. The microfabrication processes for thin-film microelectrodes are already well established for these substrates.

Currently, the greatest challenge to realizing an implant for long-term use is the chronic tissue reaction to the device. Impedance monitoring of electrode functional status coupled with a feedback mechanism for modulating the glial scar may improve the reliability of implanted microelectrodes. Future work will also integrate a telemetry system and on-board amplification and signal processing circuitry with the electrode array for a fully implantable device.

References

1. Nicolelis MAL (2001) Actions from thoughts. Nature 409:403–407.
2. Donoghue JP (2002) Connecting cortex to machines: recent advances in brain interfaces. Nature Neuroscience 5:1085–1088.
3. Taylor DM, Tillery SIH, Schwartz AB (2002) Direct cortical control of 3D neuroprosthetic devices. Science 296:1829–1832.
4. Judy JW (2001) Microelectromechanical systems (MEMS): Fabrication, design and applications. Smart Materials and Structures 10:1115–1134.
5. Nathanson HC, Newell WE, Wickstrom RA et al (1967) The resonant gate transistor. IEEE Transactions on Electron Devices 14:117–133.
6. Fan LS, Tai YC, Muller RS (1988) Integrated movable micromechanical structures for sensors and actuators. IEEE Transactions on Electron Devices 35:724–730.
7. Voldman J, Gray ML, Schmidt MA (1999) Microfabrication in biology and medicine. Annual Review of Biomedical Engineering 1:401–425.

8. Madou MJ (2002) Fundamentals of microfabrication: The Science of Miniaturization, 2nd ed: CRC.
9. Cheung KC (2007) Implantable microscale neural interfaces. Biomedical Microdevices 9:923–938.
10. Strumwasser F (1958) Long-term recording from single neurons in brain of unrestrained mammals. Science 127:469–470.
11. Nicolelis MAL, Dimitrov D, Carmena JM (2003) Chronic, multisite, multielectrode recordings in macaque monkeys. PNAS 100:11041–11046.
12. Gray CM, Maldonado PE, Wilson M (1995) Tetrodes markedly improve the reliability and yield of multiple single-unit isolation from multi-unit recordings in cat striate cortex. Journal of Neuroscience Methods 63:43–54.
13. Mechler F, Reich DS, Victor JD (2002) Detection and discrimination of relative spatial phase by V1 neurons. Journal of Neuroscience 22:6129–6157.
14. Wise KD, Angell JB, Starr A (1970) An integrated-circuit approach to extracellular microelectrodes. IEEE Transactions on Biomedical Engineering BME-17:238–247.
15. Wise KD, Angell JB (1975) A low-capacitance multielectrode probe for use in extracellular neurophysiology. IEEE Transactions on Biomedical EngineeringBME-22:212–219.
16. Najafi K, Wise KD, Mochizuki T (1985) A high-yield IC-compatible multichannel recording array. IEEE Transactions on Electron Devices 32:1206–1211.
17. Bai Q, Wise KD (2001) Single-unit neural recording with active micro-electrode arrays. IEEE Transactions on Biomedical Engineering 48:911–920.
18. Csicsvari J, Henze DA, Jamieson B (2003) Massively parallel recording of unit and local field potentials with silicon-based electrodes. Journal of Neurophysiology 90:1314–1323.
19. Blanche TJ, Spacek MA, Hetke JF (2005) Polytrodes: High-density silicon electrode arrays for large-scale multiunit recording. Journal of Neurophysiology 93:2987–3000.
20. Navarro X, Krueger TB, Lago N (2005) A critical review of interfaces with the peripheral nervous system for the control of neuroprostheses and hybrid bionic systems. Journal of the Peripheral Nervous System 10:229–258.
21. Wise KD (2005) Silicon microsystems for neuroscience and neural prostheses. Engineering in Medicine and Biology Magazine, IEEE 24:22–29.
22. Cheung KC, Lee G, Djupsund K (2000) A new neural probe using SOI wafers with topological interlocking mechanisms. Presented at 1st Annual International IEEE-EMBS Special Topic Conference on Microtechnologies in Medicine and Biology, Lyon, France.
23. Norlin P, Kindlundh M, Mouroux A (2002) A 32-site neural recording probe fabricated by DRIE of SOI substrates. Journal of Micromechanics and Microengineering 12:414.
24. Cheung KC, Djupsund K, Dan Y (2003) Implantable multi-channel electrode array based on SOI technology. Journal of Microelectromechanical Systems 12:179–184.
25. Kindlundh M, Norlin P, Hofmann UG (2004) A neural probe process enabling variable electrode configurations. Sensors and Actuators B: Chemical 102:51–58.
26. Muthuswamy J, Okandan M, Gilletti A (2005) An array of microactuated microelectrodes for monitoring single-neuronal activity in rodents. IEEE Transactions on Biomedical Engineering 52:1470–1477.
27. Muthuswamy J, Okandan M, Jain T (2005) Electrostatic microactuators for precise positioning of neural microelectrodes. IEEE Transactions on Biomedical Engineering 52: 1748–1755.
28. Campbell PK, Jones KE, Huber RJ (1991) A silicon-based, three-dimensional neural interface: Manufacturing processes for an intracortical electrode array. IEEE Transactions on Biomedical Engineering 38:758–768.
29. Rutten WLC, Frieswijk TA, Smit JPA (1995) 3D neuro-electronic interface devices for neuromuscular control: Design studies and realisation steps. Biosensors and Bioelectronics 10:141–153.
30. Normann RA, Campbell PK, Jones KE (1993) Three-Dimensional Electrode Device. U. S. P. a. T. Office, Ed. United States: The University of Utah.

31. Rutten WLC (2002) Selective Electrical Interfaces with the Nervous System. Annual Review of Biomedical Engineering 4:407–452.
32. Badi AN, Kertesz TR, Gurgel RK (2003) Development of a Novel Eighth-Nerve Intraneural Auditory Neuroprosthesis. The Laryngoscope 113:833–842.
33. Rousche PJ, Normann RA (1999) Chronic Intracortical Microstimulation (ICMS) of Cat Sensory Cortex Using the Utah Intracortical Electrode Array. IEEE Transactions on Rehabilitation Engineering 7:56–68.
34. Suner S, Fellows MR, Vargas-Irwin C (2005) Reliability of signals from a chronically implanted, silicon-based electrode array in non-human primate primary motor cortex. IEEE Transactions on Neural Systems and Rehabilitation Engineering 13:524–541.
35. Branner A, Stein RB, Normann RA (2001) Selective Stimulation of Cat Sciatic Nerve Using an Array of Varying-Length Microelectrodes. Journal of Neurophysiology 85:1585–1594.
36. Branner A, Stein RB, Fernandez E (2004) Long-term stimulation and recording with a penetrating microelectrode array in cat sciatic nerve. Biomedical Engineering, IEEE Transactions on 51:146–157.
37. House PA, Macdonald JD, Tresco PA (2006) Acute microelectrode array implantation into human neocortex: Preliminary technique and histological considerations. Neurosurgical Focus 20:E4.
38. Kovacs GTA, Storment CW, Rosen JM (1992) Regeneration microelectrode array for peripheral nerve recording and stimulation. IEEE Transactions on Biomedical Engineering 39:893–902.
39. Kovacs GTA, Storment CW, Halks-Miller M (1994) Silicon-substrate micro-electrode arrays for parallel recording of neural activity in peripheral and cranial nerves. Biomedical Engineering, IEEE Transactions on 41:567–577.
40. Mensinger AF, Anderson DJ, Buchko CJ (2000) Chronic Recording of Regenerating VIIIth Nerve Axons With a Sieve Electrode. Journal of Neurophysiology 83:611–615.
41. Akin T (1994) An integrated telemetric multichannel sieve electrode for nerve regeneration applications. Ph.D. University of Michigan, pp. 179.
42. Stieglitz T, Meyer JU (2006) Biomedical Microdevices for Neural Implants. in BioMEMS, Microsystems, G. A. Urban, Ed, pp. 71–137, Springer.
43. Motta PS, Judy JW (2005) Multielectrode microprobes for deep-brain stimulation fabricated with a customizable 3-D electroplating process. IEEE Transactions on Biomedical Engineering 52:923–933.
44. Burmeister JJ, Pomerleau F, Palmer M (2002) Improved ceramic-based multisite microelectrode for rapid measurements of L-glutamate in the CNS. Journal of Neuroscience Methods 119:163–171.
45. Burmeister JJ, Moxon K, Gerhardt GA (2000) Ceramic-based multisite microelectrodes for electrochemical recordings. Analytical Chemistry 72:187–192.
46. Day BK, Pomerleau F, Burmeister JJ (2006) Microelectrode array studies of basal and potassium-evoked release of l-glutamate in the anesthetized rat brain. Journal of Neurochemistry 96:1626–1635.
47. Pomerleau F, Day BK, Huettl P (2003) Real Time in vivo Measures of l-Glutamate in the Rat Central Nervous System Using Ceramic-Based Multisite Microelectrode Arrays. Annals of the New York Academy of Sciences 1003:454–457.
48. Hetke JF, Williams JC, Pellinen DS (2003) 3-D silicon probe array with hybrid polymer interconnect for chronic cortical recording. Presented at First International IEEE EMBS Conference on Neural Engineering, Capri Island, Italy.
49. Lefurge T, Goodall E, Horch K (1991) Chronically implanted intrafascicular recording electrodes. Annals of Biomedical Engineering 19:197–207.
50. Gilletti A, Muthuswamy J (2006) Brain micromotion around implants in the rodent somatosensory cortex. Journal of Neural Engineering 3:189.
51. Richardson J, Miller RR, Reichert WM (1993) Polyimides as biomaterials: Preliminary biocompatibility testing. Biomaterials 14:627–635.

52. Klinge PM, Vafa MA, Brinker T (2001) Immunohistochemical characterization of axonal sprouting and reactive tissue changes after long-term implantation of a polyimide sieve electrode to the transected adult rat sciatic nerve. Biomaterials 22:2333–2343.
53. Stieglitz T, Ruf HH, Gross M (2002) A biohybrid system to interface peripheral nerves after traumatic lesions: Design of a high channel sieve electrode. Biosensors and Bioelectronics 17:685–696.
54. Stieglitz T, Beutel H, Schuettler M (2000) Micromachined, Polyimide-Based Devices for Flexible Neural Interfaces. Biomedical Microdevices 2:283–294.
55. Stieglitz T, Beutel H, Meyer JU (1997) A flexible, light-weight multi-channel sieve electrode with integrated cables for interfacing regenerating peripheral nerves. Sensors and Actuators A: Physical 60:240–243.
56. Stieglitz T, Beutel H, Meyer JU (2000) "Microflex"-A New Assembling Technique for Interconnects. Journal of Intelligent Material Systems and Structures 11:417–425.
57. Ramachandran A, Schuettler M, Lago N (2006) Design, in vitro and in vivo assessment of a multi-channel sieve electrode with integrated multiplexer. Journal of Neural Engineering, 3:114–124.
58. Stieglitz T, Schuettler M, Koch KP (2005) Implantable biomedical microsystems for neural prostheses. IEEE Engineering in Medicine and Biology Magazine 24:58–65.
59. Rodriguez FJ, Ceballos D, Schuttler M et al (2000) Polyimide cuff electrodes for peripheral nerve stimulation. Journal of Neuroscience Methods 98:105–118.
60. Metz S, Holzer R, Renaud and P (2001) Polyimide-based microfluidic devices. Lab on a Chip 1:29–34.
61. Metz S, Bertsch A, Renaud P (2005) Partial release and detachment of Microfabricated Metal and Polymer Structures by Anodic Metal Dissolution. Journal of Microelectromechanical Systems 14:383–391.
62. Metz S, Jiguet S, Bertsch A, Renaud P (2004) Polyimide and SU-8 microfluidic devices manufactured by heat-depolymerizable sacrificial material technique. Lab on a Chip 4: 114–120.
63. Metz S, Bertsch A, Bertrand D, Renaud P (2004) Flexible polyimide probes with microelectrodes and embedded microfluidic channels for simultaneous drug delivery and multi-channel monitoring of bioelectric activity. Biosensors and Bioelectronics 19: 1309–1318.
64. Metz S, Trautmann C, Bertsch A, Renaud P (2004) Polyimide micro-fluidic devices with integrated nanoporous filtration areas manufactured by micromachining and ion track technology. Journal of Micromechanics and Microengineering 14:324–331.
65. Schmidt EM, McIntosh JS, Bak MJ (1988) Long-term implants of Parylene-C coated microelectrodes. Medical & Biological Engineering & Computing 26:96–101.
66. Schmidt EM, Bak MJ, Christensen P (1995) Laser exposure of Parylene-C insulated microelectrodes. Journal of Neuroscience Methods 62:89–92.
67. Takeuchi S, Ziegler D, Yoshida Y (2005) Parylene flexible neural probes integrated with microfluidic channels. Lab on a Chip 5:519–523.
68. Niklaus F, Enoksson P, Edvard K (2001) Low-temperature full wafer adhesive bonding. Journal of Micromechanics and Microengineering, 11:100–107.
69. Niklaus F, Enoksson P, Griss P (2001) Low-temperature wafer-level transfer bonding. Journal of Microelectrome-chanical Systems 10:525–531.
70. Lee K, He J, Clement R (2004) Biocompatible benzocyclobutene (BCB)-based neural implants with micro-fluidic channel. Biosensors and Bioelectronics 20:404–407.
71. Ehteshami G, Singh A, Coryell G (2003) Glial cell and fibroblast cytotoxicity study on 4026-cyclotene photosensitive benzocyclobutene (BCB) polymer films. Journal of Biomaterials Science, Polymer Edition 14:1105–1116.
72. Koeneman BA, Lee KK, Singh A (2004) An ex vivo method for evaluating the biocompatibility of neural electrodes in rat brain slice cultures. Journal of Neuroscience Methods 137:257–263.

73. Cheung KC, Renaud P, Tanila H (2007) Flexible polyimide microelectrode array for in vivo recordings and current source density analysis. Biosensors and Bioelectronics 22:1783–1790.
74. Akin T, Najafi K, Smoke RH (1994) A micromachined silicon sieve electrode for nerve regeneration applications. IEEE Transactions on Biomedical Engineering 41:305–313.
75. Hoogerwerf AC, Wise KD (1994) A three-dimensional microelectrode array for chronic neural recording. Biomedical Engineering, IEEE Transactions on 41:1136–1146.
76. Bai Q, Wise KD, Anderson DJ (2000) A high-yield microassembly structure for three-dimensional microelectrode arrays. IEEE Transactions on Biomedical Engineering 47: 281–289.
77. Olsson RH, Buhl DL, Sirota AM (2005) Band-tunable and multiplexed integrated circuits for simultaneous recording and stimulation with microelectrode arrays. Biomedical Engineering, IEEE Transactions on 52:1303–1311.
78. Chen J, Wise KD, Hetke JF (1997) A multichannel neural probe for selective chemical delivery at the cellular level. IEEE Transactions on Biomedical Engineering 44: 760–769.
79. Nicolelis MAL (2003) Brain-Machine Interfaces to Restore Motor Function and Probe Neural Circuits. Nature Reviews Neuroscience 4:417–422.
80. Kipke DR, Vetter RJ, Williams JC (2003) Silicon-substrate intracortical microelectrode arrays for long-term recording of neuronal spike activity in cerebral cortex. Neural Systems and Rehabilitation Engineering, IEEE Transactions on [see also IEEE Transaction on Rehabilitation Engineering] 11:151–155.
81. Lago N, Ceballos D, Rodriguez FJ et al (2005) Long term assessment of axonal regeneration through polyimide regenerative electrodes to interface the peripheral nerve. Biomaterials 26:2021–2031.
82. Kovacs GTA (1994) Introduction to the Theory, Design, and Modeling of Thin-Film Microelectrodes for Neural Interfaces. In Enabling Technologies for Cultured Neural Networks, DA. Stenger and TM. McKenna, Eds, pp. 121–165, Academic Press.
83. Thiebaud P (1999) Fabrication of microelectrode arrays for electrophysiological monitoring of hippocampal organotypic slice cultures by interface. In Institute of Microtechnology PhD. Neuchatel: University of Neuchatel.
84. Beebe X, Rose TL (1988) Charge injection limits of activated iridium oxide electrodes with 0.2 msec pulses in bicarbonate buffered saline. IEEE Transactions on Biomedical Engineering 35:494–495.
85. Weiland JD, Anderson DJ (2000) Chronic Neural Stimulation with Thin-Film, Iridium Oxide Electrodes. IEEE Transactions on Biomedical Engineering 47:911–918.
86. Klein JD, Clauson SL, Cogan SF (1989) Morphology and charge capacity of sputtered iridium oxide films. Journal of Vacuum Science & Technology A: Vacuum, Surfaces, and Films 7:3043–3047.
87. Weiland JD, Anderson DJ, Humayun MS (2002) In Vitro Electrical Properties for Iridium Oxide Versus Titanium Nitride Stimulating Electrodes. IEEE Transactions on Biomedical Engineering 49:1574–1579.
88. Alberts B, Bray D, Lewis J (1994) Molecular Biology of the Cell. 3rd ed. New York: Garland Publishing.
89. Gabay T, Jakobs E, Ben-Jacob E (2005) Engineered self-organization of neural networks using carbon nanotube clusters. Physica A: Statistical Mechanics and its Applications 350:611–621.
90. Zhang X, Prasad S, Niyogi S (2005) Guided neurite growth on patterned carbon nanotubes. Sensors and Actuators B: Chemical 106:843–850.
91. Gabay T, Ben-David M, Kalifa I (2007) Electrochemical and biological properties of carbon nanotube based multi-electrode arrays. Nanotechnology, vol. 18:035201.
92. Nguyen-Vu TDB, Chen H, Cassell Alan M (2006) Vertically Aligned Carbon Nanofiber Arrays: An Advance toward Electrical-Neural Interfaces. Small 2:89–94.

93. Abidian MR, Martin DC (2008) Experimental and theoretical characterization of implantable neural microelectrodes modified with conducting polymer nanotubes. Biomaterials 29:1273–1283.
94. Wang K, Fishman HA, Dai H (2006) Neural Stimulation with a Carbon Nanotube Microelectrode Array. Nano Letters 6:2043–2048.
95. McCreery DB, Agnew WF, Yuen TGH (1990) Charge Density and Charge Per Phase as Cofactors in Neural Injury Induced by Electrical Stimulation. IEEE Transactions on Biomedical Engineering 37:996–1001.
96. Abidian MR, Salas LG, Yazdan-Shahmorad A (2007) In-vivo Evaluation of Chronically Implanted Neural Microelectrode Arrays Modified with Poly (3,4-ethylenedioxythiophene) Nanotubes. Presented at Neural Engineering, 2007. CNE '07. 3rd International IEEE/EMBS Conference.
97. Abidian MR, Kim DH, Martin and DC (2006) Conducting-Polymer Na-notubes for Controlled Drug Release. Advanced Materials 18:405–409.
98. Valentini RF (2000) Nerve Guidance Channels. In The Biomedical Engineering Handbook 1, J. D. Bronzino, Ed.: CRC Press.
99. Fitch MT, Silver J (2008) CNS injury, glial scars, and inflammation: Inhibitory extracellular matrices and regeneration failure. Experimental Neurology 209:294–301.
100. Sofroniew MV (2005) Reactive Astrocytes in Neural Repair and Protection. Neuroscientist 11:400–407.
101. Fawcett JW, Asher RA (1999) The glial scar and central nervous system repair. Brain Research Bulletin 49:377–391.
102. Szarowski DH, Andersen MD, Retterer S et al (2003) Brain responses to micro-machined silicon devices. Brain Research 983:23–35.
103. Turner JN, Shain W, Szarowski DH et al (1999) Cerebral Astrocyte Response to Micromachined Silicon Implants. Experimental Neurology 156:33–49.
104. Norton WT, Aquino DA, Hozumi I et al (1992) Quantitative aspects of reactive gliosis: A review. Neurochemical Research (Historical Archive) 17:877–885.
105. Silver J, Miller JH (2004) Regeneration beyond the glial scar. Nature Review Neuroscience 5:146–156.
106. Busch SA, Silver J (2007) The role of extracellular matrix in CNS regeneration. Current Opinion in Neurobiology 17:120–127.
107. Polikov VS, Tresco PA, Reichert WM (2005) Response of brain tissue to chronically implanted neural electrodes. Journal of Neuroscience Methods 148:1–18.
108. Griffith RW, Humphrey DR (2006) Long-term gliosis around chronically implanted platinum electrodes in the Rhesus macaque motor cortex. Neuroscience Letters 406:81–86.
109. Biran R, Martin DC, Tresco PA (2005) Neuronal cell loss accompanies the brain tissue response to chronically implanted silicon microelectrode arrays. Experimental Neurology 195:115–126.
110. Stensaas SS, Stensaas LJ (1976) The reaction of the cerebral cortex to chronically implanted plastic needles. Acta Neuropathologica 35:187–203.
111. Edell DJ, Toi VV, McNeil VM (1992) Factors influencing the biocompatibility of insertable silicon microshafts in cerebral cortex. IEEE Transactions on Biomedical Engineering 39:635–643.
112. Schmidt S, Horch K, Normann R (1993) Biocompatibility of silicon-based electrode arrays implanted in feline cortical tissue. Journal of Biomedical Materials Research 27:1393–1399.
113. Henze DA, Borhegyi Z, Csicsvari J (2000) Intracellular Features Predicted by Extracellular Recordings in the Hippocampus In Vivo. Journal of Neurophysiology 84:390–400.
114. Liu X, McCreery DB, Carter RR (1999) Stability of the interface between neural tissue and chronically implanted intracortical microelectrodes. IEEE Transactions on Rehabilitation Engineering 7:315–326.
115. Ivorra A, Gomez R, Noguera N (2003) Minimally invasive silicon probe for electrical impedance measurements in small animals. Biosensors and Bioelectronics 19:391–399.

116. Williams JC, Hippensteel JA, Dilgen J (2007) Complex impedance spectroscopy for monitoring tissue responses to inserted neural implants. Journal of Neural Engineering, 4:410–423.
117. Buitenweg J, Rutten W, Willems W (1998) Measurement of sealing resistance of cell-electrode interfaces in neuronal cultures using impedance spectroscopy. Medical and Biological Engineering and Computing 36:630–637.
118. Buitenweg JR, Rutten WLC, Marani and E (2003) Geometry-based finite-element modeling of the electrical contact between a cultured neuron and a microelectrode. IEEE Transactions on Biomedical Engineering 50:501–509.
119. Merrill DR, Tresco PA (2005) Impedance characterization of microarray recording electrodes in vitro. IEEE Transactions on Biomedical Engineering 52:1960–1965.
120. Subbaroyan J, Martin DC, Kipke DR (2005) A finite-element model of the mechanical effects of implantable microelectrodes in the cerebral cortex. Journal of Neural Engineering 2:103.
121. Lee H, Bellamkonda RV, Sun W, Levenston ME (2005) Biomechanical analysis of silicon microelectrode-induced strain in the brain. Journal of Neural Engineering, 2:81–89.
122. Stice P, Gilletti A, Panitch A (2007) Thin microelectrodes reduce GFAP expression in the implant site in rodent somatosensory cortex. Journal of Neural Engineering 4:42–53.
123. Retterer ST, Smith KL, Bjornsson CS (2004) Model neural prostheses with integrated microfluidics: A potential intervention strategy for controlling reactive cell and tissue responses. Biomedical Engineering, IEEE Transactions on 51:2063–2073.
124. Mond H, Stokes K (1992) The electrode-tissue interface: the revolutionary role of steroid elution. Pacing and Clinical Electrophysiology 15:95–107.
125. Mond HG, Grenz D (2004) Implantable Transvenous Pacing Leads: The Shape of Things to Come. Pacing and Clinical Electrophysiology 27:887–893.
126. McGraw J, Hiebert GW, Steeves JD (2001) Modulating astrogliosis after neurotrauma. Journal of Neuroscience Research 63:109–115.
127. Spataro L, Dilgen J, Retterer S (2005) Dexamethasone treatment reduces astroglia responses to inserted neuroprosthetic devices in rat neocortex. Experimental Neurology 194:289–300.
128. Shain W, Spataro L, Dilgen J (2003) Controlling cellular reactive responses around neural prosthetic devices using peripheral and local intervention strategies. IEEE Transactions on Neural Systems and Rehabilitation Engineering 11:186–188.
129. Zhong Y, McConnell GC, Ross JD (2005) A Novel Dexamethasone-releasing, Anti-inflammatory Coating for Neural Implants. Presented at 2nd International IEEE EMBS Conference on Neural Engineering.
130. Wadhwa R, Lagenaur CF, Cui XT (2006) Electrochemically controlled release of dexamethasone from conducting polymer polypyrrole coated electrode. Journal of Controlled Release 110:531–541.
131. Rousche PJ, Pellinen DS, Pivin DP et al (2001) Flexible polyimide-based intracortical electrode arrays with bioactive capability. IEEE Transactions on Biomedical Engineering 48:361–371.
132. Williams JC, Holecko MM, Massia SP et al (2005) Multi-site incorporation of bioactive matrices into MEMS-based neural probes. Journal of Neural Engineering, pp. L23.
133. Agnew WF, McCreery DB, Yuen TGH et al (1993) MK-801 Protects Against Neuronal Injury Induced by Electrical Stimulation. Neuroscience 52:45–53.
134. Maher M, Pine J, Wright J (1999) The neurochip: a new multielectrode device for stimulating and recording from cultured neurons. Journal of Neuroscience Methods 87: 45–56.
135. Tatic-Lucic S, Wright JA, Tai YC et al (1997) Silicon cultured-neuron prosthetic devices for in vivo and in vitro studies. Sensors and Actuators B: Chemical 43:105–109.
136. Tooker A, Meng E, Erickson J (2004) Development of biocompatible parylene neurocages. Presented at Engineering in Medicine and Biology Society, 2004. IEMBS '04. 26th Annual International Conference of the IEEE.

137. Kim YT, Hitchcock R, Broadhead KW (2005) A cell encapsulation device for studying soluble factor release from cells transplanted in the rat brain. Journal of Controlled Release 102:101–111.
138. Kim YT, Bridge MJ, Tresco PA (2007) The influence of the foreign body response evoked by fibroblast transplantation on soluble factor diffusion in surrounding brain tissue. Journal of Controlled Release 118:340–347.
139. Richter MW, Roskams AJ (2008) Olfactory ensheathing cell transplantation following spinal cord injury: Hype or hope? Experimental Neurology 209:353–367.
140. Fairless R, Barnett SC (2005) Olfactory ensheathing cells: their role in central nervous system repair. The International Journal of Biochemistry & Cell Biology 37:693–699.

Stimulation Electrode Materials and Electrochemical Testing Methods

Andy Hung, Ira B. Goldberg, and Jack W. Judy

Abstract Neuro-stimulation can be implemented using several design choices — bipolar vs. monopolar stimulation, current vs. voltage control, and active vs. passive recharge. To ensure proper function through the desired lifetime, electrodes are typically made of titanium, platinum, or iridium. The difference between the 3 metals is primarily based on their performance in reversible oxidation/reduction mechanisms, which can be illustrated using various electrochemical techniques.

In slow cyclic voltammetry analysis, a platinum electrode can reversibly consume and release 3 mC/cm^2 of charge within the normal operating voltage range. However, for actual neuro-stimulation pulses, platinum can only safely inject 0.1 mC/cm^2, as estimated from an electrode-potential graph. Compared to platinum, *iridium* can inject 10 times the amount of charge for both neuro-stimulation and cyclic voltammetry. The greater capability is due to both the greater number of available oxidation states and utilization of bulk porous oxide.

In AC impedance, a titanium electrode exhibits the same double-layer capacitance per area as that of platinum. However, titanium suffers from irreversible buildup of a high-impedance oxide layer, which prevents sustained charge-injection usage. The irreversible oxidation can be observed using the pulse-clamp technique.

This chapter also introduces computer simulations, specifically capacitive computer models, as a method for visualizing the current-distribution pattern. The simulation shows uniform current distribution despite prominent electrode topologies. A dissolution study performed with gold electrodes confirms that the current distribution is uniform during normal usage, but exhibits severe current crowding when charge injection is increased above the safe limit.

Contents

A. Hung (✉)
Boston Scientific Neuromodulation Corporation, Valencia, CA, USA
e-mail: andy.hung@bsci.com

D.D. Zhou, E. Greenbaum (eds.), *Implantable Neural Prostheses 2*,
Biological and Medical Physics, Biomedical Engineering, DOI 10.1007/978-0-387-98120-8_7,

1 Introduction

Within a neural stimulation system, the electrodes are subjected to a harsh working environment. They may have to inject a large amount of charge, tolerate extreme local pH changes, and function for the duration of the device lifetime. To ensure longevity, the electrodes need to be thoroughly tested.

Traditionally, engineers prefer capacitive electrodes [1]. These electrodes are made from inert materials that do not react with surrounding electrolyte and therefore avoid corrosion. Examples of these are carbon electrodes for skin patches and titanium or tantalum electrodes for pacemakers. Since the electrodes do not employ oxidation reactions, they are limited in their charge-injection capability. Device designers try to compensate for this shortcoming by using larger electrodes and making the materials more porous to enhance the surface area.

In recent years, there has been a growing interest in applications that target neural tissues insulated by a thick dura layer, such as in spinal cord and retina. The large amount of charge injection that is required for these applications can only be safely provided by platinum and iridium electrodes. These two materials have substantial charge storage capacity within the water window (the voltage range over which water electrolysis does not occur), thereby allowing for a much higher charge-injection capability.

Titanium, platinum, and iridium materials have been shown to be well tolerated by the tissue without concerns related to toxicity. Platinum is a widely used material because of its durability and its relative ease of manufacture. Although iridium can inject even higher amounts of charge than platinum, its high mechanical strength leads to difficulties in machining or molding. Instead, it is usually sputtered or electroplated onto the electrode surface, adding complexity to the manufacturing process. Titanium is often used as a casing for the electronics, but suffers from relatively poor charge-injection capability. In this chapter, the electrochemical properties of these three materials are illustrated using various testing methods.

2 Electrode Reactions

2.1 Electrode Interface

Neural stimulation devices depend on the ability to excite the neural cells with an electric-voltage gradient. The gradient is established by using electrodes that pass current through the tissue. Within the implant device itself, electrical current is carried by the electrons. However, within tissue or saline, this current is carried by the movement of charged ions such as Na^+, Cl^-, and $H_2PO_4^-$. Devoid of dissolved salt, pure deionized water exhibits a very low conductance. The function of the electrode is to convert the charge transfer between electrons in the circuits to ions in the tissue. It does so through two main mechanisms: a capacitive double-layer charging/discharging and Faradaic processes. Conceptually, this is represented by a circuit with two conduction pathways. Faradaic pathway is a complex impedance that can have both reversible (pseudo-capacitive) and irreversible (conductive) properties (Fig. 1).

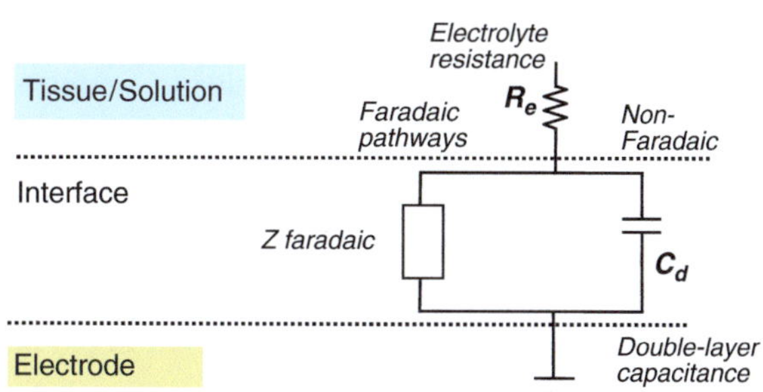

Fig. 1 Schematic showing the two main charge-injection pathways of the electrode interface

2.2 Electrical Double Layer

An electrical double layer is formed whenever a metal is submerged in a liquid. The charges on the metal can attract positive or negative ions, resulting in an ion concentration at the interface region that differs from that in the bulk solution. When a submerged metal is charged, the charges in the metal and oppositely charged ions in the liquid attract each other and concentrate near the electrode-liquid interface. The small separation distance (about 20 nm) between charges in the electrode and in the electrolyte results in high capacitance, typically around 400 nF/mm^2. The electric double-layer charging is fast acting and carries the majority of the current immediately after the start of a neural stimulation pulse. The double-layer capacitance is not constant, but increases with higher electrode potentials or higher ion concentration levels. This is due to the increased amount of ions that attract to and accumulate at the electrode surface [2].

2.3 Reversible Metal Oxidation/Reduction

Traditionally, all the noncapacitive charge-injection mechanisms are grouped together into the "Faradaic" pathway. This includes both the reversible and irreversible chemical reactions. They are grouped together because the distinction between the two can be unclear in some instances. The reversible oxidation/reduction reaction plays an important role for charge injection and it is part of what gives each metal its unique electrochemical properties.

By changing the electrode potential, the metal can be made to react with the surrounding water molecules and ionic species. A common reaction is the formation of metal oxide. When a metal is deprived of electrons (i.e., positively biased), it starts to attract negatively charged ions. With sufficient driving force (in the form of electrical bias), the metal may react with oxygen in water molecules to form an oxide. On the other hand, when a platinum group metal is supplied with abundance of electrons, it can bind to hydrogen atoms to form a reduced hydride form.

The electrical bias at which a particular oxidation-reduction reaction occurs is expressed as the standard potential ($E°$) of a reaction. This standard potential is measured in relation to the voltage at which hydrogen is oxidized ($H_2 \rightarrow 2H^+ + 2e^-$) in acid with 1 M activity. In actual experiments, the reaction voltage is typically measured by comparing the value to a reference electrode (e.g., Ag/AgCl). For example, the Ag/AgCl electrode has a constant potential of 0.2 V when compared to the standard hydrogen electrode (SHE). The voltage measured with AgCl can then be subtracted by 0.2 V to arrive at the standard potential $E°$ of the reaction. The measured reaction voltage is dependent on the composition of the surrounding electrolyte. For platinum submerged in 1 M sulfuric acid, the reaction voltage for hydrogen gas formation is at 0 V. However, in physiological saline (pH 7), the H_2 gas formation occurs at −0.83 V.

In saline, platinum oxidation starts at a standard potential E° >0.2 V. The oxidation reaction can be reversed by negatively biasing the electrode (Fig. 2) (Table 1). At a given voltage, E, the ratio of oxidized [O] to reduced [R] species can be described by Nernst equation:

$$E = E^0 + \frac{RT}{nF}\ln\frac{[O]}{[R]} \text{ or } \frac{[O]}{[R]} = e^{\frac{nF}{RT}(E-E^0)}$$

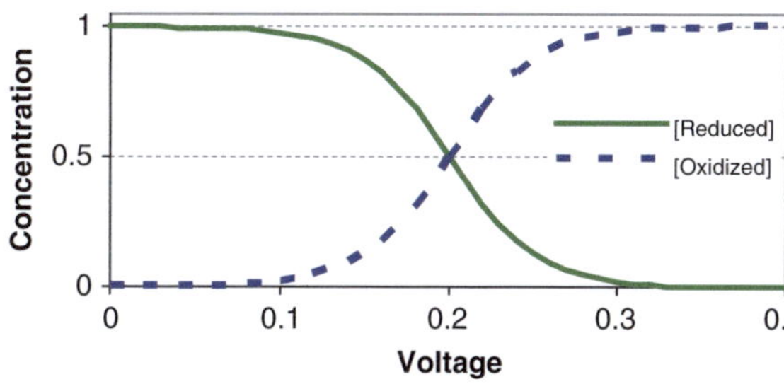

Fig. 2 Equilibrium concentration of oxidized vs. reduced species as function of voltage

Table 1 Constants for Nernst equation

$R = 8.3\ ^{J}/_{K\cdot mol}$
$T = 298\ ^\circ K$
$F = 9.65 \times 10^4\ ^{C}/_{mol}$
$R \cdot T / F = 0.0257$ V
n= number of e^- transferred
E= electrode voltage

2.4 Irreversible Chemical Reaction

If the electrode potential is further increased near the electrolysis potential, it will eventually start to decompose water to form oxygen or hydrogen gas. The gas quickly diffuses away, making the reaction chemically irreversible. The irreversible reaction pathway is then dominated by the electrolysis reaction.

In simplified electrode models, the irreversible reaction pathway is typically represented by a resistor R_i. The resistor denotes that charge is consumed and not stored. However, the pathway actually behaves more like a diode. The reaction-current equation is described approximately by:

$$i = i_0(e^{(V_E - V_{act})\cdot\frac{-n\cdot F}{R\cdot T}}),$$

When the electrode voltage (V_E) exceeds the reaction potential (V_{act}), the current increases exponentially with voltage.

Even at voltages below that of water electrolysis, other undesirable reactions can occur [3]. Two common examples are dissolution of metals by formation of metal chloride and formation of hydrogen peroxide.

While these reactions are typically irreversible, a small portion of them may be reversed in the neuro-stimulation setting. This is because neuro-stimulation typically uses short pulses that are charge balanced. When the current polarity is reversed, the H_2 and O_2 gases still adsorbed on the electrode surface may be dissociated [4]. Likewise, dissolved metal chloride can be electroplated back onto the metal surface.

2.5 *Electrolyte Resistance*

The resistance of an electrolyte, or the tissue surrounding the electrode, is represented by a simple resistor R_e. In saline, ions can move freely, resulting in low R_e value of around 50 Ω·cm. For a 10-mm^2 electrode implanted in biological tissue, the resistance is typically around 500∼1500 Ω. Certain tissues (e.g., muscle fascia) can also have a small capacitive component, making the impedance slightly frequency dependent. It should be noted that while all the other electrode capacitance and resistance parameters scale with an r^2 relationship, the electrolyte resistance is inversely proportional to the electrode radius r. The resistance relationship for a hemispherical electrode can be derived by integrating the electrolyte resistance of a hemispherical volume from the electrode surface to infinity. Each slice of the hemispherical shell has an area of $2{\cdot}\pi{\cdot}r^2$, and a thickness dr.

$$R_s = \int \frac{\rho}{A} dl = \int_r^{\infty} \frac{\rho}{2\pi \cdot r^2} dr = \frac{\rho}{2\pi \cdot r}$$

With increasing distance from the electrode surface, the resistance decreases by r^2. Therefore, the R_s is dominated by the solution resistance at the surface of the electrode. For a planar disk electrode, $R_s = {}^{\rho}/_{4r}$ [5].

3 Common Electrochemical Tests and Electrode Materials

3.1 *Cyclic Voltammetry*

The occurrence of oxidation and electrolysis reactions can be best observed with the slow cyclic voltammetry technique. In this analysis, the electrode voltage is slowly ramped up or down (typically at <0.1 V/s), and the measured current is plotted in relationship to the voltage (Fig. 3). The slow voltage ramp allows oxidation/reduction to equilibrate, resulting in more well-defined peaks.

If the measured system is just a simple resistor, the current would be linearly proportional to the voltage, as dictated by $\mathrm{I} = {}^{\mathrm{V}}/_{\mathrm{R}}$. If the system were a pure capacitor, the current would be a constant value ($I = C{\cdot}{}^{\mathrm{dV}}/_{\mathrm{d}t}$) (Fig. 4).

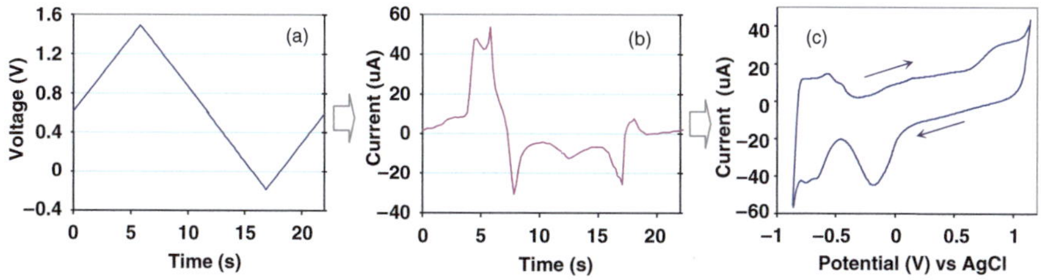

Fig. 3 In cyclic voltammetry, (**a**) electrode voltage is ramped up or down, and (**b**) the current is measured. (**c**) The result is plotted as current vs. voltage

Fig. 4 (**a**) current voltage graph of a simple resistor, and (**b**) a capacitor

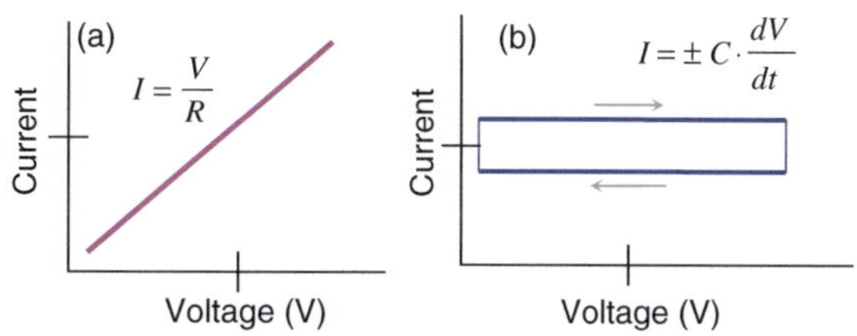

If the electrode surface is oxidized or reduced, there would be a surge of current at the corresponding reaction potential. As seen in Fig. 5, the concentrations of the oxidized and reduced species change near the reaction potential. The resulting current is proportional to the rate of change in the concentration: $I = {}^{d[O]}/_{dt}$.

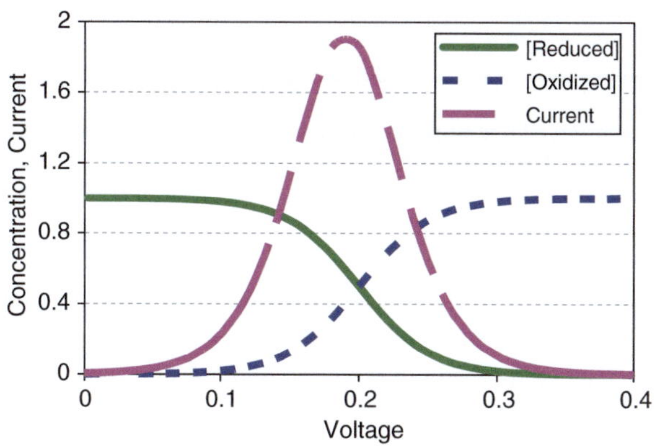

Fig. 5 The concentration of the reduced vs. oxidized species, and the current as a result of voltage

To obtain a clear CV graph, it is preferable to use electrodes larger than 3 mm^2. Using a larger electrode allows for higher levels of oxidation-reduction currents, which can be more clearly recorded over the parasitic capacitance and resistance current levels.

3.2 Platinum

The cyclic voltammogram for a smooth, polycrystalline platinum electrode submerged in saline is shown in Fig. 6. The range of the voltage scan is limited by the generation of oxygen gas at $E°$ of +1.23 V (Fig. 6-1), and hydrogen gas at -0.83 V (Fig. 6-4). This is observed in the curve by the sharp exponential increase in the current when the electrode is near the hydrolysis voltage. The voltage range in which the electrode can be operated without undergoing hydrolysis is referred to as the "water window."

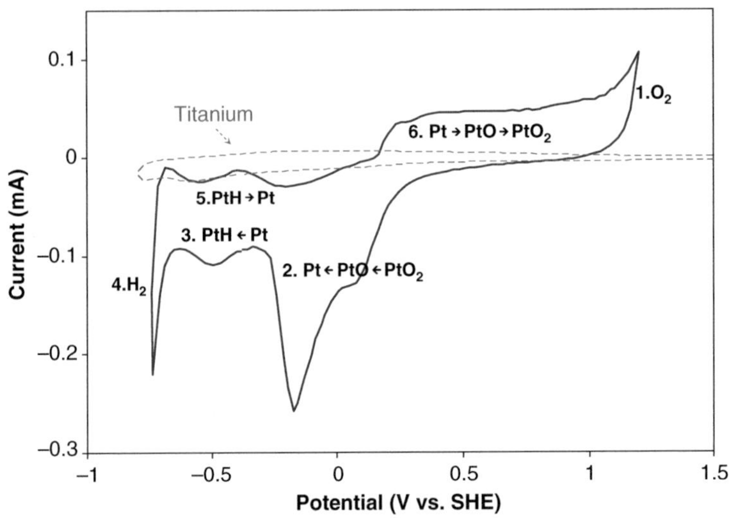

Fig. 6 Cyclic voltammetry of 4-mm^2 platinum, measured in 0.1-M NaCl in open air (21% oxygen) and a scan rate of 0.1 V/s. The marked points indicate hydrogen released (*5*), hydrogen plated (*3*), oxidation of platinum (*6*), reduction to Pt (*2*), O_2 generation (*1*), and H_2 Generation (*4*)

Within the water window, platinum is observed to undergo several oxidation changes. By removing electrons from platinum (i.e., applying a positive voltage), platinum can be induced to react with oxygen in the water, forming platinum oxide. On the positive sweep, the oxidation reaction starts at 0.2 V (Fig. 6-6) and continues to occur even up to values of 1.23 V. Each platinum atom can be bound to a maximum of 2 oxygen atoms, resulting in a +4 state. Platinum oxide can be made very thick through repeated anodic high-current pulses [6]. On the negative sweep, the oxide is converted back to elemental platinum when the voltage falls below 0.2 V (Fig. 6-2). At more negative voltages, adding excess electrons to platinum can induce hydrogen atoms to bond (Fig. 6-3). Unlike oxygen, hydrogen is only loosely bound to the platinum surface (Table 2).

Table 2 Platinum reactions, potential listed in reference to standard hydrogen electrode [7]

Reaction	E°
$H_2\,(g)\,2\,OH^- \leftrightarrow 2H_2O + 2\,e^-$	−0.83
$2H_2O \leftrightarrow O_2\,(g) + 4H + +4e^-$	1.23
$H_2O_2 \leftrightarrow O_2 + 2H^+ + 2e^-$	0.70
$PtH_2 + OH^- \leftrightarrow PtH + H_2O + e^-$	−0.6
$PtH_2 + OH^- \leftrightarrow Pt + H_2O + e^-$	−0.5
$Pt + H_2O \leftrightarrow PtO + 2H + +2e^-$	0.25
$PtO + H_2O \leftrightarrow PtO_2 + 2H^+ + 2e^-$	0.3
$Pt + 6Cl^- \leftrightarrow PtCl_6^{2-}\,(aq) + 6e^-$	0.73~1.4

3.3 Titanium

Titanium injects relatively little charge within the operating voltage range. Observation of the reaction potentials shows that only one reversible oxidation reaction occurs within the voltage range, with the oxidation state change between +4 and +3.

$$2TiO_2\,(s) + 2H^+ + 2e^- \rightarrow Ti_2O_3\,(s) + H_2O \quad E^\circ = -0.56$$

With repeated voltage cycling, titanium does not exhibit a charge-injection increase due to "activation" (see iridium). However, the voltage at which oxygen evolution occurs will gradually shift higher. This is due to the buildup of a resistive oxide layer. The increased voltage drop across the oxide requires a higher electrode potential to achieve the hydrolysis current (Fig. 7).

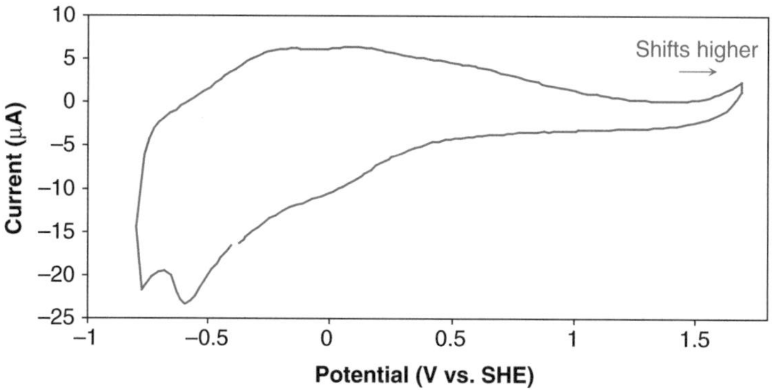

Fig. 7 Cyclic voltammogram of 4-mm² titanium in 0.1 M saline

3.4 Iridium

Compared to platinum, iridium can inject an even greater amount of charge through oxidation reactions for the same electrode area, because each individual iridium

Table 3. Iridium reactions [11]

$Ir^{+3} + 4H_2O = IrO_4^{-2} + 8H^+ + 3e^-$	$E^\circ = 1.448 - 0.4576pH + 0.0197 \log (IrO_4^{-2}/Ir^{+3})$
$Ir + 2H_2O = IrO_2 + 4H^+ + 4e^-$	$E^\circ = 0.926 - 0.0391pH$
$IrO_2 + 2H_2O = IrO_4^{-2} + 4H^+ + 2e^-$	$E^\circ = 2.037 - 0.01182pH + 0.0293 \log(IrO_4^{-2})$

atom can be bound to a maximum of 4 oxygen atoms to +8 state (Table 3). The iridium surface can also be "activated" with repeated voltage cycling in saline [8–10]. This "activation" process creates a thick layer of porous iridium oxide that further enhances charge injection. Because the activated iridium utilizes the three-dimensional bulk material instead of just the surface, the resulting effective surface area is many times that of the topological area. Cyclic voltammetry analysis shows that for the as-deposited state, the iridium surface can deliver about the same amount of charge as platinum. However, after 400 cycles, the iridium delivers almost 10 times the charge within the same voltage range. Because of the many reaction potentials, the individual current peaks cannot be easily distinguished in the graph (Fig. 8).

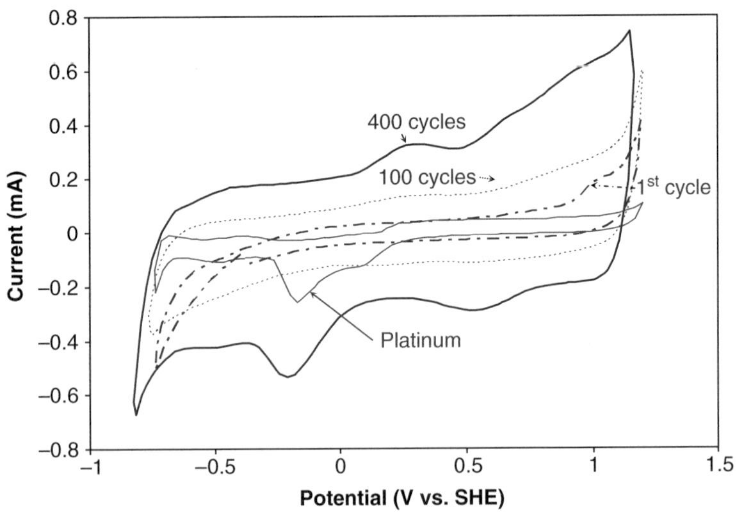

Fig. 8 Cyclic voltammogram for 4-mm^2 iridium electrode in saline. Current increase is observed with prolonged voltage cycling

By integrating the area within the C-V graph, the iridium charge injection is calculated to be 30 mC/cm^2 [12]. However, not all of this charge is available during neural stimulation. Because of the many oxidation-state changes, iridium charge injection requires many milliseconds to equilibrate. This is the result of slow mass transport of oxygen into the bulk of iridium material. While typical slow cyclic voltammetry takes 1 minute for a complete sweep, neural stimulation pulses are much shorter, typically less than 1 ms. For the short pulses used in neural stimulation, the practical charge-injection limit for iridium is 1 mC/cm^2.

3.5 *Effect of Protein*

Protein is known to reduce corrosion for platinum-stimulation electrodes [13]. Studies performed with a quartz-crystal microbalance further show that protein can competitively bind to the platinum surface and prevent oxidation. This study is performed by depositing platinum on a piece of piezo crystal with a natural harmonic frequency of 10 MHz [14]. Voltage cycling the platinum in liquid causes the platinum to oxidize. The additional oxygen mass lowers the harmonic frequency. Each type of amino acid reduces the increased mass by a varying degree. The proteins cause the C-V curve to lose the distinct peaks for platinum and, instead, to show peaks corresponding to the oxidation of the different amino acids.

3.6 *Impedance Measurement*

An impedance measurement can be used as a standardized test for quickly estimating the roughness and porosity of a surface. It is performed by applying a small ac voltage signal (10 mV) at various frequencies (typically 10 kHz–0.1 Hz) and using the measured electrode current to calculate its impedance [15–16]. The small-signal perturbation used in the impedance minimizes the oxidation and hydrolysis reactions. The electrode impedance is thus dominated by the double-layer capacitance C_d, which is in series with the electrolyte resistance R_e. This forms a high-pass filter with cutoff frequency at:

$$f = \frac{1}{2 \cdot \pi \cdot R_e \cdot C_d}$$

At very high frequencies (>10 kHz), the capacitor C_d shunts all the current, and the resulting electrode impedance is approximately the electrolyte resistance R_e. At very low frequencies (~0.1 Hz), C_d becomes an open circuit, resulting in a total electrode impedance of $R_e + R_i$. At intermediate frequencies, the impedance is governed by the capacitance:

$$Z = \frac{1}{j \cdot w \cdot C}, \text{ with } C \infty A$$

Therefore, an electrode with a rougher surface would exhibit a larger surface area, and thus a higher capacitance. As seen in Fig. 9, three similarly sized electrodes made of titanium, platinum, and iridium exhibit similar impedance values when they are initially submerged. However, after the iridium electrode is subjected to repeated voltage cycles at the full voltage range of the water window (as in cyclic voltammetry), its impedance at lower frequencies decreases by up to 10 fold [17].

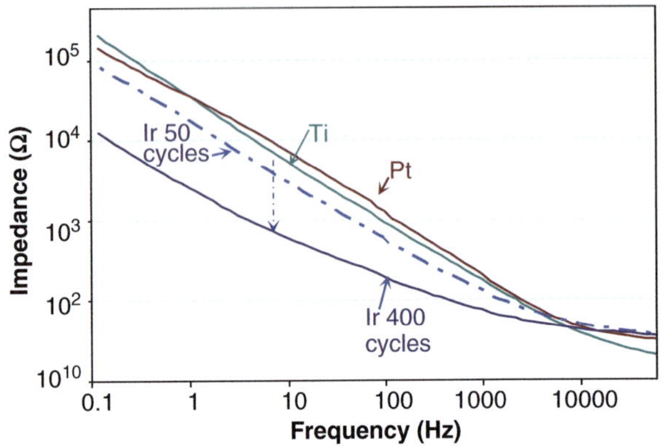

Fig. 9 Impedance spectroscopy of titanium, platinum, and iridium electrodes; all are around 2.5 mm^2 in size. Smaller electrodes are used here so the electrode impedances are above that for connection wires

4 Neural Stimulation Settings Which Affect Charge Injections

4.1 Basis of Neural Stimulation

Figure 10 depicts an electrode injecting current into the surrounding tissue, with the target neuron nearby. Current passing through the semipermeable cell membrane establishes a voltage gradient across the membrane. If the transmembrane voltage increases by a value that exceeds a threshold (typically 15 mV above the

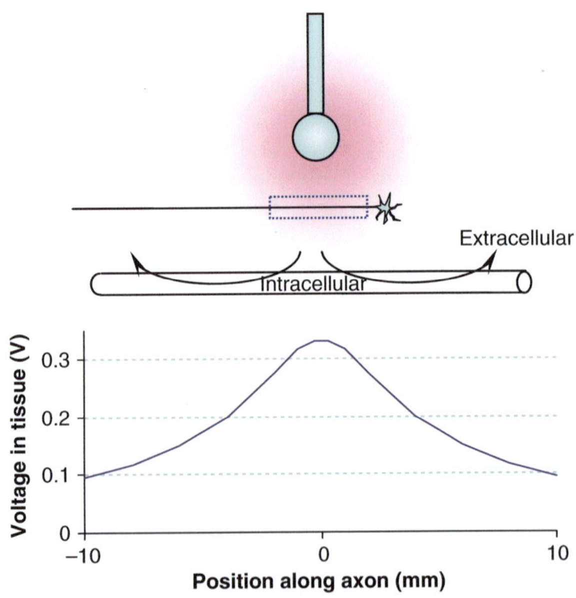

Fig. 10 An electrode placed near the target neuron, establishing a voltage gradient in tissue

resting state), the voltage change activates a cascade of ion channels that leads to cell excitation.

Neural excitation can be best achieved with an electric field that is nonhomogeneous (i.e., near the edge of an electrode). Calculations using a conducting-cable model show that stimulation efficacy is proportional to the gradient of the electric field along the neural membrane (alternatively this can be calculated as the second spatial derivative of the electric potential) [18]. The nonhomogeneous field increases the current flux into the cell membrane at the region of the highest potential. Current then travels down the axon and exits at regions of the membrane nearby.

4.2 Components of a Neural Stimulation System

In order to provide the stimulation pulses to the electrode, the stimulation system typically contains custom circuitry, communication modules, and a power supply (battery). These components are housed in hermetic casing, traditionally made of titanium (Fig. 11).

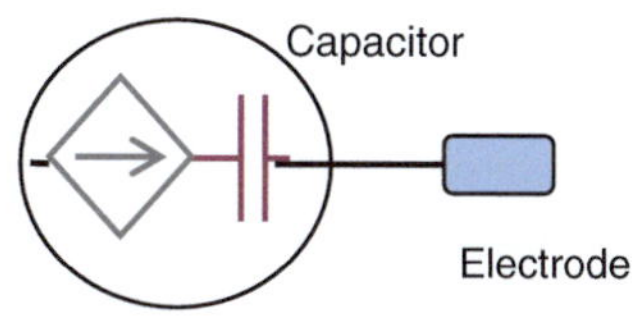

Fig. 11 Schematic showing output capacitor

To guarantee safety, a capacitor is placed at the output terminal of the electronics. In the event of electronics failure, in which the output is shorted to one of the supply voltages, the output capacitor is crucial in preventing a continuous DC charge injection. The output capacitor is also useful in implementing charge balance, as can be seen in the monophasic stimulation example (Section 4.3).

4.3 Monophasic Voltage Stimulation

Monophasic voltage stimulation is the most basic neuro-stimulation scheme. During stimulation, the circuit outputs a constant-voltage pulse (Fig. 12a). Current is injected through the output capacitor and to the electrode to stimulate the tissue. The current level is limited by the tissue resistance. After the pulse, the circuit output returns to ground. The voltage stored across the output capacitor and electrode interface drives the reaction in reverse and helps to recover the charge.

4.4 Constant Voltage vs. Constant Current

Voltage-controlled systems rely on the tissue resistance to limit the current. Due to the buildup of voltage across the capacitor and electrode interface, the remaining voltage drop across the tissue decreases over the pulse duration. This is reflected in

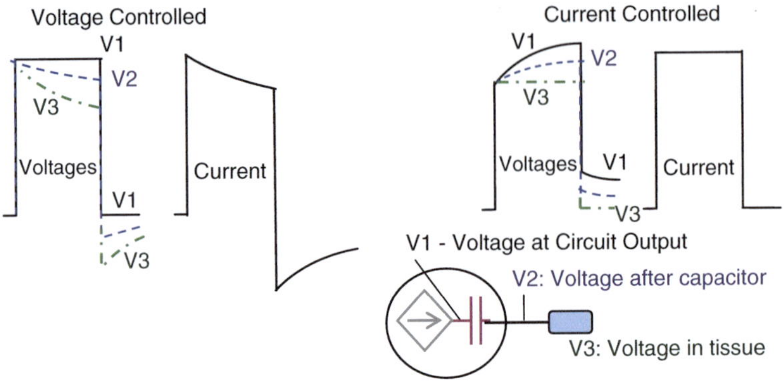

Fig. 12 Voltage control vs. current control

a nonuniform current level. For certain applications, the initial current spike may actually stimulate other neuron populations that respond to shorter pulses.

The current-controlled scheme avoids this problem by automatically increasing the output voltage to maintain constant current (Fig. 12b). Another benefit for using current control is the ability to use multiple electrodes to establish a constructive current profile. Multiple electrodes cannot be easily implemented with voltage control because the impedance between electrodes can vary depending on the relative distance, leading to uncertainty in the current. The drawback of current control is the increased complexity in the circuit design. Additionally, the p and n type output transistors need to be calibrated to improve charge balance.

4.5 Analysis of Constant-Current Electrode-Voltage Waveform

A diagram of the electrode-voltage waveforms for constant-current stimulation is illustrated in Fig. 13. A constant-current pulse is injected via the counter electrode, and returned through the test electrode (connected to ground). With the onset of constant-current injection, current passing through the electrolyte (or tissue) resistance (R_{Ω}) would result in a voltage drop (Fig. 13a). For some high-current stimulation applications, this resistive voltage drop can exceed 10 V. Because the current level is constant for the pulse duration, the *iR* voltage drop also stays constant and can easily be subtracted from the measured voltage. The resistive voltage is not part of the electrode potential because it occurs in the tissue, and is therefore independent of the reactions on the electrode surface. The injected current also charges the double-layer capacitance (C_d), resulting in a voltage ramp (Fig. 13b). As the electrode voltage increases, metal oxidation/reduction or other electrode reactions may start to occur. Current injection then becomes divided between the double-layer charging and Faradiac reactions including metal oxidation, thus decreasing the voltage-ramping speed (Fig. 13c). Because metal oxidation stores charge reversibly like a capacitor, it is sometimes termed as 'pseudo-capacitance.' If there is a

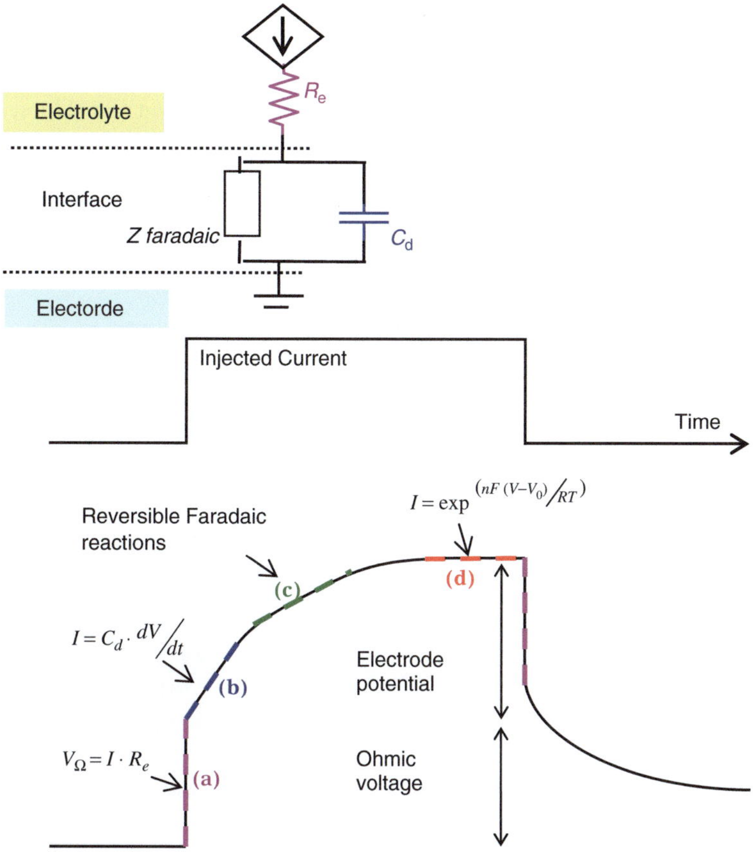

Fig. 13 Voltage waveform generated by a constant-current pulse through the electrode-electrolyte interface

sufficient increase in electrode potential (~1 V), electrolysis occurs and thus shunts the majority of the current (Fig. 13d). If the electrode is continuously operated in the electrolysis regime, undesirable byproducts are formed and corrosion accelerates [19]. Since the electrode current becomes an exponential function of the voltage when in the electrolysis-dominated regime, a small increase in voltage can cause a large increase in current. This characteristic can be used to identify the onset of hydrolysis.

4.6 Current Control with Passive Recharge

A commonly used neuro-stimulation scheme is to inject charge with a constant-current pulse, then recover the charge by clamping the circuit output back to ground voltage. In the passive recharge phase, the circuit operates similarly to the recharge phase in the monophasic mode, relying on the voltage stored in the capacitor and electrode to drive the recharge. Passive recharge conserves energy because it utilizes the voltage potential already stored in the capacitor to drive the recharge current. However, it suffers from slower recharge, with some high-current stimulation applications taking more than 10 ms to recover. Due to the limited driving strength of the return pulse, the change in oxidation state may not be completely reversed. Furthermore, leakage current in the output capacitor accumulates over time and as a result the anode typically exhibits signs of corrosion due to the passive recharge (Fig. 14).

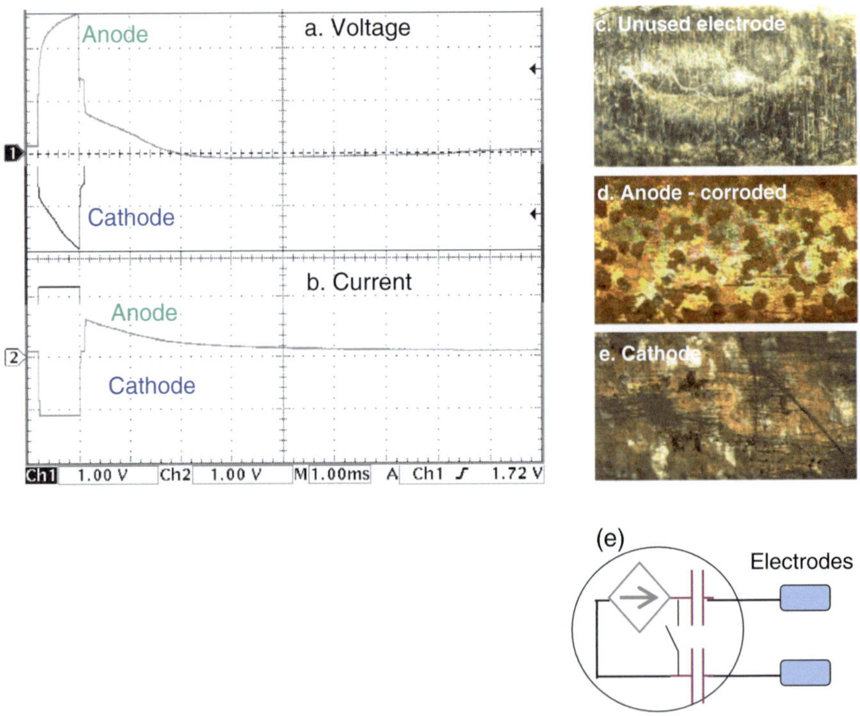

Fig. 14 Passive recharge observed in the electrode (**a**) voltage and (**b**) current waveforms, the resulting corrosion morphology (**b–d**), and (**e**) schematic of constant-current injection with passive discharge shorting

4.7 Active Recharge

For high-frequency stimulation, active recharge can be used to accelerate the recharge speed. A circuit designed to provide active recharge would actively drive both the cathodic and anodic phases. However, at the end of both active phases, the circuit still needs to incorporate a "shorting" phase, as done in passive recharge.

The shorting phase discharges any residual charge between the two active phases and prevents the bias voltage located at the electronics output (before capacitor) from slowly drifting away from ground.

It is important to note that active recharge may not reduce the likelihood of electrode corrosion. During the recharge phase, the cathode is positively biased and tends to maintain the positive bias at the end of the stimulation sequence, leaving the electrode susceptible to corrosion.

4.8 Cathodic Bias

For platinum electrodes, corrosion can be best minimized with a slight cathodic bias, rather than completely balancing the charge [20]. This is because the platinum electrode is more susceptible to corrosion when positively biased, increasing the likelihood for oxidation and dissolution by chloride.

To implement a cathodic bias, a slightly higher amount of charge is injected in the cathodic pulse than the anodic pulse. This ensures that the oxidation that occurs during stimulation can be completely reversed, rather than continuing to build up in successively thicker oxide layers. Since the cathodic bias scheme requires charge imbalance, the output capacitor cannot be used. The imbalance in charge injection must be supplied by a third, sacrificial anode, which is needed to provide sufficient material for corrosion to occur over the device lifetime. While this is technically feasible, the concept of intentionally not balancing charge has not gained wide usage.

4.9 Bipolar vs. Monopolar Stimulation

Many devices are designed to operate in either bipolar or monopolar modes, allowing the user to select the most-effective mode.

In bipolar mode, current is passed between two electrodes that are close to each other, resulting in a current pathway that is concentrated between the two electrodes. This leads to localized stimulation in the tissue immediately surrounding the electrodes.

In applications where the target tissue may be located further away or protected by the high-impedance of the dura, monopolar stimulation can be used to increase the stimulation distance. By using a return electrode that is near the implant casing, one creates an electrical gradient that has increased spatial coverage (Fig. 15).

5 Other Measurement Techniques

Cyclic voltammetry and impedance analysis were previously introduced as standard methods for studying the electrochemical properties of the electrode materials. However, these techniques cannot accurately predict the electrode behaviors when

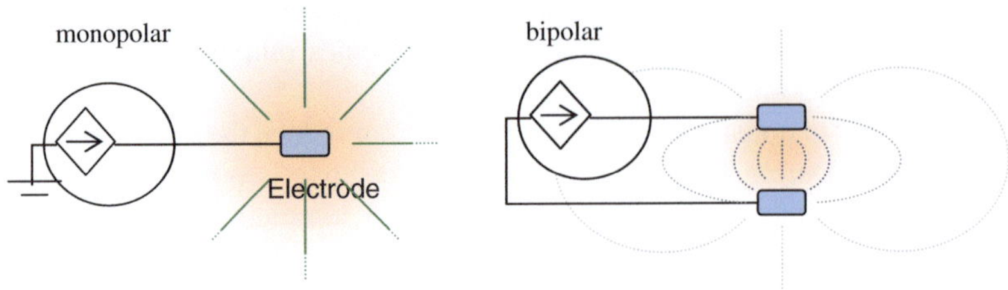

Fig. 15 Monopolar and bipolar electrode configurations

used in neuro-stimulation applications. Tests that are appropriate for this unique environment are introduced in the following sections. The charge-injection limit of an electrode can be quickly estimated by measuring electrode potential as a function of charge injection. Implementing a pulse-clamp is a new technique that has the potential for measuring the charge-storage characteristics. Finally, computational methods can be used to reveal detailed information describing the charge-injection mechanism.

5.1 Electrode-Potential Measurement

A quick method to estimate the safe charge-injection limit of the electrode is to measure the electrode potential as a function of the injected charge [21]. At low charge-injection levels, the electrode potential increases linearly with increasing charge injection as the double-layer capacitance charges. However, with the onset of electrolysis, any additional charge is consumed by hydrolysis and the electrode potential reaches a plateau.

The test is typically conducted by setting a stimulation protocol with a constant pulse width. The current amplitude is then gradually increased as successive measurements are recorded. This sequence is used to avoid the effect of electrode activation. To minimize electrode dissolution, a cathodic pulse measurement of the same magnitude and duration should be made following each anodic pulse measurement.

As shown in Fig. 16, the platinum electrode potential reaches a broad maximum at a current injection of about 0.15 mC/cm^2. At greater charge-injection values, the electrode potential appears to gradually decrease. This is likely due to the fact that the electrode experiences surface cleaning and roughening over the duration of the experiment.

Compared to platinum, iridium exhibits a much higher charge-injection limit at about 0.8 mC/cm^2. The iridium curve displays a very gradual flattening of the slope. This indicates that with increased charge injection and electrode potential, iridium reaches a greater oxidation state and increased charge-injection effectiveness.

An electrode with a fresh titanium surface shows similar curvature to that of a platinum electrode at low charge densities. However, at charge densities greater than 0.1 mC/cm^2, the electrode voltage continues to increase and does not plateau due

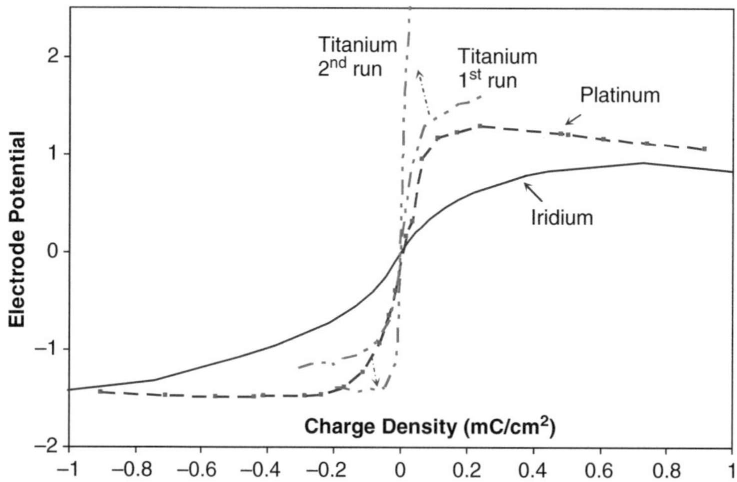

Fig. 16 Electrode potential for a titanium, platinum, and iridium electrode, measured at the end of 1-ms-long pulse. Titanium measurement is repeated to show drastic increase in electrode potential

to the increased thickness of the resistive oxide layer. When the titanium is immediately tested a second time, the electrode voltage is substantially higher due to the thick oxide layer. The diode-like property of the titanium oxide results in an asymmetrical curvature [22]. For cathodic charging, the curve plateaus at -0.05 mC/cm^2. For the anodic pulsing, the electrode potential continues to increase past 4 V.

5.2 *Pulse Clamp*

Pulse clamping is a relatively new technique that promises to provide more information about the charge-storage characteristics of neural stimulation electrodes. Pulse-clamp circuits operate similarly to a typical neural stimulation system. The difference lies in the additional circuit components that precisely measure the current level and use positive feedback to greatly accelerate the discharge [23, 24]. A typical pulse-clamp sequence starts with the working electrode set to 0 V (or any voltage) against the counter electrode or the reference electrode in most cases. During testing, the instrument switches to current mode and injects the working electrode with a constant-current cathodic or anodic pulse. Then, the instrument clamps the test electrode potential back to the initial voltage. In this state, all of the charge stored on the working electrode is discharged and detected by the instrument. The charge that is not recovered due to Faradaic loss can thus be calculated. If the electrode behaves as a pure resistor, then no stored charge will be recovered. Conversely, if the electrode behaves as an ideal capacitor, then all of the injected charge will be recovered and measured. The recovered current will exhibit an exponential time decay, with a time constant of $\tau = R_e \cdot C_d$, where R_e is the ohmic electrolyte resistance and C_d is the electrode capacitance.

When using pulse clamping to compare electrodes of different materials, the difference in discharge time can affect the measurement results. For example, iridium discharges more slowly than platinum, leading to higher measured charge loss. Similarly, larger electrodes also discharge slower because they exhibit greater double-layer capacitance. To facilitate comparison, the discharge time can be reduced by increasing the positive feedback for the slower electrode.

The pulse-clamp result can be displayed as the fraction of charge lost divided by the charge injected (in the constant-current pulse) (Fig. 17). At very low charge density, platinum electrodes behave like ideal capacitors. Utilizing only double-layer capacitance, the stored charge is quickly and completely removed during the discharge phase. With increased charge injection, oxidation-reduction reactions begin to occur. The oxidation reaction is typically not completely reversed during discharging. This is observed in the curve as higher fraction not recovered.

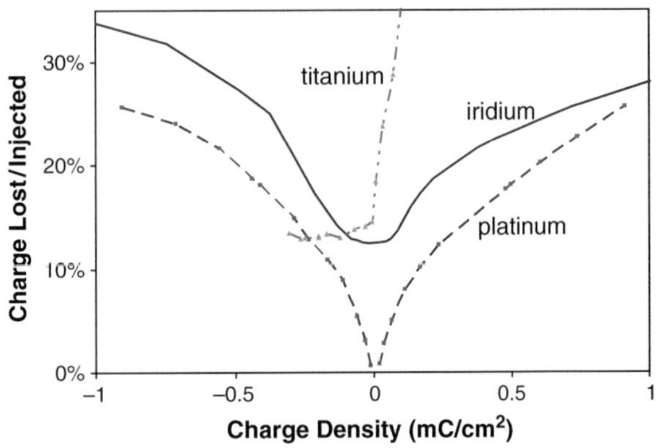

Fig. 17 Pulse-clamp data of titanium, platinum, and iridium electrodes

Accurately measuring iridium electrodes can be difficult because iridium utilizes bulk oxide with slow reaction kinetics, meaning the discharge time can be greater than 30 ms. Depending on the data-acquisition method, part of the discharge current may be truncated resulting in high level of measured charge loss.

Titanium discharges very quickly, relying almost completely on double-layer charge injection. Unlike platinum or iridium, titanium does not exhibit symmetrical behavior for cathodic and anodic pulses. As a result, the anodic charge injection is almost completely consumed in the formation of oxide. Repeating the experiment produces nearly identical results, despite the fact that the electrode potential has increased.

5.3 Computer Simulation

A well-designed model of the electrode interface can greatly stream-line the process of designing new systems and help visualize the reaction dynamics. However, creating a realistic model of the interface is difficult, due to its complex nonlinear behavior. If one were to create a detailed model that can simulate cyclic voltammetry, one would need a set of equivalent elements for each individual reaction that occurs at different potentials [25–27]. For example, according to Table 2, modeling platinum would require at least 7 equivalent pathways, with each pathway containing information about the reaction speed, potential, and charge capacity. However, it is possible to use a much simplified model if one just needs an accurate computer representation for a particular stimulation condition.

Because neuro-stimulation occurs at a relatively fast time scale, the reactions do not have time to equilibrate. To simulate the condition of many reactions occurring concurrently, one just needs the average impedance during the entire pulse duration. Typically, good simulation can be obtained using a second-order model similar to Fig. 1. The drawback is that because the electrode behavior is dependent on the stimulation conditions, this model may lose accuracy when applied to different pulse width and amplitude settings.

Computer simulations have long been used to reflect a reactive electrode surface, such as when a high charge injection results in high level of electrolysis current [28– 31]. These simulations employ a resistive network, and all illustrate significant increases in current density at convex protrusions (Fig. 18).

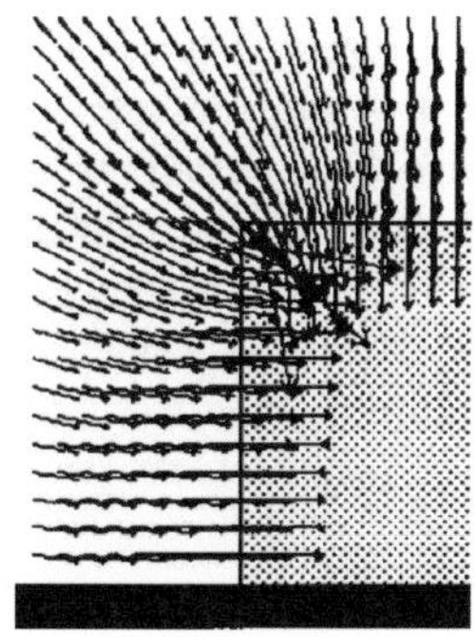

Fig. 18 Resistive models reflect the current-crowding pattern that occurs in hydrolysis regime or with DC bias [30]

In normal usage, electrodes should avoid reaching electrolysis potentials [32], and instead rely on the double-layer capacitance and reversible oxidation/reduction mechanisms. To simulate the electrode behavior in this regime, capacitive elements are needed. Because the voltage of the capacitive elements is the result of charge accumulation over time, a time-stepping computation is necessary to track the behavior of these capacitive elements through the pulse duration. SPICE and ANSYS are shown as two examples of software that can be used to perform this time-stepping simulation. The parameters are obtained from curve-fitting SPICE simulation to recorded pulse-clamp data, using 1-ms pulse width and charge density of 0.1 mC/cm^2 [33].

Performing the time-domain simulation with ANSYS illustrates the typical current pathway that for capacitive electrode surfaces (Fig. 19). Areas of higher current are observed in the solution near the sharp convex edges. However, at the electrode-solution interface, the current density is low and uniform. This is shown in the short arrows near the electrode surface that are perpendicular to the surface. The increased current density in the solution is the result of current having to navigate around the protruding portions of the electrode. Compared to the resistive model, the capacitive electrode model shows a very different current pattern. This uniform current distribution on the electrode surface is reflected in dissolution testing in Section 5.4.

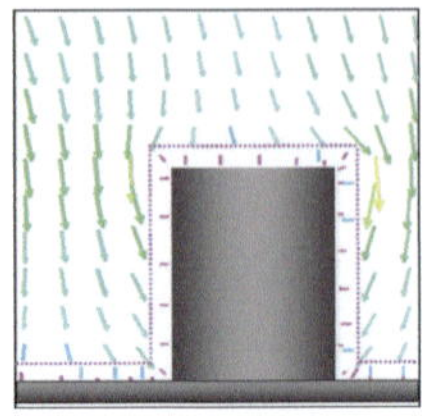

Fig. 19 ANSYS-simulation of a nonflat electrode operated in capacitive regime, 50 μs after the start of current injection. Vectors indicate the direction and magnitude of current

The same simulation can be performed using SPICE to illustrate the current contribution from different pathways. Because SPICE lacks meshing capability, a custom C-script is required to generate a netlist of nodes that describe the two-dimensional space of the electrode surface and surrounding electrolyte. Figure 19 shows the simulated current densities for the protruding and recessed portions of the electrode. At the beginning of the pulse, there is a brief 20-μs duration in which higher current concentrates on the convex protrusions (Fig. 20a). This initial current spike is required to charge the capacitance and establish proper voltage bias. The voltage gradient is established after 20 μs, and the current density becomes uniform, with less than 2% variation on the entire electrode surface.

As the electrode voltage increases during the pulse, a proportionally higher amount of current is injected through the metal-oxidation pathway, rising to more

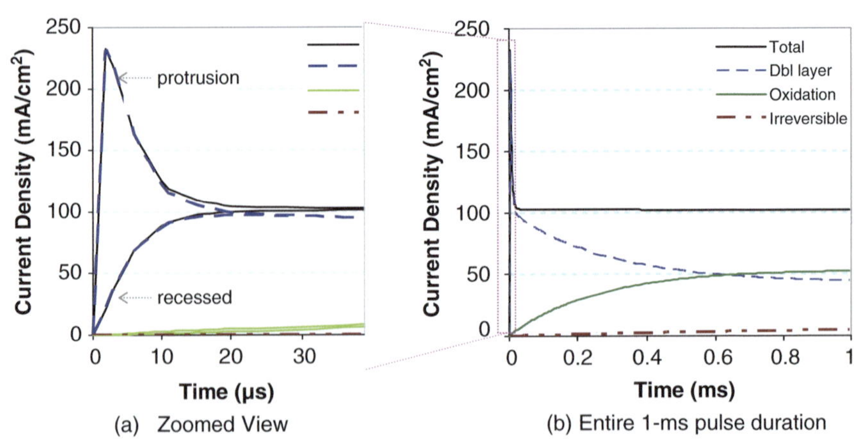

Fig. 20 Contribution of charge injection from different mechanisms. The expanded view in (**a**) shows the first 50 μs

than 50% at the end of the 1-ms pulse. The electrolysis current remains low, contributing less than 5% of the total current at 1 ms.

5.4 Dissolution Testing

Dissolution testing can be performed to illustrate the current distribution at different charge-injection settings. To demonstrate the effect of extreme surface topology, gold microposts (10-μm wide and tall) are electroplated on top of a gold electrode surface [34]. Gold is used in this test because it dissolves readily during the anodic pulse and is redeposited on the metal surface during the cathodic pulse. The deposition pattern is a good indication of the current distribution.

The morphology of surface redeposition shows a pattern of uniform current density when the electrode is injected with 30 μC/cm^2, well within its charge-injection limit. However, at 2 mC/cm^2, increased current density is observed at sharp convex edges, indicating a dissolution-dominated mode of charge injection at the higher current level (Fig. 21).

In general, long-term dissolution testing is required to ensure the lifetime of an electrode. The various techniques introduced in this chapter are useful in estimating the electrode performance at different usage conditions, but they cannot be substituted for the dissolution test. The test is generally conducted with the electrode submerged in buffered saline, with oxygen purged from the saline. The stimulation is set to the maximum pulse width and current amplitude used in clinical setting. The frequency may be further increased to speed up the experiment.

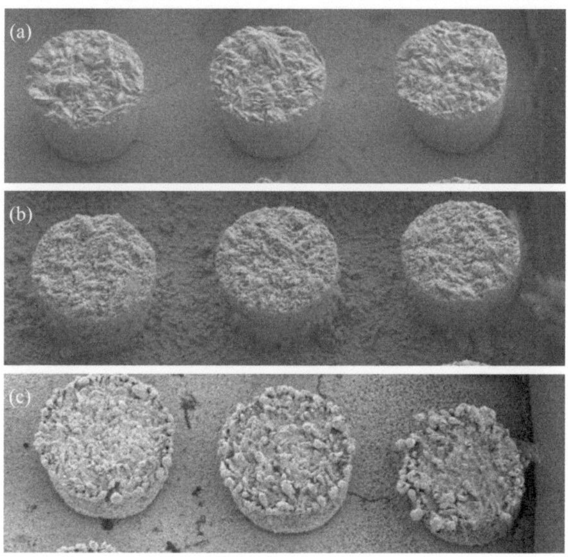

Fig. 21 SEM image of an array of gold microposts (**a**) as fabricated, (**b**) subjected to 30 μC/cm^2 dissolution for 2 weeks and (**c**) 2 mC/cm^2 dissolution

5.5 *Inductively Coupled Plasma (ICP)*

The most effective method of measuring the dissolved ion is with ICP equipment. The test solution containing dissolved metal is nebulized into the chamber. A radio-frequency field generates plasma and the optical spectrum of the plasma is analyzed for emission from various elements.

Most ICP equipments are not optimized for analyzing samples with high carbon contents. If protein is added to the test solution, it may quickly leave a carbon deposit on the optics. For these samples, one should instead use inductively coupled plasma-mass spectrometry (ICP-MS). An ICP-MS ionizes the sample in the plasma chamber and then uses mass spectrometry to detect the different elements. While ICP has sensitivity of parts-per-billion range, ICP-MS can detect in the parts-per-trillion range.

6 Summary

This chapter introduced several methods for analyzing the electrochemical properties of neuro-stimulation electrodes. Cyclic voltammetry is a generic technique that shows the oxidation/reduction potentials of a material, thereby revealing important electrode behavior properties. Furthermore, the impedance spectra and the electrode potential can reflect the surface area and the charge-injection limit of the electrode, respectively. Lastly, to determine their functional lifetimes, electrodes must be subjected to long-term dissolution testing.

The computer simulations described in this chapter are very simplistic models, with parameters obtained by fitting to actual stimulation waveforms. This type of simulation can only reflect the electrode behavior within a narrow range of charge-injection settings. Nevertheless, with better understanding of the materials, future models might be able to predict the electrode behavior at all charge-injection conditions, and even predict the rate of corrosion.

References

1. Rose TL, Kelliher EM, Robblee LS (1985) Assessment of capacitor electrodes for intracortical neural stimulation. *J Neurosci Meth* 12 (3):181–193
2. Bard AJ, Faulkner LR (1980) *Electrochemical methods: fundamentals and applications.* John Wiley & Sons, New York
3. Brummer SB, McHardy J, Turner MJ (1977) Electrical stimulation with Pt electrodes: trace analysis for dissolved platinum and other dissolved electrochemical products. *Brain Behav Evol* 14:10–22
4. Morton SI, Daroux MI, Mortimer JT (1994) The role of oxygen reduction in electrical stimulation of neural tissue. *J Electrochem Soc* 141:122
5. Newman J (1966) Resistance for flow of current to a disk. *J Electrochem Soc* 113:501
6. Tremiliosi-Filho G, Jerkiewicz G, Conway BE (1992) Characterization and significance of the sequence of stages of oxide film formation at platinum generated by strong anodic polarization. *Langmuir* 8:658

7. Pikelny AY (2003) Mechanism of platinum deterioration under stimulation pulses polarization. IFESS
8. Petit MA, Plichon V (1997) Anodic electrodeposition of iridium oxide films. *J Electroanal Chem* 444:247–252
9. Weiland JD, Anderson DJ, Humayun MS (2002) In vitro electrical properties for iridium oxide versus titanium nitride stimulating electrodes. *IEEE Trans Biomed Eng* 49:1574–1579
10. Wessling B, Besmehn A, Mokwa W et al. (2007) Reactively sputtered iridium oxide. *J electroch Soc* 154:f83–f89
11. Pourbaix M (1974) *Atlas of electrochemical equilibria in aqueous solutions.* Pergamon Press, Brussels
12. Beebe X, Rose TL (1988) Charge injection limits of activated iridium oxide electrodes with 0.2 ms pulses in bicarbonate buffered saline. *IEEE Trans Biomed Eng* 35:6
13. Robblee LS, McHardy J, Marston JM, Brummer SB (1980) Electrical stimulation with Pt electrodes V The effect of protein on Pt dissolution. *Biomaterials* 1(3):135–139
14. Hibbert DB, Weitznera K, Tabor B, Carter P (2000) Mass changes and dissolution of platinum during electrical stimulation in artificial perilymph solution. *Biomaterials* 21(1): 2177–2182
15. Macdonald JR (1992) Impedance spectroscopy. *J Annals Biomed Eng* 20:289–305
16. Norlin A, Pan J, Leygraf C (2005) Investigation of electrochemical behavior of stimulation/sensing materials for pacemaker electrode applications I Pt, Ti, and TiN coated electrodes. *J Electrochem Soc* 152 (2):J7–J15
17. Meyer RD, Cogan SE, Nguyen TH et al. (2001) Electrodeposited iridium oxide for neural stimulation and recording electrodes. *IEEE Trans Neural Sys Rehab* 9:2
18. Durand DM (1999) Electrical stimulation of excitable tissue. *Biomedical Engineering Handbook.* CRC, Boca Raton, FL
19. Robblee LS, McHardy J, Agnew WF et al. (1983) Electrical stimulation with Pt electrodes VII, Dissolution of Pt electrodes during electrical stimulation of the cat cerebral cortex. *J Neurosci Methods* 9(4):301–308
20. Scheiner A, Mortimer JT, Roessmann U (1990) Imbalanced biphasic electrical stimulation: muscle tissue damage. *Ann Biomed Eng* 18(4): 407–425
21. Rose TL, Robblee LS (1990) Electrical stimulation with Pt electrodes VIII Electrochemically safe charge injection limits with 0.2 ms pulses. *IEEE T BioMed Eng* 37:1118
22. Campbell SA, Kim HS, Gilmer DC et al. (1999) Titanium dioxide (TiO2)-based gate insulators. *IBM J Res Dev* 43(3):383–392
23. Bonner M, Daroux M, Crish T et al. (1993) The pulse-clamp method for analyzing the electrochemistry on neural stimulating electrodes. *J Electrochem Soc* 140:2740
24. Hung A, Zhou D, Greenberg R et al. (2007) Pulse-clamp technique for characterizing neural-stimulating electrodes. *J Electrochem Soc* 154(9):479–486
25. Moorhead ED, Stephens MM (1990) A finite element Galerkin/B-spline (GBS) numerical model of electrochemical kinetics, transport, and mechanism for multi-geometry working electrodes II: A study of quasi-reversible linear sweep voltammetry. *J Electroanal Chem* 282(1):1–26
26. Speiser B (1996) Numerical simulation of electroanalytical experiments: recent advances in methodology. *Electroanal Chem* 19:1–108
27. Alden J (1998) *Computational electrochemistry*, PhD Thesis, Oxford University, http://compton.chem.ox.ac.uk/john/Thesis/
28. Feldberg SW (1972) *Electrochemistry, calculations, simulations and instrumentations.* Marcel Dekker, New York
29. Goldberg IB, Bard AJ (1972) Resistive effects in thin electrochemical cells: digital simulations of current and potential steps in thin layer electrochemical cells. *Electroanal Chem* 38:313
30. Ferrigno R, Brevet PF, Girault HH (1997) Finite element simulation of the chronoamperometric response of recessed and protruding microdisc electrodes. Electrochimica Acta 42(12):1895–1903

31. Mcintyre CC, Grill WM (2001) Finite element analysis of the current-density and electric field generated by metal microelectrodes. Ann Biomed Eng 29(3):227–235
32. Rubinstein JT, Spelman FA, Soma M (1987) Current-density profiles of surface mounted and recessed electrodes for neural prostheses. *IEEE Trans Biomed Eng* 34 (11):864–875
33. Ksienski DA (1992) A minimum profile uniform current-density electrode. *IEEE Trans BioMed Eng* 39(7):682
34. Brummer SB, Turner MJ (1977) Electrical stimulation with Pt electrodes: II-estimation of maximum surface redox limits. *IEEE Trans BioMed Eng* 24:440
35. Hung A (2005) Techniques for sensing the current distribution and charge storage of high-density neuroelectrodes. PhD Thesis UCLA
36. Hung A, Zhou D, Greenberg R et al. (2005) Dynamic simulation and testing of the electrode-electrolyte interface of 3-D stimulating microelectrodes. *IEEE Neural Eng,* March 16–19

Conducting Polymers in Neural Stimulation Applications

David D. Zhou, X. Tracy Cui, Amy Hines, and Robert J. Greenberg

Abstract With advances in neural prostheses, the demand for high-resolution and site-specific stimulation is driving microelectrode research to develop electrodes that are much smaller in area and longer in lifetime. For such arrays, the choice of electrode material has become increasingly important. Currently, most neural stimulation devices use platinum, iridium oxide, or titanium nitride electrodes. Although those metal electrodes have low electrode impedance, high charge injection capability, and high corrosion resistance, the neural interface between solid metal and soft tissue has undesilable characteristics. Recently, several conducting polymers, also known as inherently conducting polymers (ICPs), have been explored as new electrode materials for neural interfaces. Polypyrrole (PPy), polyaniline (PANi), and poly(3,4-ethylenedioxythiophene) (PEDOT) polymers may offer the organic, improved bionic interface that is necessary to promote biocompatibility in neural stimulation applications. While conducting polymers hold much promise in biomedical applications, more research is needed to further understand the properties of these materials. Factors such as electrode impedance, polymer volume changes under electrical stimulation, charge injection capability, biocompatibility, and long-term stability are of significant importance and may pose as challenges in the future success of conducting polymers in biomedical applications.

This chapter looks into the current research and challenges for conducting polymers and their applications in neural stimulation electrodes.

Contents

D.D. Zhou (✉)
Second Sight Medical Products, Inc., Sylmar Biomedical Park, Sylmar, CA 91342, USA
e-mail: dzhou@2-sight.com

D.D. Zhou, E. Greenbaum (eds.), *Implantable Neural Prostheses 2*,
Biological and Medical Physics, Biomedical Engineering, DOI 10.1007/978-0-387-98120-8_8,

1 Introduction

Implantable neural prostheses have been widely used to improve or restore main functions of nervous systems for patients with neural damage. Some common neural prostheses include cochlear implants [1–2], spinal-cord stimulators [3–6], and deep-brain stimulators [7–10]. Novel neural prostheses, such as retinal prostheses [11–12] and brain–machine interfaces [13–14], with higher resolution and site specificity are being actively investigated. These devices require larger numbers of microelectrodes patterned in a very small area, more sophisticated circuit designs, and longer lifespans.

In a retinal prosthesis, vision for patients blinded by macular degeneration or retinitis pigmentosa is recovered through electrical stimulation of remaining retinal neurons [11, 15–16]. The retinal prosthesis is designed to detect and convert light into electrical stimuli, which are then delivered to the retina, bypassing damaged photoreceptor layers, to evoke a visual response [11, 17–20]. Many technical challenges exist before the retinal prosthesis can become clinically successful including image processing, power and data telemetry, microelectronics, interconnection, and packaging. This chapter addresses factors specifically relating to microelectrode arrays.

The demand for high-resolution stimulation in neural prosthetic devices requires the development of high performance, high density microelectrode arrays. For such arrays, the choice of electrode materials has become increasingly important. Stimulation electrodes serve as the interface between electronic devices and neural systems. Since high levels of charge injection and electric fields are applied to functional and responsive tissues and electrode materials, it is critical that damage to tissues or nerves, as well as to electrode materials, from chronic stimulation be minimized or avoided [21].

Recently, several conducting polymers have been explored as new electrode materials for neural interface. Polypyrrole, polythiophene, and their derivatives can be electrochemically polymerized from aqueous solution and deposited on the neural electrodes [22–28]. Bioactive molecules, such as cell-adhesion peptides, ECM proteins, growth factors, etc., can be incorporated into the polymer as dopants to promote neuronal growth and binding to the electrodes [22–25, 29–30]. Poly(3,4-ethylenedioxythiophene) (PEDOT) presents properties especially promising for chronic neural interfaces. PEDOT has a dioxyethylene bridging group across the 3- and 4-positions of the hetero-ring, which blocks the possibility of coupling and, consequently, renders superior electrochemical stability [31]. When electrochemically deposited onto the microelectrode sites of neural probes, PEDOT, along with dopant polystyrenesulfonate (PSS), was found to be electrochemically stable after hundreds of cyclic voltammetric scans. PEDOT/PSS coating decreases the impedance modulus by almost two orders of magnitude [24–25].

Most studies on conducting polymers focus on improving the neural interface or neural electrode for recording. For example, high quality acute neural recordings were obtained through electrodes with the PEDOT-incorporated peptide sequence DCDPGYIGSR in acute animal tests [24–25]. More recent work demonstrated an improvement in chronic recording performance in neural probes coated with PEDOT/PSS [32–33]. Our particular interest is in the material used for the chronic stimulating electrodes. There are similarities in the requirements for a neural recording electrode and a neural stimulating electrode, such as low impedance, biocompatibility, and stability. However, one important requirement for a neural stimulation electrode is the charge-injection capacity. During neural stimulation, either current or voltage is passed through the electrode for charge delivery, while in neural recording, an electrode is typically operated through a very high impedance amplifier with virtually no passing current. This poses a challenge to a neural stimulation electrode to maintain low impedance, biocompatibility, and stability while delivering adequate charges for prolonged periods of time. In this chapter, a brief introduction on the developments of conducting polymers for neural interfaces will be presented. Some preliminary work exploring conducting polymers as stimulation electrodes will be reviewed. The challenges in the development of stimulation electrodes using conducting polymers, especially for chronic stimulation, will be discussed.

2 Neural Stimulation and Electrode Materials

Most neural stimulation applications use a biphasic, charge-balanced, cathodic-first-current pulse. The cathodic phase is believed to be more effective in exciting cells or neurons than the anodic phase. After the cathodic phase, electrodes are biased negatively. A lagging anodic phase will remove such negative bias and keep the electrode voltage near neutral. This biphasic method minimizes electrode polarization by the charge-balancing second phase, which cancels out cathodic bias and maximizes charge delivery. During each phase, electrodes are subjected to charges that may exceed their charge-delivery capabilities. Electrode materials used for chronic stimulation are required to maintain high charge delivery capability [34].

2.1 Charge Transfer Processes During Stimulation

Electrical stimulation of biological tissue with metal electrodes requires the flow of ionic charge in the biological tissue. This flow of charge can be induced by both

Faradaic and capacitive mechanisms. The Faradaic mechanism of charge injection involves electron transfer across the electrode-tissue interface. One such example is the charge injection through metal oxidation and reduction mechanisms. Metal oxidation involves binding the metal to oxygen, forming a new oxidation state or metal oxide binding with oxygen to form a higher oxidation state. For some metal oxides, charge stored can be recovered through reversible reduction reactions. Such reversible metal oxidation and reduction processes can be used for safe neural stimulation. If charge density is too high for a given electrode, irreversible electrochemical reactions such as metal corrosion or dissolution, gas evolution, or production of toxic chemical reaction products can occur. Such induced harmful electrochemical reactions during Faradaic charge transfer not only can cause electrode damage, but also can cause tissue or nerve damage [35–38].

The capacitive mechanism involves charging or discharging of a layer of solution, known as the electrochemical double layer, whose composition is different from the bulk solution. Ions in the tissue are attracted or repelled by charge on the electrode to produce pulses of ionic current. There is no charge transfer across the electrode–electrolyte interface. There is a tendency for charged species to be attracted to or repelled from the metal–solution interface. This gives rise to a separation of charge and the electrochemical double layer. As a result of the variation of the charge separation with the applied potential, the electrochemical double layer has an apparent capacitance, known as the "double-layer capacitance" [39]. Double-layer capacitance is the result of ionic accumulation at the electrode surface and does not involve any electrochemical reaction. Charging or discharging the double-layer capacitance is an ideal mechanism of charge injection because no electrochemical reactions can occur in the electrode–tissue interface [35].

For a given electrode, capacitance, C, is calculated according to the equation:

$$C = \varepsilon\varepsilon^{\circ}\ (A/d)$$

where ε is the dielectric constant of the solvent, ε° the permissivity in vacuum, A the electrochemical surface area of the electrode surface, and d the thickness of the dielectric layer.

To achieve large charge storage capacity for a given material and device design, the real surface area of the electrode is a key factor: the higher the surface area, the higher the capacitance. For high density microelectrode arrays used in neural stimulation, the geometric surface area is often limited by the application, and an effective way to increase the electrochemical surface area without enlarging array size is to increase the surface roughness of the electrode.

2.2 Electrodes for Neural Stimulation Implants

In neural prosthetics, the demand for high performance, high density microelectrodes is increasing as more and more neural prosthetic devices are developed [40]. The choice of electrode materials has become a key factor for the success of such neural prostheses. The electrodes must be made smaller to increase spatial resolution and at the same time deliver adequate charge without generating irreversible electrochemical reactions.

Neurological stimulation requires high quality, electrochemically stable and biocompatible electrodes. These electrodes must have low electrode impedance, high charge injection capability, and high corrosion resistance [35]. Current devices typically use platinum, iridium oxide, or titanium nitride electrodes to inject the necessary current to elicit a neural response. These materials are chosen because of their ability to inject large amounts of charge with negligible electrode degradation. To ensure longer device lifetimes, most devices incorporate electrodes with a large surface area and operate at well below the charge injection limit of the electrode material to avoid unnecessary dissolution and gas evolution, which could potentially damage biological cells.

Platinum is an electrode material widely used in neural stimulation. Its safe stimulation limit, defined by the amount of charge applied before hydrolysis and gas evolution, ranges from 0.1 to 0.35 mC/cm^2 [41–43]. For effective stimulation, the electrode must transfer sufficient charge to exceed the threshold potential for neuronal depolarization. For retinal electrode arrays in retinitis pigmentosa (RP) patients, the disk electrodes are 500 μm in diameter and the injection charge required for neuronal depolarization is 0.4–0.6 μC per phase of a biphasic-stimulating pulse [44]. For many stimulation applications that require a high density electrode array in a very small region, electrode size should be no larger than 100 μm in diameter. The charge density required for effective stimulation with such small smooth Pt electrodes will be 1 mC/cm^2 or higher, which exceeds the safe stimulation limit [12, 19]. Hydrogen/oxygen evolution due to water hydrolysis induced by stimulus will alter the pH, causing metal corrosion and possible tissue damage in the electrode–tissue interface [45–47]. Therefore, smooth Pt is not a suitable candidate for these applications [33, 47].

Iridium oxide (IrOx) and titanium nitride (TiN) have been reported to hold higher safe stimulation limits of 0.9–4 mC/cm^2 [12]. These two materials, in a recent study, were directly compared on silicon-based electrode arrays [48]. TiN is more microfabrication friendly than IrOx, but has higher impedance. As a result, the stimulator compliance voltage needed to drive the necessary current through the TiN electrode will need to be high. Iridium oxide, when activated, has much lower impedance, reducing the amount of power necessary. In addition, the electrode size could also be reduced for better spatial resolution. Few studies have been done to evaluate the biocompatibility of these materials. TiN has been used as a pacemaker electrode material clinically. However, exposure of TiN to retinal cell culture increased cell death [49]. Iridium oxide has been shown to be biocompatible in several studies on cortical implants [50]. However, chronic aggressive stimulation resulted in degradation of iridium and adverse tissue response [50–51].

Conducting polymers offer some advantages over metal electrodes. Conducting polymers are unique materials, which have the ability to conduct electricity, but are organic in nature. In contrast to metal materials, soft conducting polymers may provide an improved bionic interface between the rigid electronic devices and the soft, amorphous biological systems [52–53]. A conducting polymer's surface and structure can be readily modified to improve charge injection capability and promote biocompatibility. Such advantages make it an attractive candidate for further research as coatings for neural stimulation electrodes.

3 Conducting Polymers

The early work on conducting polymers, also known as inherently conducting polymers (ICPs), started from polymerization of acetylene [54]. It was expected that such conjugated structures, which possess an extended π-conjugation along the polymer backbone, would exhibit electronic conductivity. However, the conductivity of polyacetylene, a charcoal-like black powder, was found to be fairly low; it was at best semiconducting (in the 10^{-7} S/cm range) [55]. The pioneer work by Heeger, MacDiarmid, and Shirakawa led to the discovery of doping of conducting polymers [56, 57]. They oxidized polyacetylene with iodine vapor, increasing its electrical conductivity by up to 11 orders of magnitude [54]. The doping process, adopted from terminology used for semiconductors, is an oxidation or reduction process. Upon oxidation or reduction, electrons are removed from or inserted into the polymer backbone producing a cation or anion. Electron transfer along the conjugated p- or n-molecular orbital backbone, coupled with the motion of charge carriers in the material, provides conductivity to doped conducting polymers [58].

3.1 Various Conducting Polymers

Many conducting polymers and their derivatives or composites have been investigated. Early conducting polymers, such as polyacetylene, suffer from instability in air and were difficult to prepare [59]. Extensive research has yielded several promising polymers and their derivatives, including polyanilines, polypyrroles, and polythiophenes [52, 59], specifically for biomedical applications [54, 60].

3.1.1 Polypyrrole

Polypyrrole (PPy) is one of the most studied conducting polymers due to its aqueous compatibility, low oxidation potential of the monomer, ease of preparation, and high conductivity [61–62]. The structure of PPy is shown in Fig. 1. Electropolymerizaion of polypyrrole involves the formation of pyrrole cation radicals, dimer radicals, and

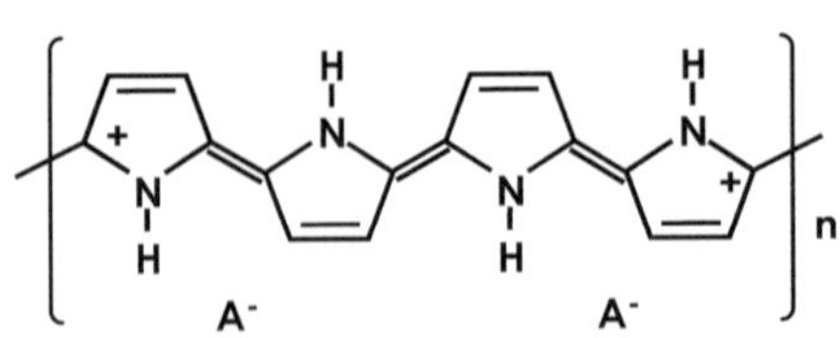

Fig. 1 The p-doped polypyrrole (PPy) structure. The dopant anions (A^-) are required to balance the charge and maintain electroneutrality

coupling of the dimers and monomers [58]. Depending on oxidation stages, one or more electrons can be removed for each polymer section during oxidation. The polymer backbone is positively charged. The dopant anions (A^-) are required to balance the charge and maintain electroneutrality. During reduction, the anions, if small enough, are expelled from the polymer.

The physical, chemical, and electrical properties of PPy can be easily modified by various doping agents and preparation conditions. There are many available dopant ions for the generation of good quality deposited polymer films [63–64]. Two of the most common dopants that are codeposited with PPy are polystyrenesulfonate (PSS) and sodium dodecylbenzenesulfonate (NaDBS). PPy/PSS and PPy/NaDBS polymers have been used in many applications ranging from actuators and neural scaffolds to neural electrode coatings [22, 53, 61, 65]. It is reported that electrode materials, electrolyte solution, deposition methods (current- or potential-controlled deposition), deposition time, and solution temperature during electrochemical polymerization affect both the structure and electroactivity of PPy films [66].

3.1.2 Polyaniline

There are many different forms of polyaniline (PANi) that exist based on their oxidation states ranging from fully reduced to the fully oxidized. These oxidation states are affected by pH and are interchangeable from the protonation and/or oxidation reactions. It is reported that the most conductive form among various redox forms is the emeraldine salt as shown in Fig. 2 [67].

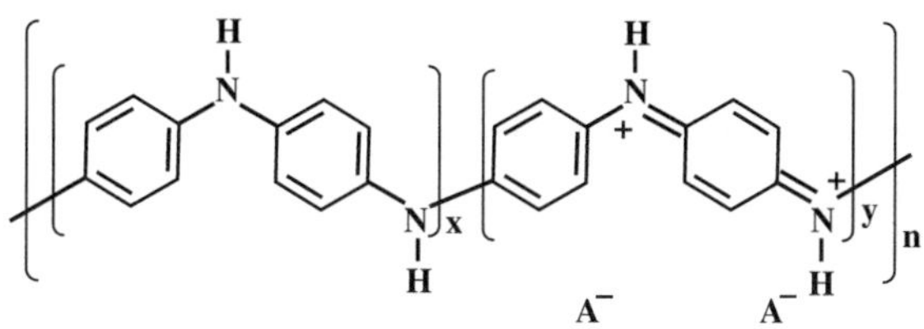

Fig. 2 One of the PANi structures – emeraldine salt. x and y are the degrees of polymerization and x=y=0.5 for emeraldine salt. This is the most conductive form among various PANi redox forms

Polyaniline and its derivatives have been widely used for biosensing applications based on their crystallinity. For example, PANi and its derivatives were found responding to the saturated alcohol vapors by undergoing a change in resistance [54]. The change in resistance of the polymers on exposure to different alcohol vapors was attributed to their chemical structure, chain length, and dielectric nature.

3.1.3 PEDOT

Poly(3,4-ethylenedioxythiophene) (PEDOT) is formed from polymerization of bicyclic monomer 3,4-ethylenedioxythiophene (EDOT). The structure of PEDOT is shown in Fig. 3. Due to its narrow energy gap of the highest occupied molecular orbital (HOMO) to the lowest unoccupied molecular orbital (LUMO) [59, 68],

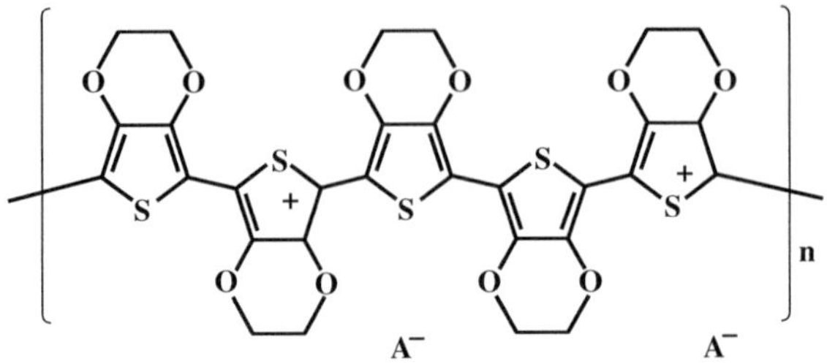

Fig. 3 The p-doped PEDOT structure. The oxidant A^- served as the charge balancing counter-ion to the PEDOT during polymerization

PEDOT exhibits improved conductivity and thermal stability relative to many other polyheterocycles, such as the polypyrroles or polythiophenes [31, 62, 69–70]. In addition, PEDOT has a dioxyethylene bridging group across the 3- and 4-positions of the heterocycle preventing α-β' coupling, and as a result is more electrochemically stable than PPy [24–25, 32]. Most PEDOT polymers studied are p-doped for better stability while only a few are n-doped [71]. Similar to PPy, many doping agents are available for the polymerization of PEDOT. One such doping agent is sodium polystyrenesulfonate (PSS). The oxidant PSS served as the charge balancing counter-ion to the PEDOT during polymerization. EDOT monomers have poor solubility in aqueous solution, usually less than 10 mM at room temperature. PSS is a water-soluble polymer and also functions as a good dispersant for aqueous PEDOT [59].

Another approach to increase the solubility of EDOT monomers is to append a water-soluble side chain onto the monomer rings. For example, the addition of a hydroxymethylated group ($MeOH^-$) onto EDOT results in a significant increase of aqueous solubility up to 0.1 M. Xiao et al. have reported that thin films of PEDOT-MeOH/PSS were easily deposited on the gold electrodes of a neural probe with 20 mM EDOT-MeOH and an applied potential of 0.8–0.9 V vs. SCE [62]. The resulting films from the complex of PEDOT-MeOH/PSS are more uniform compared to PEDOT/PSS coatings. This indicates that water-soluble groups such as MeOH-modified monomers improved current distribution during electrochemical polymerization.

Often, the colors of PEDOT are reported to be sky blue or deep blue [59, 62]. Actually, the coating color of PEDOT films produced electrochemically depends on plating conditions and film thicknesses. The charge used in the deposition appears to affect the film color significantly for a given polymer/dopant system. In the potentiometric-controlled mode, it was found that the coating color is dependent on plating time at a given current density or is dependent on current density at a given plating time. Generally speaking, conducting polymer films deposited with lower charge appear lighter in color, primarily because of the reduced thickness [33]. Figure 4 shows some typical colors for PEDOT/PSS films electrochemically deposited on the Pt electrodes. The plating is carried out at a fixed plating time of 60 s at room temperature with variable plating current densities, ranging from 20 to 160 $\mu A/cm^2$. The color changes from light blue at 20 $\mu A/cm^2$, light purple at 40 $\mu A/cm^2$, dark purple at 80 $\mu A/cm^2$ to dark brown at 160 $\mu A/cm^2$.

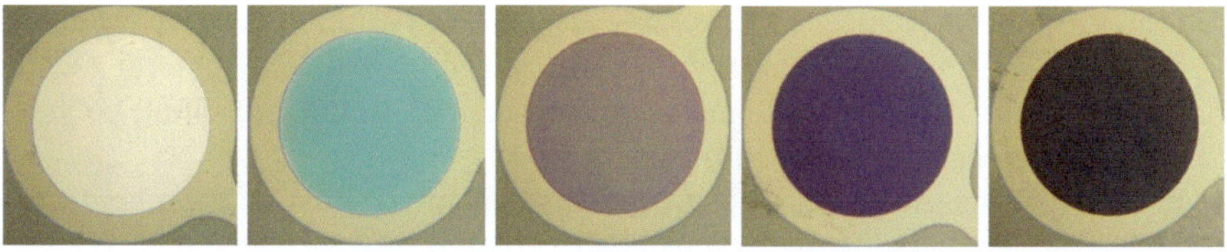

Fig. 4 Some typical colors for PEDOT/PSS films electrochemically deposited on the Pt electrodes. The PEDOT was coated potentiometrically for 60 s at room temperature with variable current densities. From *left* to *right*: (**a**) noncoated Pt surface, (**b**) light blue at 20 μA/cm^2, (**c**) light purple at 40 μA/cm^2, (**d**) dark purple at 80 μA/cm^2 to (**e**) dark brown at 160 μA/cm^2

3.2 Methods of Preparing Conducting Polymers

Conducting polymers can be polymerized by chemical and electrochemical methods. Chemical synthesis is typically used to produce large quantities of material, which involves mixing a strong oxidizing agent with a monomer solution [72–73]. The resulting polymers are usually in powder or very thick films. For example, nanometer-sized PEDOT/PSS particles were chemically synthesized with the oxidant iron(III) p-toluenesulfonate [59]. During the polymerization reaction, the radicals formed from oxidized monomer conjugated into the neutral polythiophene backbone. The neutral polymer backbones are subsequently doped by further oxidation and incorporated with some counter-ions for maintaining charge balance [58, 74].

Electrochemical polymerization is widely utilized during the investigation of conducting polymers for neural probes and for neural stimulation. The advantages in electrochemical polymerization are obvious. The technique is established and simple. It allows accurate controls over thickness and morphology of the resulting polymer coating on the target electrode surface. Very thin and well-defined coatings can be deposited over microelectrodes in a neural probe or microelectrode arrays [63–64]. In addition to lateral growth, vertical growth of polymer coatings provides flexibility in the three-dimensional design of electrodes [53, 75–76].

Electrochemical polymerization can be carried out by current-controlled (potentiometric) or voltage-controlled (amperometric or cyclic voltammetric) methods. Although both methods produced quality coatings, the potentiometric method appears easier to control. When a constant current is applied to the electrochemical cell in a potentiometric mode, the charge used for polymer deposition is simply determined by time and the film grows at a steady rate. However, when a constant voltage is used in an amperometric mode, the resulting electrode current varies and depends on many factors including the base electrode material, polymer systems, and plating conditions. The calculation for the charge used for the deposition is not straight forward and the film growth is less steady.

Unlike metal depositions where metal ions are reduced and deposited on the cathode, most electrochemical polymerizations are an anodic process. Positive (anodic) current or voltage is applied to the working electrode to perform oxidation reactions. Monomers at the working electrode surface undergo oxidation to form radical cations that react with other monomers or radical cations, forming insoluble polymer chains on the electrode surface [60].

In an amperometric process, electrochemical polymerization on a working electrode (WE) is subjected to anodic potentials, versus a reference electrode (RE), in a three-electrode cell. However, in some studies, a cell voltage, i.e., a working-electrode voltage versus a counter electrode (CE), has been used in a simple two-electrode cell [77]. In addition to constant voltage, a scanned voltage using cyclic voltammetry can also been used for electropolymerization. It is reported that the films produced by cyclic voltammetry are thicker than those produced potentiostatically at a potential equal to the upper limit of the cyclic voltammetry [78]. Typically, a working electrode voltage of +0.6 V to +1.0 V versus Ag/AgCl is used for most depositions in a three-electrode cell. Overoxidation of the polymer occurs when higher anodic voltages are used. A typical three-electrode electrochemical cell for electropolymerization is shown in Fig. 5.

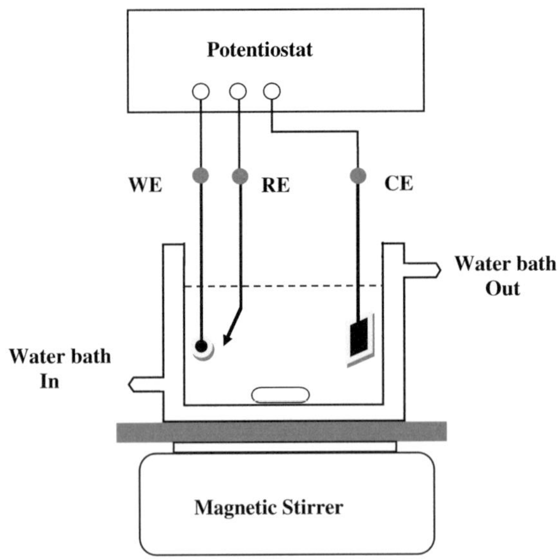

Fig. 5 A typical three-electrode electrochemical cell for electropolymerization of conducting polymers on microelectrodes. A magnetic stirrer is used for solution agitation while water bath is used to control temperature of a water-jacketed glass cell. A potentiostat is used to control current or voltage for electrochemical polymerizations and measurements

In a potentiometric process, a wide range of current or charge densities are reported. A typical current density of 0.5 mA/cm^2, or 30 mC/cm^2, is a good starting point for exploring optimal plating conditions. It was found that the use of current densities greater than 2 mA/cm^2 resulted in the formation of an overoxidized polypyrrole film with reduced conductivity [79].

The surface morphology and mechanical properties of polypyrrole coatings using different electrodeposition methods have been compared by Herrasti et al. [80]. Polypyrrole films have been deposited on steel by three electrochemical methods: constant potential, constant current, and cyclic voltammetry. The deposition was conducted in a three-electrode cell in 0.1 M $LiClO_4$ and 0.5 M pyrrol. The working electrode was a steel sheet. A graphite electrode was used as the auxiliary electrode while an Ag/AgCl electrode was used as the reference electrode. The deposition was controlled with a charge of 2 C/cm^2, and the thickness of the resulting films were approximately 5–6 μm. The morphology of all the deposits is shown in Fig. 6.

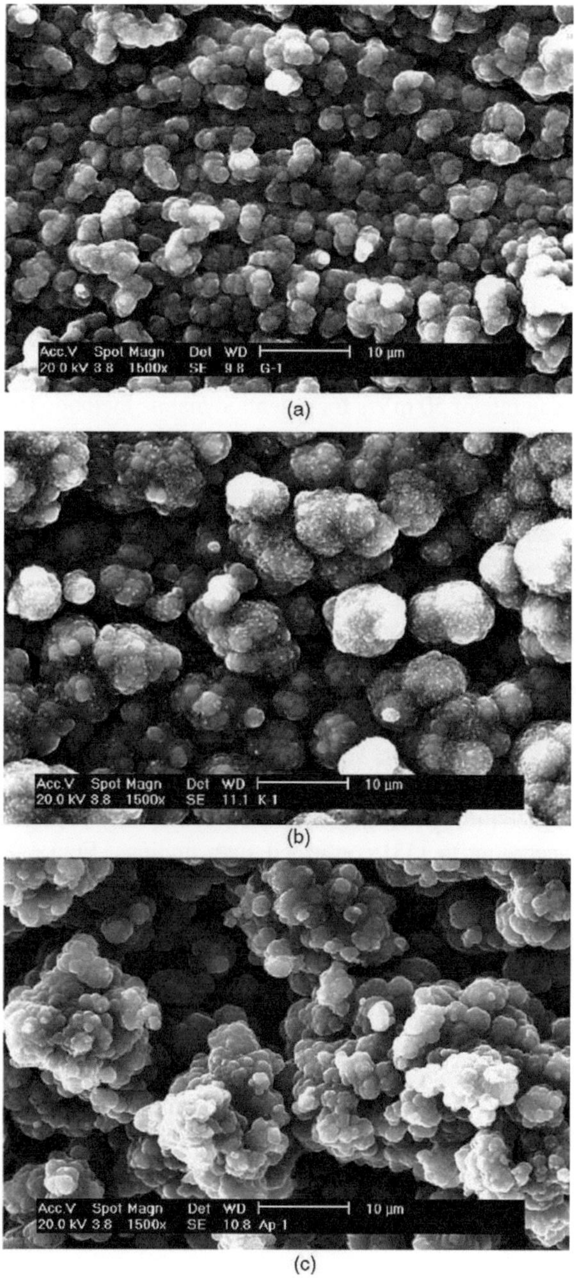

Fig. 6 SEM micrographs of the polypyrrole films prepared by three electrochemical methods in oxidized state (**a**) constant potential (0.7 V vs. Ag/AgCl) for 500 s, (**b**) constant current (5 mA/cm^2 for 400 s), and (**c**) cyclic voltammetry (–0.2 V to 0.7 V at 50 mV/s, 30 cycles) (Reproduced from [80], with permission from Elsevier)

It seemed that all the deposits were of a granular type. However, the differences in the grain size and porosity of the deposits were clear, and they were dependent on the type of deposition method used. The grain size and porosity increase from samples obtained at constant potential, to constant current, to cyclic voltammetry. The authors attributed the differences in polypyrrole morphology to the deposition rates. They speculated that in a constant-potential method, the polymerization has a higher initial growth rate and the polymer deposits are more compact and harder due to the smaller grain size. In a constant-current method, the deposits obtained have larger grain sizes due to a more uniform growth rate throughout the deposition process. In a cyclic voltammetry method, variable deposition rates and the oxidation and reduction process of the polymer during its growth might contribute to a more rough and porous structure.

In contrast to PPy, the trend in morphology changes due to deposition methods was different for the polyaniline films. Choi and Park [81] reported that polyaniline films prepared by cyclic voltammetry were smoother and more compact compared to the films obtained by the constant-potential method. PANi films were deposited on Pt electrodes in a solution containing 0.030 M aniline and 1.0 M $HClO_4$ at +1.0 V for constant voltage method, and at scanned voltages of –0.1 V to +1.15 V with a scan rate of 50 mV/s for potentiodynamic method. The authors pointed out that the aging effects and the degradation reaction might lead to more compact films during the potentiodynamic cycles. During reverse scans to the voltage lower than 0.7 V, the degradation reaction produces soluble products such as benzoquinone and/or *p*-aminophenol, which would reduce the amount of the deposited polymer. Such dissolution reaction had less impact on the film growths when the constant voltage was maintained at +1.0 V.

Many factors affect the quality of the polymer coating including film morphology, adhesion, conductivity, impedance, and surface activity. In addition to the deposition methods and deposition charge discussed above, other important factors include solution temperature, pH, nature of the supporting electrolyte, dopant and doping levels, surface condition of bare electrode, and concentration of the monomer solution. For instance, the surface topographies of PPy can be controlled through electrochemical polymerization temperature [53]. Work reproduced according to [53] confirmed such findings. Figure 7 shows surface morphology changes when the plating solution temperatures are varied. During the deposition of PPy/NaDBS coatings, smooth and dense morphology is obtained at the room temperature. However, when the solution temperature is lowered to 4°C, rough and porous surface features can be obtained while other conditions are kept the same.

When temperatures are varied, in addition to the surface-morphology differences, current distribution over the electrode surface also seems to be affected. By comparison, surface coating is more uniform when plated at a lower temperature, while surface plated at room temperature shows clear edge effects, as shown in Fig. 8. This indicates a better current distribution is achieved when polymerization is carried out at a lower temperature. It is reported that more uniform surface coating could give higher conductivity [59].

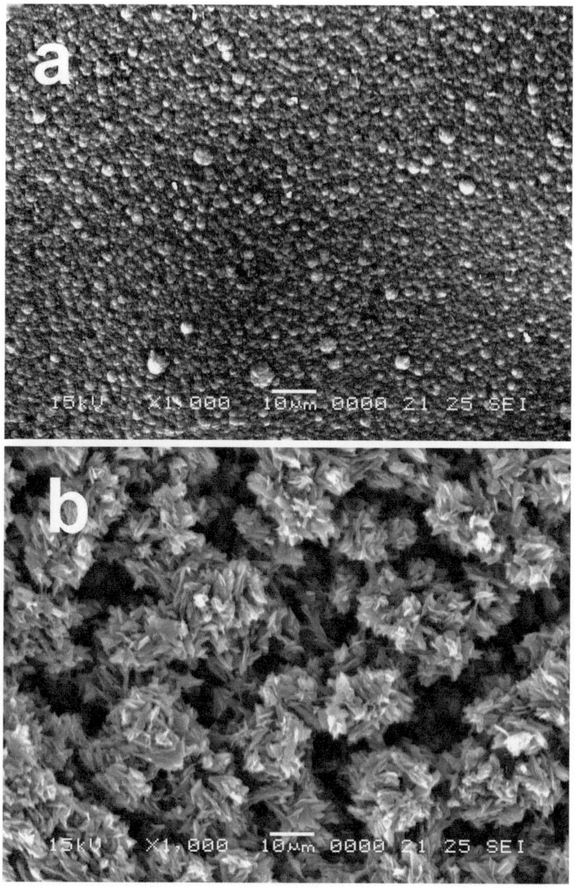

Fig. 7 PPy/NaDBS surfaces on Pt electrodes polymerized at two temperatures. (**a**) Smooth and dense morphology at 25°C. (**b**) Rough and porous morphology at 4°C

Surface modification and preparation on substrate have been reported to improve coating adhesion. For instance, premodification of the polyimide surface with self-assembled alkylsilane monolayers induced anisotropic lateral growth of PPy films along the substrate surface and enhanced film adhesion [76]. In this work, PPy film was electrochemically deposited onto Pt microelectrodes on a polyimide substrate and the PPy-coated microelectrode was used in external stimulation of cultured cardiac myocytes.

In addition to chemical and electrochemical synthesis, other polymerization methods such as photochemical synthesis, emulsion, and pyrolysis have been reported [58, 63]. In a recent study, a self-assembled monolayers (SAMs) method has been used to deposit thiolated poly(alkylthiophene)s and functionalized alkanethiols [82]. The authors claimed that resulting polymers have low impedance at 1 kHz and are more robust and better controlled in their composition than existing electrodeposited conductive polymer coatings.

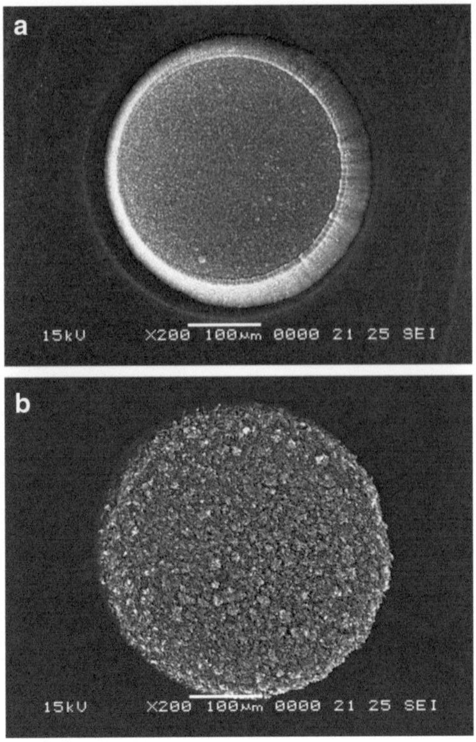

Fig. 8 Edge effects at different plating temperatures. PPy/NaDBS-coated surface on Pt shows clear edge effects when plated at room temperature (**a**), while the coating is more uniform when plated at a lower temperature (**b**)

3.3 Biomedical Applications of Conducting Polymers

There has been a large effort focusing on the development of conducting polymers for biomedical engineering applications such as biosensing [83–84], tissue engineering [85], drug delivery devices [77, 86–87], and neural electrode arrays [24]. A comprehensive review of conducting-polymer properties specific to biomedical engineering applications is provided by Schmidt and coworkers [60].

Conducting polymers are promising for the applications of sensors such as electrochemical sensors, biosensors, and gas sensors [88–90]. In sensor applications, conducting polymers are used either as sensing elements like electrodes and sensing membrane or as matrices for immobilizing sensing materials [54]. Physical adsorption, entrapment, and covalent bonds have been explored to incorporate sensing materials into sensor systems. The sensing materials can be incorporated simply by mixing them with the solution of dopant and monomers during electrochemical polymerization. For example, PPy has been used to trap biosensing materials such as glucose oxidase and electron transfer mediators such as ferrocene for sensitive glucose sensors. In this case, biomolecules are mixed with monomers and dopant during electrochemical polymerization [60]. Optical property changes with oxidation states of conducting polymers have been found suitable for construction of pH sensors. pH sensors based on PPy [91] and PANi [82] have been reported with fast reversible color response upon pH change [54, 89]. The PANi-based optical pH sensors can be exposed in air for over 1 month without any deterioration in sensor performance [92].

Recently, PPy has been used to construct a reference electrode for an RuO_2 thin-film pH sensor [93]. The solid-state miniaturized PPy-coated ITO (indium tin oxide) reference electrode showed improved properties such as lower drift rate and narrower hysteresis width in comparison to a commercial Ag/AgCl electrode.

Conducting polymers have been explored as biomaterials to replace metals or modify metal surfaces in tissue engineering. Localized electrical stimulation has been shown to promote tissue repair, cell growth, bone regrowth, and wound healing [52, 58].

The investigations of conducting polymers for improving neural interfaces have attracted much attention recently [94–97]. Studies indicated that polypyrrole, polythiophene, and their derivatives can be electrochemically polymerized from aqueous solution and deposited on the neural electrodes [23–24]. Rough surface morphology, increased surface area, and low impedance of conducting polymer coatings improved neural recoding performances [24, 32]. Furthermore, the neural probes can be modified with biologically functional macromolecules such as peptides, proteins [29, 98, 99], polysaccharides, and living cells [100–101]. These macromolecules can be incorporated by mixing with the solution of dopant and monomers during electrochemical polymerization. In one study, Wadhwa et al. [77] coated gold electrodes with polypyrrole and dexamethasone by electrochemical codeposition. Dexamethasone is an anti-inflammatory drug and can reduce inflammation in the Central Nervous System (CNS) to inhibit astro-glial sheath formation. Localized and controlled release of Dex from PPy coatings may reduce such sheath formation around implanted micromachined neural electrode arrays for chronic recordings.

Polypyrrole has been explored as an implantable microelectrode array for stimulating the nervous system [102]. PPy patterns were prepared by the micropatterning method using photochemical reaction of oxidative polymerization agents. A mixture of iron (III) chloride ($FeCl_3$) and aqueous solution of polyvinyl alcohol (PVA) was coated on a plastic film as the oxidative polymerization agent. $FeCl_3$ is known to be photochemically reactive and will not cause polymerization by irradiation of UV light. The coated agent was exposed under UV light with a patterned mask. The PPy patterns were prepared on the nonirradiated surface after the contact with pyrrole vapor. However, neither electrochemical test data nor electrode performance were reported by the authors.

Recently, we reported work on the evaluation of PEDOT for neural stimulation applications [33]. PEDOT was electrochemically deposited on thin-film Pt electrodes of stimulation electrode arrays to evaluate its properties for chronic stimulation. The coated electrodes demonstrated much lower impedance than thin-film Pt due to the high surface area and high ion conductivity across the film. The PEDOT film also presents intrinsic redox activity, which contributes to the low impedance as well as a much higher charge storage capacity. The charge injection limit of PEDOT electrodes was found to be 2.3 mC/cm^2, comparable to IrOx, and much higher than thin-film Pt. Under biphasic stimulation, the coated electrodes exhibited lower voltage and linear voltage excursion. Well-coated PEDOT electrodes were stable under chronic stimulation conditions, suggesting that PEDOT is a promising electrode material for chronic neural stimulation applications.

4 Challenges of Conducting Polymers for Chronic Neural Stimulation

Chronic neural stimulation requires charge injection through electrode/tissue interface at a high frequency for a prolonged period of time. This poses unique challenges to conducting polymers used as neural stimulation electrodes. Neural stimulation requires electrodes that are biocompatible and have low impedance and high charge delivery capability. In addition, the conducting polymer coating should have good adhesion on the electrode surface. For chronic stimulation applications, it is critical that these properties are stable for a prolonged period of time to achieve long-term quality performance.

4.1 Electrode Impedance

A unique feature for conducting polymers is their π-bonded conjugated polymer backbone with alternating single and double bonds. The mobile charge carriers formed during the doping process move along the polymer backbone and between adjacent chains, generating conductivity [58, 60]. While good conductivity is necessary to reduce the dc resistance of an electrode, having a low ac impedance (i.e., low resistance and high capacitance over a wide range of frequencies) is more important for a stimulation electrode. Electrode impedance is related to interfacial surface area between the electrode and electrolyte with impedance decreasing as surface area is increasing. Electrochemical Impedance Spectroscopy (EIS) is a useful tool to characterize electrode impedance.

Many have reported that electrode impedance is lowered after coating with conducting polymers. The reduction in impedance presents two features: reduction in impedance modulus and reduction in phase angles. For electrodes coated with PEDOT of different thicknesses, a significant decrease of impedance modulus ranging from 1 to 3 orders of magnitude can be seen mostly at the lower frequency region (1–500 Hz) [24, 100, 103]. More impedance reduction was observed in electrodes coated with PEDOT nanotubes and PEDOT nanotubes combined with hydrogel scaffolds [104–105]. A lesser decrease in impedance at higher frequencies is obtained. For example, after deposition of PEDOT/PSS, the impedance modulus of the Pt electrode decreased 2–3 orders of magnitude at the frequency range of 1–1000 Hz, but only 1–2 orders of magnitude at higher frequencies of 1–10 kHz, as seen in Fig. 9.

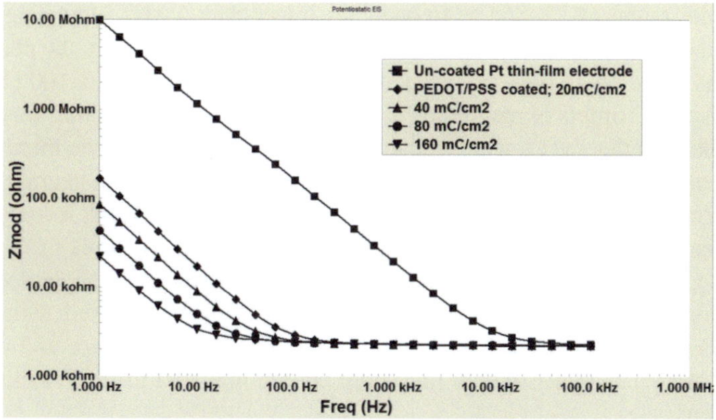

Fig. 9 Comparison of the impedance modulus of the Pt electrode and PEDOT-coated Pt electrodes. The PEDOT/PSS was coated potentiometrically at room temperature with variable polymerization charge densities from 20, 40, 80 to 160 mC/cm^2. All 4 PEDOT-coated electrodes have reductions in impedance modulus at the frequency range of 10 kHz or lower while significant reductions appear at 1 kHz or lower

Conducting polymer coatings also reduced the phase angles in impedance spectroscopy. For example, coating with PEDOT/PSS dramatically decreases the phase angles as shown in Fig. 10. The phase plot of the impedance spectroscopy measured in phosphate-buffered saline (PBS) reveals phase angles transition from low to high at the very high frequency of 63 kHz for a bare Pt electrode. The phase angles

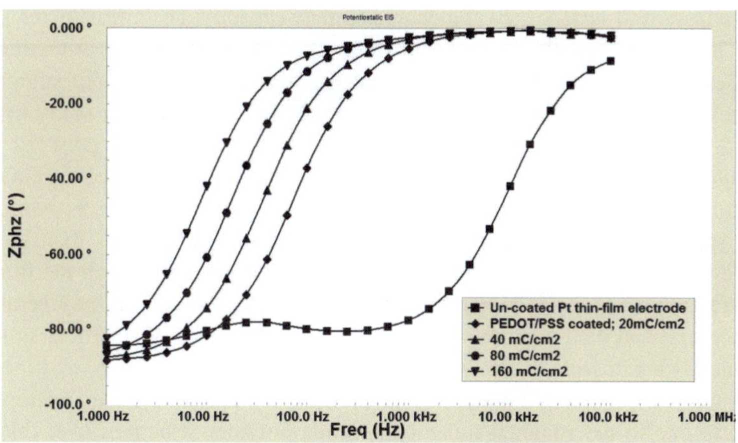

Fig. 10 Comparison of the impedance phase angles of the Pt electrode and PEDOT-coated Pt electrodes. The PEDOT/PSS was coated potentiometrically at room temperature with variable polymerization charge densities from 20, 40, 80 to 160 mC/cm^2. All four PEDOT-coated electrodes have reductions in phase angle at the frequency range of 100 Hz–10 kHz while significant reductions appear around 1 kHz

were in the range of 77–85° at frequencies lower than 1 kHz indicating that the bare Pt electrode is primarily functioning as a capacitor. In contrast, the phase angle transitions from low to high occurred at very low frequencies of 10–100 Hz, with a reduction of 1–2 orders of magnitude after conducting polymer coating was applied. Lower phase angles over a wide frequency range make the electrode more resistive as opposed to capacitive (low capacitance), which is desirable for a neural recording electrode or a stimulation electrode since this will provide an overall lower electrode impedance. Both ionic- and electrical-conductive porous structures of conducting polymer coatings are attributed to impedance modulus and phase angle deductions. Richardson-Burns et al. [100] have reported that electrodes coated with PEDOT polymerized in direct contact with neural tissue had low impedance and increased charge storage capability, which may help charge injection in in vivo stimulation applications.

Many equivalent circuit models have been proposed in EIS data analysis [104, 106]. Most researchers use an equivalent circuit of RWC to model the impedance spectra of coated electrodes [24, 29, 107]. An RWC model consists of a series connection of a solution resistance (R), a finite-length Warburg impedance (W or Z_D), and an electrode capacitance (C) [107]. The solution resistance is independent of the film thickness or roughness and is dominated by the geometric surface area of the microelectrodes for a given electrolyte. The finite Warburg diffusion impedance is characterized by a diffusional time constant, a diffusional pseudocapacitance, and a diffusion resistance [108]. The electrode capacitance is dominated by the surface area, which depends on surface roughness for a given polymer.

The RWC model was successfully applied to analyze the relationship of coating thickness and roughness. For example, on PEDOT/PSS-coated Pt electrodes, the electrode capacitance increases with the increase of film thickness, which suggests an increased surface area and charge storage capacity. The diffusion resistance, which is a characteristic of the deposited film, increases as the film thickness increases. These observations suggest that the film growth starts out smooth and becomes rough and porous as it gets thicker. Surface morphology study by SEM verified this hypothesis. Figure 11 shows the PEDOT coatings that grew for different lengths of time on the thin-film Pt electrode. The thinnest film shows featureless morphology. As the film grows, more topographic features, such as bumps and pores, start to appear. Thicker films have larger bumps and pores than the thinner ones. The open porous structure allows fast ion transport through the film. The good electronic and ionic conductivity of PEDOT suggests that it may be a better interface material than smooth metal electrodes for converting electrical signals to the ionically conductive tissue [33].

Although the coating impedance is reduced by increasing the coating layer thickness due to the rougher and more porous polymer structure, over-thick and over-porous structures appear fragile and are susceptible to delamination. Therefore, an optimized coating thickness, giving the best coating impedance, stability, and charge capacity, needs to be found [59].

As neural stimulation and neural response are high frequency in nature, impedance behavior at high frequencies is of particular interest for stimulation. At

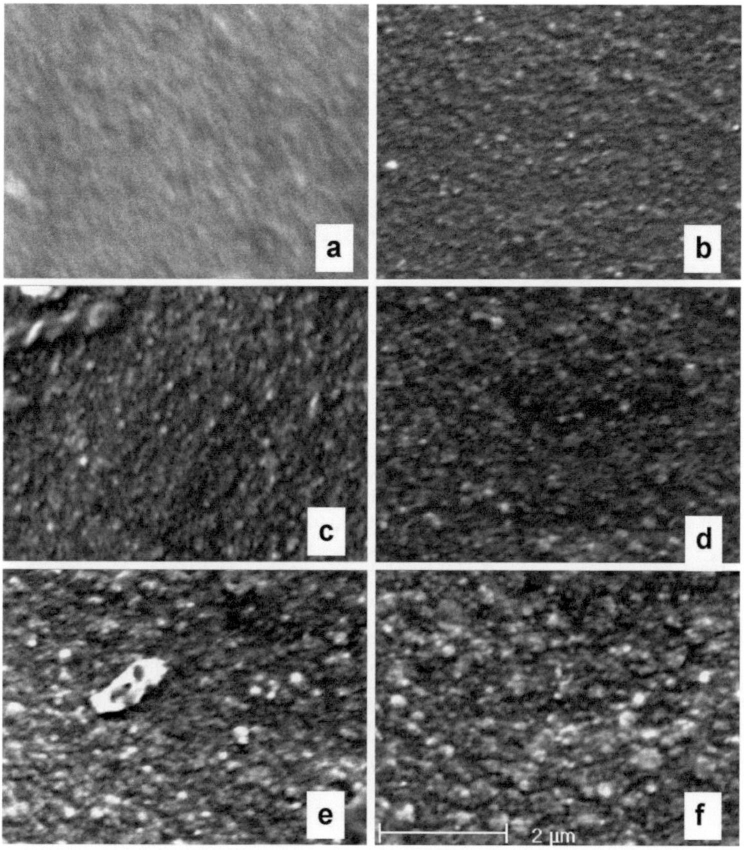

Fig. 11 Surface morphology of the PEDOT coatings on Pt electrodes. Films were grown for (**a**) 10, (**b**) 50, (**c**) 100, (**d**) 190, (**e**) 300, and (**f**) 400 s at 0.64 mA/cm^2 (Reproduced from [33], with permission from IEEE)

the clinically relevant frequency of 1 kHz or higher, the electrode impedance in Fig. 12 is nearly identical for electrodes coated under different charges. Therefore, high charge storage capacity from very thick coatings may not be fully accessed during fast stimulation. For a very porous thick structure, high frequency stimulation current cannot access deep pores due to time restraints in mass transport and *iR* drop in the deep pores.

A neural stimulation system typically uses constant current to deliver charges required to excite cells or neurons; however, this often causes voltage excursion to exceed 1 V on the stimulation electrode. In comparison, impedance spectroscopy traditionally operates in the 5–50 mV range, typically 10 mV, without dc bias for simplicity in data interpretation [34, 109]. However, the small potential without dc bias does not accurately reflect the performance at high potentials that are involved in oxidation or reduction reactions for conducting polymers. In fact, electrode

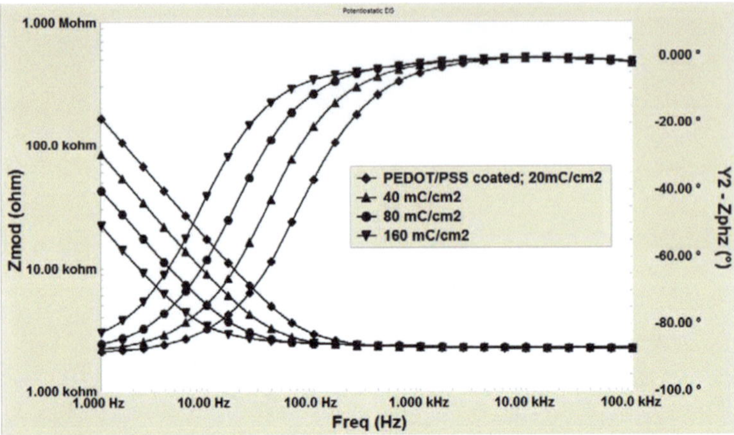

Fig. 12 At the clinically relevant frequency of 1 kHz, the electrode impedance modulus is nearly identical for electrodes coated with different polymerization charges (i.e., different thicknesses), while the phase angles have a slight decrease with the increase of coating charges

impedance is greatly affected by dc biases, especially at the lower frequency range [107]. Examples of Nyquist plots of a PPy/NaDBS-coated Pt electrode under different dc biases are shown in Fig. 13. For PPy/NaDBS-coated electrodes, the electrode impedance decreased as dc bias increased from 0 V bias to 0.6 V vs. Ag/AgCl. At high dc bias, electrode impedance is lower and the Warburgh impedance, which

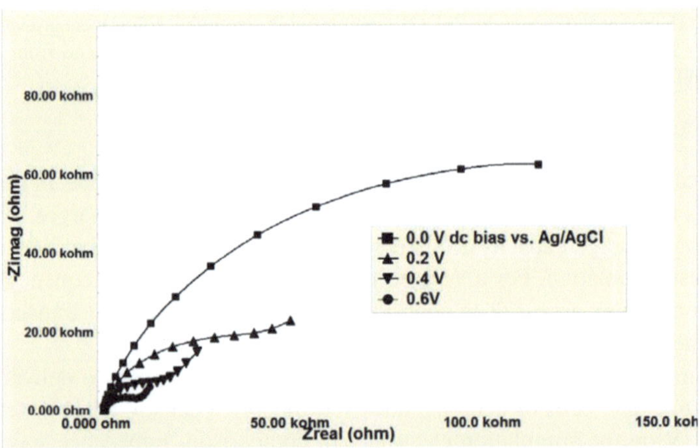

Fig. 13 Impedance spectra changes with dc biases. EIS measurements were made on a PPy/NaDBS-coated Pt electrode in PBS within a frequency range of 100 kHz to 10 Hz. The ac bias was 10 mV. The electrode impedance decreased when the dc bias increased toward anodic voltages

indicates mass transfer and diffusion controlled electrode reactions, dominates impedance spectra.

4.2 Polymer Volume Changes Under Electrical Stimulation

One unique property of conducting polymers is their ability to bind and expel molecules or ions in response to an electrochemical potential. During the doping process, i.e., oxidation (for p-doped polymers), the neutral polymer backbone is positively charged to form mobile charge carriers. To maintain electroneutrality, some counter-ions are required to diffuse into the polymer during charging and diffuse out during neutralization [58]. Such movement of anionic species in the polymer structure caused significant change in polymer volume [80, 110].

The volume change in PPy films has been studied by Otero et al. [110]. The experiments were carried out on free-standing PPy film under a wide potential range of –0.6 V up to 0.8 V. It was observed that the polymer length increased with the oxidation state of the polymer, where a linear relationship was found between charge consumed and film extension. Polymers swell during oxidation and shrinks during reduction [80, 110]. It was determined that the amplitude of change in the polymer length was of 12.5 ± 0.5 μm for a free-standing PPy film of 17×12.5 mm and such length variation is associated with the change of the polymer volume when it is oxidized or reduced.

Herrasti et al. [80] compared the structure changes of PPy films due to oxidation states. As can be seen in Fig. 14, the structure of a reduced polymer is much more compact with smaller grain size and dense films due to the loss of counter-ions and the reduction of its volume.

The volume change under reduction and oxidation was linked to the film's hardness and the Young's modulus. The hardness of the reduced samples is always greater than oxidized samples due to the more compact structure from the loss of counter-ions and the reduction of its volume [80]. The Young's modulus increased with the oxidation state of the polymer to achieve a plateau at a high oxidized state of the polymer. When the PPy is oxidized, a swelling of the polymer caused a stretching of the polymer chains. Hence, a reduction in the number of degrees of freedom of the polymer chains (from a folded to a stretched conformation) means a reduction in the elastic properties of the polymer film, increasing the Young's modulus of the polymer [110].

The volume change in conducting polymers has been linked to electrode impedance changes. It is reported that poly-3-octylthiophene (POT) films under increasing oxidation potentials decreased in charge transfer resistance. Such reduction in impedance was linked to film swelling, i.e., polymer film volume increases up to 25–30% during oxidation. Such film swelling caused an easier uptake of counter-ions into compact structures of the film and, thus, decreased the energy barrier for injection of anions into the film [106].

The volume change of conducting polymers under different bias or electrical stimulation has been explored for the development of artificial muscles and actuators

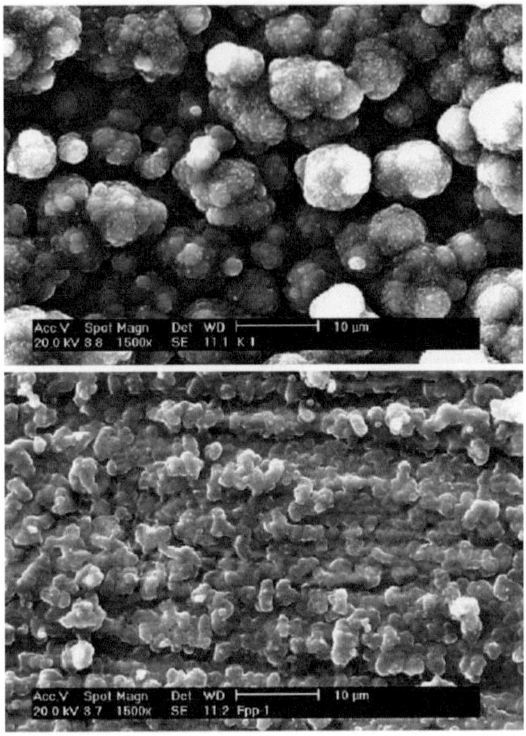

(a)

(b)

Fig. 14 Comparison of morphology of polypyrrole films in the oxidized (**a**) and reduced (**b**) states. The film in the reduced state has a smaller grain size and a more compact structure than that in its oxidized state. The polypyrrole films were prepared by the constant-current method at 5 mA/cm^2 for 400 s (Reproduced from [80], with permission from Elsevier)

[74]. The flow of a current through a film of conducting polymer with well-defined three-dimensional shape in the electrolyte causes electrochemical reactions in the film, producing a volume change. This change of volume can be transformed into a mechanical stress for the construction of artificial muscles [110, 111, 112].

The volume change of conducting polymers under electrical control, i.e., electro-chemo-mechanical properties, has been used in drug delivery applications [60]. In one study, Xu et al. [113] reported a device using electrochemically deposited thin PPy films on evaporated gold to release drugs from reservoirs under a low electrode voltage. Research reported by Abidian and coworkers described the application of PEDOT nanotube for the electrochemically controlled release of a drug [86, 104]. Under the cyclic potential stimulation, the drug-incorporated nanotube undergoes a fast swelling and deswelling process caused by ion and water movement in and out of the film.

The change in volume for conducting polymer coatings may be potentially a disadvantage for their applications in neural stimulations. During neural stimulation, the electrode is pulsed between its oxidation and reduction states, especially under high charge densities. If such cycling in redox states will indeed induce significant volume changes, the coating adhesion may be compromised, causing coating delamination and breakdown. This poses a challenge for the application of conducting polymers as stimulation electrode materials.

4.3 Charge Injection Capability

Electrochemical processes including oxidation and reduction reactions can be characterized by cyclic voltammetry (CV). A typical cyclic voltammogram of a PEDOT-coated electrode in comparison to a Pt electrode is shown in Fig. 15a. PEDOT undergoes reversible oxidation and reduction reactions as indicated by the anodic and cathodic current peak on the CV curve. The peak potential indicates the voltage at which the reaction takes place and the enclosed area of the curve is proportional to the charge storage capacity (CSC), i.e., charge passed during one CV cycle. Clearly the PEDOT film has a significantly higher CSC than the smooth thin-film Pt electrode. The CSC of PEDOT film is 10 times higher for the thinner films,

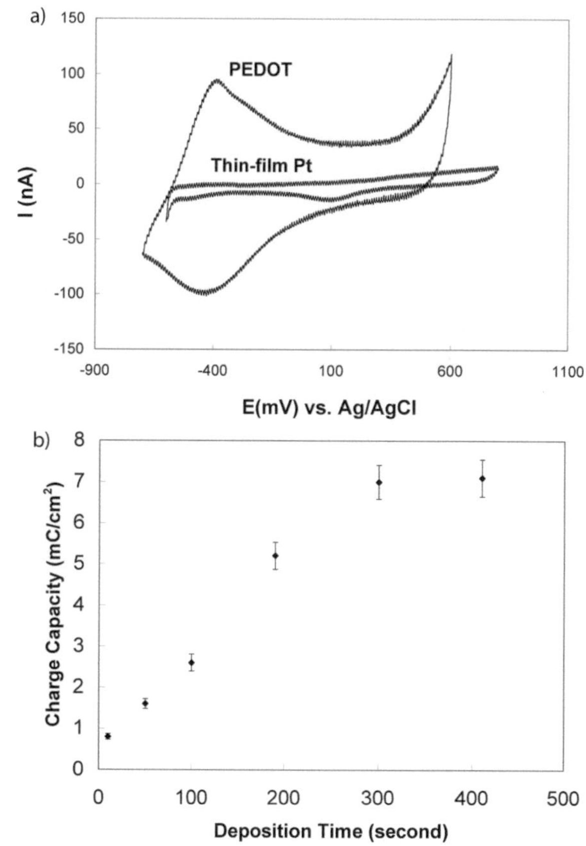

Fig. 15 (**a**) Cyclic Voltammograms (CV) of PEDOT-coated and uncoated Pt electrodes. The CV scan rate was 100 mV/s. (**b**) The charge capacity as a function of deposition time (Reproduced from [33], with permission from IEEE)

and 20 times higher for the thicker films. From CV measurements, PEDOT electrodes show current peaks due to Faradaic reactions and high capacitance current due to double-layer charging/discharging, similar to IrOx electrodes. This suggests that the charge injection of PEDOT electrodes, like IrOx, is through both Faraday (redox reactions) and non-Faraday (double-layer capacitance) processes [33]. The charge storage capacity of PEDOT films increased with deposition charge as shown in Fig. 15b. This is expected because thicker films are rougher and present more electro-active species. The relationship is linear for thinner films. However, the linearity falls off for thicker films, which suggests lower doping levels and more defects for the thicker films.

Cyclic voltammetry, used to characterize conducting polymer coatings, does not operate at the same time scales or voltage amplitudes as those used for neural stimulation; it may be inadequate for investigating true electrode dynamics. Typical neural stimulation uses constant-current pulses of 0.1–1 ms in duration, where voltages often exceed 1 V on stimulation electrodes. In comparison, CV has a typical sweep rate of 10–100 mV/s or 0.5–5 min/cycle for a potential range from –0.6 V to +0.8 V, and thus is ineffective in reflecting the reaction kinetics of neural stimulation. CV at a faster sweep rate would be more appropriate for characterizing stimulation electrodes. However, at faster scan rates of 1–20 V/s, charging current from electrode capacitance dominates the total current response causing distortion of voltammograms. This makes it difficult to analyze these voltammograms in detail [114].

Most neural stimulation applications use a biphasic, cathodic-first current pulse. The voltage response (voltage excursion) of an electrode is a direct indicator of its charge injection capability. For a given pulse current, the electrode that shows lower electrode voltage presents higher charge injection capability. A low voltage with linear voltage excursion is favorable for neural stimulation because it has less effect from irreversible electrochemical reactions. These reactions may produce harmful by-products and/or cause electrode corrosion.

Voltage excursions were measured under pulse stimulation currents to evaluate the PEDOT electrode charge injection capacity [47]. The electrodes were stimulated in PBS under a charge balanced, cathodic-first, biphasic pulse current at 1 mC/cm^2 at 50 Hz. The voltage excursion curves are shown in Fig. 16. Because of the lower impedance and higher charge capacity of the PEDOT polymer, voltage across the electrode under stimulation is lower than the thin-film Pt electrode and the excursion profile is linear. Based on voltage excursion under pulse stimulation currents, the charge injection limit or charge injection capacity (Q_{inj}) of a stimulation electrode can be determined. Q_{inj} is defined as the maximum cathodic charge that resulted in cathodic or anodic electrode voltage to exceed 0.6 V versus Ag/AgCl. The average charge injection limit measured on the PEDOT/PSS electrodes was 2.3 mC/cm^2 $\pm$0.6 ($\pm$ StdV, n=5). This charge injection capacity is comparable to the reported values for IrOx and TiN [12, 48] and much higher than that of smooth thin-film Pt. It meets the requirement for retinal prosthesis [44]. Care needs to be taken in in vivo stimulations to not exceed the threshold charge density that induces neural damage [33].

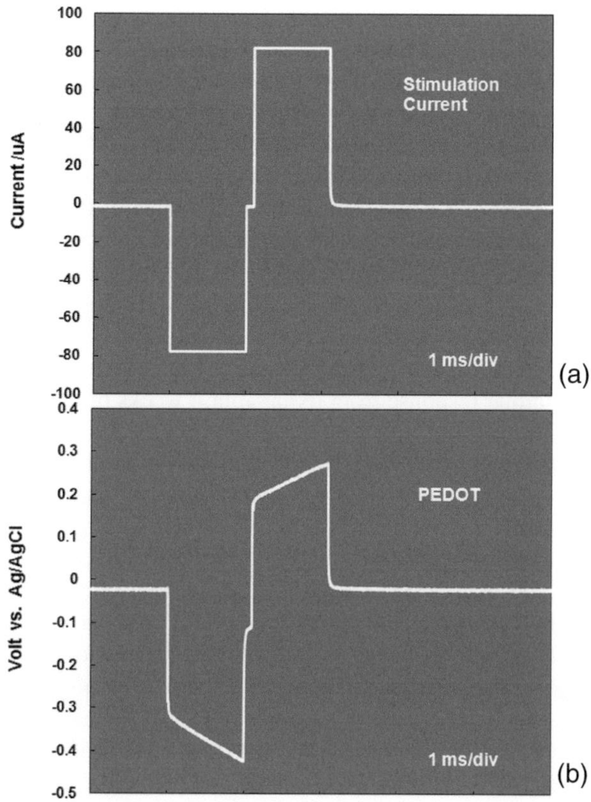

Fig. 16 Current and voltage waveforms of a PEDOT-coated Pt electrode under pulse stimulation (**a**). The electrode was stimulated in PBS under a charge-balanced, cathodic-first, biphasic pulse current at 1 mC/cm^2 at 50 Hz; (**b**). Voltage excursion was measured under pulse stimulation currents to evaluate the PEDOT electrode charge injection capacity

The charge injection capacity (Q_{inj}) determined from voltage excursions of pulsed current stimulation is often much smaller than the charge storage capacity (CSC) determined by CV measurements. For PEDOT-coated electrodes, Q_{inj} is about 10–100 times smaller than their CSC values. This suggests the full charge capacity of conducting polymer coatings cannot be fully utilized during stimulation. As noted above, the current cannot reach deep pores of porous polymer coatings under faster pulse stimulations.

4.4 Biocompatibility

Biocompatibility of conducting polymers has been studied for biomedical applications including implantable medical devices [53, 61, 76]. Biocompatible polymer coatings will provide a good integration of electrodes and surrounding tissues. It will reduce tissue response and ensure longer performance of the electrodes [52].

The biocompatibility of polypyrrole has been assessed in vitro and in vivo as an effective guidance channel for the regeneration of nervous tissue and as a method of conducting current to enhance the repair of a nerve [115]. A PPy coating of 25 μm was electrochemically deposited on a platinum wire. In vitro, the cell responses from L929 mouse fibroblast and neuro2a neuroblastoma cells to the polymer coatings were evaluated under a constant current. The results indicated that the polypyrrole was cytocompatible in vitro if prepared by appropriate extraction techniques. After polymerization, extraction in methanol for a period of 1 week was carried out to remove residual electrolyte. Some evidence of toxicity was evident when a current of 1 mA was applied across the polymer for periods up to 96 h. In vivo results show there was only minimal response from tissue after 4 weeks of implantation in a rat model.

Most biocompatibility studies of conducting polymers check cell growths in vitro and inflammation response in vivo for low cytotoxicity and good biocompatibility. The tests last typically from a few days to several weeks [58, 116]. In one study by Wang et al., the biocompatibility of PPy with nerve tissue was evaluated for 6 months in vitro and in vivo [117]. Both PPy powders synthesized chemically and PPy film synthesized electrochemically on the indium tin oxide conductive borosilicate glass were evaluated. Tests for acute toxicity, subacute toxicity, cell viability, hemolysis, allergens, and micronuclei were carried out according to ISO 10993 and ASTM F1748–82 standards. Their results indicated that PPy has a good biocompatibility with rat peripheral nerve tissue. They observed better growth of the Schwann cells on the surface of PPy-coated glass as compared with those of bare glass control group.

George et al. [53] examined the effect of surface roughness of PPy coatings on biocompatibility. Free-standing PPy films of $2 \times 3 \times 0.25$ mm with different surface roughness and doped with NaDBS or PSS were electrochemically deposited at different temperatures. The electrochemically deposited PPy films were implanted into rats' cerebral cortexes for a period of 3–6 weeks with Teflon as a control. The intensity and extent of inflammatory responses (e.g., gliosis) demonstrated that PPy implants were biocompatible and performed similar to or better than the inert and unreactive Teflon. The PPy/NaDBS sample electrodeposited at 4°C, which had a rougher surface, had less peak gliosis at 3 weeks than the same sample type at 25°C ($p < 0.05$). This suggests that the rougher surface induces a less inflammatory response and provides better implant integration with the surrounding tissue.

Several studies showed PANi had no inflammatory responses in a rat model for both the conductive emeraldine salt (E-PANi) and nonconductive emeraldine base (PANi) [116, 118]. Wang et al. also reported that noncovalent or covalent modifications on PANI surface improved biocompatibility [119].

The biocompatibility of PEDOT has not been thoroughly studied. However, in vitro and in vivo data so far suggest that it is compatible with cultured cells and brain tissue and does not release any substance that elicits toxicity [24]. Furthermore, the peptide sequence DCDPGYIGSR can be incorporated in the PEDOT coating to promote cell attachment. High quality acute neural recordings were obtained through PEDOT/DCDPGYIGSR-coated electrodes in acute animal tests [24]. More recent work demonstrated an improvement in chronic recording performance in neural probes coated with PEDOT/PSS [32]. However, biocompatibility of polymer coatings under neural stimulation conditions in longer term implantation has yet to be determined.

4.5 Long-Term Stability

Another important parameter for the application of conducting polymer in neural stimulation is chronic stability. While many factors can affect their long-term stability, three major factors are redox reversibility, stability of dopants, and mechanical adhesion.

Similar to metal electrode materials, conducting polymers stimulation requires reversible reactions driven under stimulation currents to ensure low voltage excursion and reduced production of by-products. Reports suggest that conducting polymers, especially PPy, are susceptible to irreversible oxidation [120]. Overoxidation of conducting polymers was reported to cause permselectivity changes, loss of conductivity, and de-doping [58, 121–122].

Conducting polymers require incorporated dopants to maintain conductivity and stability. During biphasic neural stimulation, electrodes are cycled between reduction and oxidation stages. Such redox reactions result in dynamic doping and de-doping processes. A study carried out by Farrington and Slater [120] indicated that overoxidized PPy films displayed permselectivity favoring cation transport, but excluding anions. Since conducting polymers, such as PPy and PEDOT, used as stimulation electrodes have positively charged polymer backbones, anions (A^-) of dopants are required to balance the charge and maintain electroneutrality. Excluding anions from polymer films leads to de-doping, causing conductivity loss. In one study, PPy/PSS was found to retain only 5% of its original electrochemical activity after polarization at 0.4 V for 16 h [31].

Leaching of dopants, especially small dopants, and/or exchanging dopants with other ions compromises the stability of conducting polymers. In one study, PPy-coated electrodes suffered breakdown under cyclic potential stimulation [77]. The PPy-coated electrodes were used for the electrochemically controlled release of dexamethasone. Cracks on the original smooth film developed after ~30 CV cycles. Under the cyclic potential stimulation, the film with the drug had undergone a fast swelling and deswelling process caused by ion and water movement in and out of the film, causing its delamination and breakdown.

In neural stimulation, conducting polymers are usually electrochemically deposited on metal seed layers such as Pt and Au. For such noncovalently coated polymer layers, the adhesion to metal is critical for the stability of long-term applications. In a recent study, the chronic behavior of PEDOT-coated electrodes was examined [33]. Thirty electrodes from three electrode arrays were stimulated and evaluated for stability. It was found that the well-coated PEDOT electrodes were stable for 2 weeks of electrical stimulation, suggesting the possibility of using such material for chronic stimulation. The thickness of the PEDOT coatings, which is approximately proportional to deposition time, has direct effect on the mechanical failure observed. It is found that thicker films have lower impedance and larger charge capacity. However, they showed more mechanical failures such as cracking and delamination under chronic stimulation, possibly due to the higher stress imposed on the film. Thinner films (plated for 100 s or less) have better adhesion and appear physically stable. However, the charge capacity is not enough to handle stimulation charge density of 1 mC/cm^2.

Physical degradation, including cracks and delamination, in PEDOT coatings has been observed as the major mode of failure in long-term stimulation. As the number of stimulation cycles increased, the electrode impedance increased, which often times coincided with film cracking and delamination. Figure 17 shows representative plots of electrode impedance changes over time during chronic stimulation for thin-film Pt, stable, and delaminated PEDOT coatings, respectively. Stable PEDOT has much lower impedance than uncoated Pt electrode and the impedance remained unchanged over the 2-week stimulation period. Unstable coatings showed an abrupt increase in impedance at the point of failure and the impedance value became close to that of Pt electrodes suggesting delamination of the film and exposure of the Pt.

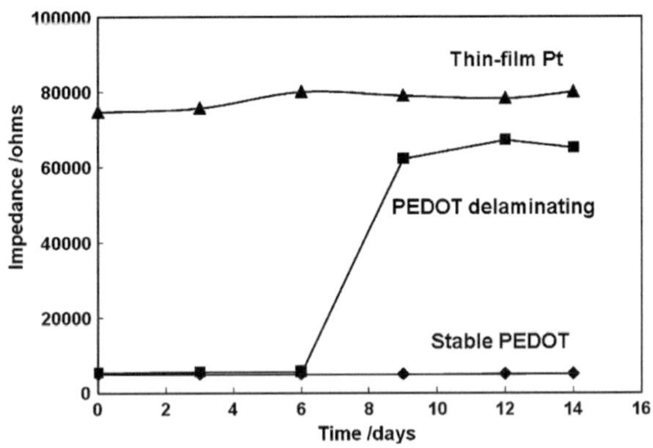

Fig. 17 Electrode impedance changes over time during chronic stimulation of thin-film Pt electrodes. Delamination of PEDOT coatings from a Pt electrode caused the impedance value to increase to close to that of Pt electrode (Reproduced from [33], with permission from IEEE)

When examined under SEM right after failure point, all the coatings that had unstable impedance showed different degrees of cracking and delamination as seen in Fig. 18.

In addition to determining the reversibility and CSC of polymer coatings, CV measurements can also reveal the stability of the polymer coating. Cyclic voltammetry showed some interesting changes of film properties under continuous CV cycles. Figure 19 shows voltammogram changes recorded in a continuous cyclic voltammetry on a PPy/NaDBS-coated Pt electrode. As the number of stimulation cycles increased, the separation between redox peaks enlarged, i.e., the anodic peak moved to more positive potentials while the cathodic peak to more negative

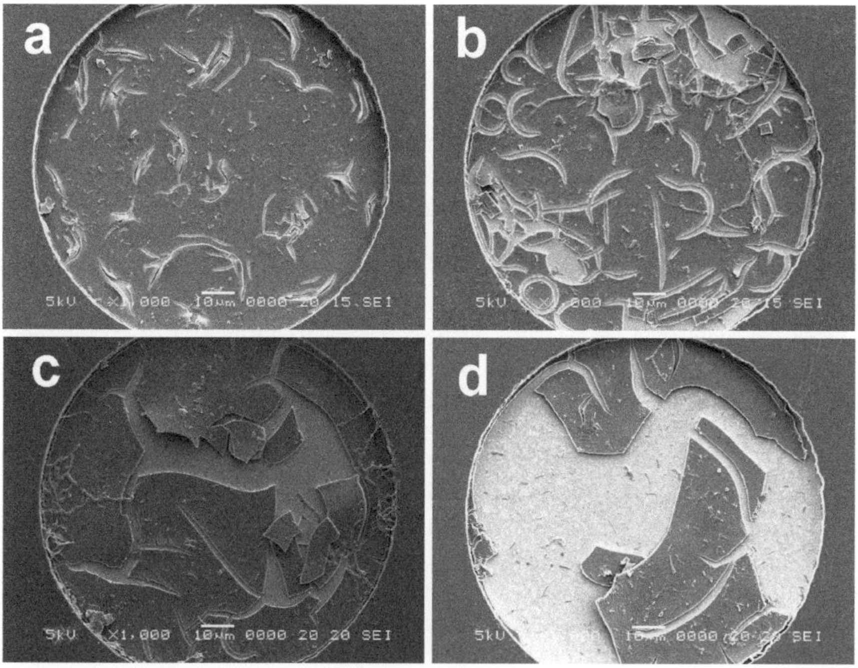

Fig. 18 SEM micrographs of electrode surfaces with cracking and delamination during 14 days of pulse stimulation under a charge density of 350 μC/cm^2. From (**a**) to (**d**), the cracks and delamination increased with the pulse stimulation time

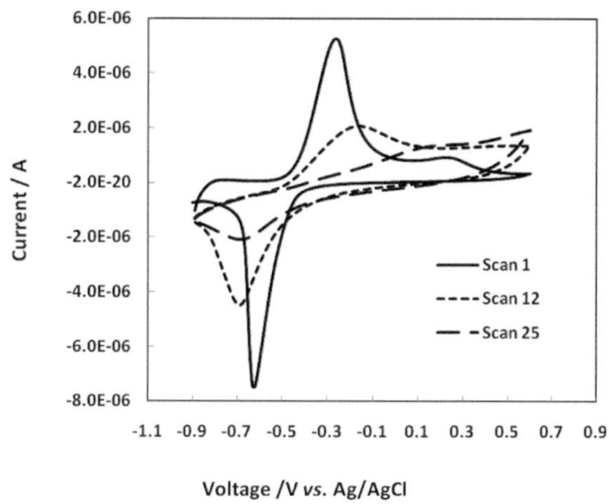

Fig. 19 Peak current and potential changes in cyclic voltammograms of a PPy/NaDBS-coated Pt electrode. Voltammogram changes were recorded in a continuous cyclic voltammetry in PBS with a scan rate of 50 mV/s

potential. The peak currents diminish with CV cycle increases. Larger peak separations suggest higher energy barriers required for oxidation and reduction reactions or poor reversibility. Lower peak currents indicate the reduction of electroactivity and loss of charge storage capacity. In contrast, PEDOT-coated electrodes are much more stable under CV conditions for at least hundreds of cycles [24–25]. Changes in CV were observed on some PEDOT-coated electrodes following chronic pulse stimulations [33].

Efforts have been made to improve the adhesion of conducting polymer to metal substrates. One strategy is to covalently attach the polymer to the electrode surface. A series of thiol-modified pyrroles have been synthesized and tested as an intermediate monolayer to link the gold electrode surface and the polypyrrole film [123–126]. However, characterizations of the films revealed no evidence that pyrrole moieties in the monolayer actually covalently linked to the overlying polypyrrole film. Slight improvements in adhesion by thiol-modified pyrrole were found to be most likely due to the roughened surface by chemical treatment, adsorption of pyrrole onto the carbonaceous layer, or coupling of the polypyrrole with decomposition products of the monolayer. Another, more effective approach is roughening the surface of the metal substrates by electroplating [24] or chemical etching [127]. PPy/PSS deposited on neural electrodes that were preprocessed with a fuzzy electroplated gold layer showed electrochemical properties and morphologies similar to those deposited on smooth-sputtered gold electrodes, but with significantly improved adhesion to the substrate [24–25]. In an actuator study, electroplating a thin layer of 1000 angstroms of Au in a low concentration plating bath consistently improved lifetimes to at least 10,000 cycles (on an unmodified substrate, delamination unpredictably occurred between 1 cycle and 20,000 cycles, with almost half the samples failing before 3000 cycles) [127]. In the same study, chemical etching was found to produce a roughened surface with undercut morphology that presumably strengthened the interlock between the polypyrrole film and the substrate. This effect was found to be sensitive to the grain structure of the gold substrate [127]. Cogan et al [94] investigated strategies for obtaining adherent PEDOT on metal electrodes. Their results suggested that adhesion layers of sputtered iridium oxide and electrodeposited gold were most effective in providing stable PEDOT coatings on Pt-Ir microelectrodes.

5 Conclusions

Significant progress in the development of conducting polymers for biomedical applications has been made in the past decades. Conducting polymers have been explored for tissue engineering, biosensing, actuator, drug delivery, and neural probe applications. The unique features of conducting polymers with metallic conductivity and organic polymer properties have made them very promising for the modification of neural probes.

To date, few studies evaluating conducting polymers as stimulation electrode materials have been reported. More research is needed to further understand charge storage and charge injection mechanisms in conducting polymer-coated stimulation electrodes. Although high conductivity is required, low impedance at higher frequencies is critical for neural stimulation and neural prosthetic applications.

The challenge lies in depositing less stress, low impedance, high charge capacity, and thin polymer coatings for high density microstimulation electrode arrays. Efforts should be made to minimize overoxidation and improve polymer/metal adhesion. Perhaps the most challenging area in the application of conducting polymers as neural stimulation electrode materials is their chronic stability. Although many challenges remain, conducting polymers appear to be a favorable candidate for improving bionic interfaces between neural prosthetic devices and biological environments in neural stimulation applications.

References

1. Middlebrooks JC, Bierer JA, Snyder RL (2005) Cochlear implants: The view from the brain. Curr Opin Neurobiol 15:488–493
2. Clark GM (2006) The multiple-channel cochlear implant: The interface between sound and the central nervous system for hearing, speech, and language in deaf people—a personal perspective. Philos Trans R Soc Lond B 361:791–810
3. Giszter SF (2008) Spinal cord injury: Present and future therapeutic devices and prostheses. Neurotherapeutics 5:147–162
4. North RB (2008) Neural interface devices: Spinal cord stimulation technology. Proc IEEE 96:1108–1119
5. Cameron T (2004) Safety and efficacy of spinal cord stimulation for the treatment of chronic pain: A 20-year literature review. J Neurosurg 100:254–267
6. Rise MT (2000) Instrumentation for neuromodulation. J Arc Med 31:237–247
7. Perlmutter JS, Mink JW (2006) Deep brain stimulation. Annu Rev Neuro. 29:229–257
8. Diamond A, Jankovic J (2005) The effect of deep brain stimulation on quality of life in movement disorders. J Neurol 76:1188–1193
9. Bittar RG, Kar-Purkayastha I, Owen SL et al (2005) Deep brain stimulation for pain relief: A meta-analysis. J Clin Neurosci 12:515–519
10. Volkmann J (2004) Deep brain stimulation for the treatment of Parkinson's disease. J Clin Neurophysiol 21:6–17
11. Humayun MS, Weiland JD, Fuji GY et al (2003) Visual perception in a blind subject with a chronic microelectronic retinal prosthesis. Vision Res 43:2573–2581
12. Weiland JD, Liu WT, Humayun MS (2005) Retinal prosthesis. Annu Rev Biomed Eng 7:361–401
13. Schwartz AB (2004) Cortical neural prosthetics. Annu Rev Neurosci 27:487–507
14. Friehs GM, Zerris VA, Ojakangas CL et al (2004) Brain-machine and brain-computer interfaces. Stroke 35:2702–2705
15. Zrenner E (2002) Will retinal implants restore vision? Science 295(5557):1022–1025
16. Maynard EM (2001) Visual prostheses. Annu Rev Biomed Eng 3:145–168
17. Greenberg RJ (2000) Visual prostheses: A Review. Neuromodulation 3(3):161–165
18. Weiland JD, Humayun MS (2008) Visual prosthesis. Proc IEEE 96:1076–1084
19. Margalit E, Maia M, Weiland JD et al (2002) Retinal prosthesis for the blind. Surv Ophthalmol 47:335–356
20. Rizzo J, Wyatt J, Humayun M et al (2001) Retinal prosthesis: An encouraging first decade with major challenges ahead. Ophthalmology 108(1):13–14
21. Butterwick AF, Vankov A, Huie P et al (2006) Dynamic range of safe electrical stimulation of the retina SPIE Proceedings, Ophthalmic Technologies XVI, SPIE 6138:1–8
22. Cui XY, Hetke JF, Wiler JA et al (2001) Electrochemical deposition and characterization of conducting polymer polypyrrole/PSS on multichannel neural probes. Sens Actuators A93: 8–18

23. Cui X, Lee VA, Raphael Y et al (2001) Surface modification of neural recording electrodes with conducting polymer/biomolecule blends. J Biomed Mater Res 56:261–272
24. Cui XY, Martin DC (2003) Electrochemical deposition and characterization of poly(3,4-ethylenedioxythiophene) on neural microelectrode arrays. Sens Actuators B Chem 89: 92–102
25. Cui, XY, Martin DC (2003) Fuzzy gold electrodes for lowering impedance and improving adhesion with electrodeposited conducting polymer films. Sens Actuators A: Phys 103: 384–394
26. Xiao YH, Cui XY, Martin DC (2004) Electrochemical polymerization and properties of PEDOT/S-EDOT on neural microelectrode arrays. J Electroanal Chem 573:43–48
27. Yang JY, Martin DC (2004) Microporous conducting polymers on neural microelectrode arrays II. Physical characterization. Sens Actuators A Phys 113A:204–211
28. Kim DH, Abidian M, Martin DC (2004) Conducting polymers grown in hydrogel scaffolds coated on neural prosthetic devices. J Biomed Mater Res 71A:577–585
29. Cui XY, Wiler J, Dzamann M et al (2003) In vivo studies of polypyrrole/peptide coated neural probes. Biomaterials 24:777–787
30. Kim DH, Sequerah C, Hendricks JL et al (2007) Effect of immobilized nerve growth factor (NGF) on conductive polymers electrical properties and cellular response. Adv Funct Mater 17:79–86
31. Yamato H, Ohwa M, Wernet W (1995) Stability of polypyrrole and poly(3,4-ethylenedioxythiophene) for biosensor application. J Electroanal Chem 397:163–170
32. Ludwig KA, Uram JD, Yang JY et al (2006) Chronic neural recordings using silicon microelectrode arrays electrochemically deposited with a poly(3,4-ethylemedioxythiophene) (PEDOT) film. J Neural Eng 3:59–70
33. Cui X, Zhou D (2007) Poly (3,4-ethylenedioxythiophene) for Chronic neural stimulation. IEEE Trans Neural Syst Rehabil Eng 15:502–508
34. Cogan SF (2008) Neural stimulation and recording electrodes. Annu Rev Biomed Eng 10:14.1–14.35
35. Robblee LS, Rose TL (1990) Electrochemical guidelines for selection of protocols and electrode materials for neural stimulation. In: Agnew WF, McCreery DB (eds) Neural Prostheses: Fundamental Studies, pp. 25–66. Englewood Cliffs, NJ, Prentice Hall
36. McCreery DB, Agnew WF, Yuen TG et al (1990) Charge density and charge per phase as cofactors in neural injury induced by electrical stimulation. IEEE Trans Biomed Eng 37:996–1001
37. Brummer SB, Turner MJ (1977) Electrochemical considerations for safe electrical stimulation of the nervous system with platinum electrodes. IEEE Trans Biomed Eng 24:59–63
38. Merrill DR, Bikson M, Jefferys J (2005) Electrical stimulation of excitable tissue: Design of efficacious and safe protocols. J Neurosci Met 141:71–198
39. Bard A, Faulkner L (1980) Electrochemical Methods. Chapter 1, Wiley, New York
40. Rodger DC, Fong AJ, Li W (2008) Flexible parylene-based multielectrode array technology for high-density neural stimulation and recording. Sens Actuators B: Chem 132:449–460
41. Brummer SB, Robblee LS, Hambrecht FT (1983) Criteria for selecting electrodes for electrical stimulation: Theoretical and practical considerations. Ann NY Acad Sci 405:159–171
42. Robblee LS, McHardy J, Agnew WF (1983) Electrical stimulation with Pt electrodes. VII. Dissolution of pt electrodes during electrical stimulation of the cat cerebral cortex. J Neurosci Methods 9:310–318
43. Rose TL, Robblee LS (1990) Electrical stimulation with pt electrodes. VIII. Electrochemically safe charge injection limits with 0.2 ms pulses. IEEE Trans Biomed Eng 37:1118–1120
44. Humayun MS, de Juan E Jr, Weiland JD (1999) Pattern electrical stimulation of the human retina. Vision Res 39:2569–2576
45. Huang CQ, Carter PM, Shepherd PK (2001) Stimulus induced pH changes in cochlear implants—an in vitro and in vivo study. Ann Biomed Eng 29:791–802

46. Mortimer JT, Kaufman D, Roessman U (1980) Intramuscular electrical stimulation tissue damage. Ann Biomed Eng 8:235–244
47. Hu Z, Zhou D, Greenberg R et al (2006) Nanopowder molding method for creating implantable high-aspect-ratio electrodes on thin flexible substrates. Biomaterials 27: 2009–2017
48. Weiland JD, Anderson DJ, Humayun MS (2002) In vitro electrical properties for iridium oxide versus titanium nitride stimulating electrodes. IEEE Trans Biomed Eng 49: 1574–1579
49. Guenther E, Troger B, Schlosshauer B et el (1999) Long-term survival of retinal cell cultures on retinal implant materials. Vision Res 39:3988–3994
50. Weiland JD, Anderson J (2000) Chronic neural stimulation with thin-film, iridium oxide electrodes. IEEE Trans Biomed Eng 47:911–918
51. Cogan SF, Guzelian AA, Agnew WF et al (2004) Over-pulsing degrades activated iridium oxide films used for intracortical neural stimulation. J Neurosci Methods 137:141–150
52. Wallace G, Spinks G (2007) Conducting polymers – bridging the bionic interface. Soft Matter 3:665–671
53. George PM, Lyckman AW, LaVan DA et al (2005) Fabrication and biocompatibility of polypyrrole implants suitable for neural prosthetics. Biomaterials 26:3511–3519
54. Adhikari B, Majumdar S (2004) Polymers in sensor applications. Prog Polym Sci 29: 699–766
55. Ramakrishnan S (1997) Conducting polymers from a laboratory curiosity to the market place. Resonance 2:48–58
56. Shirakawa H, Louis EJ, MacDiarmid AG (1977) Synthesis of electrically conductive organic polymers:Halogen derivatives of polyacetylene (CH)x, J Chem Soc Chem Commun 16: 578–580
57. Alan G. MacDiarmid AG (2002) Synthetic metals: A novel role for organic polymers. Synth met 125:11–22
58. Ateh DD, Navsaria HA, P. Vadgama P (2006) Polypyrrole-based conducting polymers and interactions with biological tissues. J R Soc Interface 3:741–752
59. Simpson J, Kirchmeyer S, Reuter K (2005) Advances and applications of inherently conductive polymer technologies based on poly(3,4-ethylenedioxythiophene) AIMCAL Fall Technical Conference and 19th International Vacuum Web Coating Conference, Session 5: Advances in Technology, October 16–20, 2005, Myrtle Beach, South Carolina
60. Guimard NK, Gomez N, Schmidt CE (2007) Conducting polymers in biomedical engineering. Prog Polym Sci 32:876–921
61. Schmidt CE, Shastri VR, Vacanti JP et al (1997) Stimulation of neurite outgrowth using an electrically conducting polymer. Proc Natl Acad Sci 94:8948–8953
62. Xiao YH, Cui XY, Hancock JM, et al (2004) Electrochemical polymerization of poly(hydroxymethylated-3,4-ethylenedioxythiophene) (PEDOT-MeOH) on multichannel neural probes. Sens Actuators B Chem 99:437–43
63. Kumar D, Sharma RC (1998) Advances in conductive polymers. Eur Polym J 34:1053–1060
64. Inzelt G, Pineri M, Schultze JW et al (2000) Electron and proton conducting polymers: Recent developments and prospects. Electrochim Acta 45:2403–2421
65. Kang HR, Jo NJ (2006) Solid-state conducting polymer actuator based on electrochemically-deposited polypyrrole and solid polymer electrolyte. High Perform Polym 18:665–678
66. Skotheim TA, Elsenbaumer RL, Reynolds JR (1997) Handbook of Conducting Polymers, Vols. 1–2, Marcel Dekker, New York
67. Mottaghitalab V, Xi B, Spinks GM et al (2006) Polyaniline fibers containing single walled carbon nanotubes: Enhanced performance artificial muscles. Synth Met 156:796–803
68. Aziz EF, Vollmer A, Eisebitt S et al (2007) Localized charge transfer in a molecularly doped conducting polymer. Adv Mater 19: 3257–3260
69. Dietrich M, Heinze J, Heywang G et al (1994) Electrochemical studies on poly(3,4-ethylenedioxithiophene) and related materials. J Electroanal Chem 369:87–92

70. Morvant MC, Reynolds JR (1998) In situ conductivity studies of poly(3,4-ethylenedioxythiophene). Synth Met 92:57–61
71. Hillman AR, Daisley SJ, Bruckenstein S (2008) Ion and solvent transfers and trapping phenomena during n-doping of PEDOT films. Electrochim Acta 53:3763–3771
72. Armes SP (1987) Optimum reaction conditions for the polymerization of pyrrole by iron (III) chloride in aqueous solution. Synth Met 20:365–371
73. Duchet J, Legras R, Demoustier-Champagne S (1998) Chemical synthesis of polypyrrole: Structure-properties relationship. Synth Met 98:113–122
74. Otero TF, Sansinena JM (1995) Artificial muscles based on conducting polymers. Bioelectroch Bioener 38:411–414
75. LaVan DA, George PM, Langer R (2003) Simple, three-dimensional microfabrication of electrodeposited structures. Angew Chem 42:1262–1265
76. Nishizawa M, Nozaki H, Kaji H (2007) Electrodeposition of anchored polypyrrole film on microelectrodes and stimulation of cultured cardiac myocytes. Biomaterials 28: 1480–1485
77. Wadhwa R, Lagenaur CF, Cui XT (2006) Electrochemically controlled release of dexamethasone from conducting polymer polypyrrole coated electrode. J Control Release 110:531–541
78. Martins NCT, Silva T, Montemor MF et al (2008) Electrodeposition and characterization of polypyrrole films on aluminum alloy 6061-T6. Electrochimica Acta 53:4754–4763
79. Lewis TW, Spinks GM, Wallace GG et al (2001) Investigation of the applied potential limits for polypyrrole when employed as the active components of a two-electrode device, Synth Met 122:379–385
80. Herrasti P, Diaz L, Ocon P et al (2004) Electrochemical and mechanical properties of polypyrrole coatings on steel. Electrochim Acta 49:3693–3699
81. Choi S, Park S (2002) Electrochemistry of conductive polymers XXVI. Effects of electrolytes and growth methods on polyaniline morphology. J Electrochem Soc 149: E26–E34
82. Widge A, Jeffries-El M, Cui X (2007) Self-assembled monolayers of polythiophene conductive polymers improve biocompatibility and electrical impedance of neural electrodes. Biosens and Bioelectron 22:1723–1732
83. Malhotra BD, Chaubey A, Singh SP (2006) Prospects of conducting polymers in biosensors. Analytica Chimica Acta 578:59–74
84. Ramanavicius A, Kausaite A, Ramanavicien A (2005) Polypyrrole-coated glucose oxidase nanoparticles for biosensor design. Sens Actuators B 111–112:532–539
85. Schmidt CE, Leach JB (2003) Neural tissue engineering: Strategies for repair and regeneration. Annu Rev Biomed Eng 5:293–347
86. Abidian MR, Kim DH, Martin DC (2006) Conducting-polymer nanotubes for controlled drug release. Adv Mat 18:405–409
87. Thompson BC, Simon E. Moulton SE, Ding J (2006) Optimising the incorporation and release of a neurotrophic factor using conducting polypyrrole. J Control Release 116: 285–294
88. Teles FRR, Fonseca LP (2008) Applications of polymers for biomolecule immobilization in electrochemical biosensors. Mater Sci Eng, C. doi:10.1016/j.msec.2008.04.010
89. Lange U, Roznyatovskaya NV, Mirsky VM (2008) Conducting polymers in chemical sensors and arrays. Anal Chim Acta 614:1–26
90. Rohwerder M, Michalik A (2007) Conducting polymers for corrosion protection: What makes the difference between failure and success? Electrochim Acta 53:1300–1313
91. de Marcos S, Wolfbeis OS (1996) Optical sensing of pH based on polypyrrole films. Analytica Chimica Acta 334:149–153
92. Jin Z, Su Y, Duan Y (2000) An improved optical pH sensor based on polyaniline. Sens Actuators B: Chem 71:118–122

93. Liao YH, Chou JC (2008) Comparison of polypyrrole-conducting polymer and Ag/AgCl reference electrodes Used for ruthenium dioxide pH electrode. J Electrochem Soc 155: J257–J262
94. Cogan SF, Peramunage D, Smirnov A et al (2007) Polyethylenedioxythiophene (PEDOT) coatings for neural stimulation and recording electrodes. Mater Res Soc Meet, (Abstr. QQ2.7), Nov. 26–30, 2007, Boston
95. Innis PC, Moulton SE, Wallace GG (2007) Biomedical applications of inherently conducting polymers (ICPs) in Skotheim T (ed) Conjugated Polymers: Processing and applications, Handbook of Conducting Polymers, 3rd edn. CRC press, New York
96. Green RA, Poole-Warren LA, Lovell NH (2007) Novel Neural interface for vision prosthesis electrodes: Improving electrical and mechanical properties through layering. Proceedings 3rd International IEEE/EMBS Conference on Neural Engineering, pp. 97–100
97. Green RA, Lovell NH, Wallace GG et al (2008) Conducting polymers for neural interfaces: Challenges in developing an effective long-term implant. Biomaterials 29:3393–3399
98. Stauffer WR, Bi B, Cui XT (2007) Conducting polymer based neurochemical release system. In Biomedical Engineering Society Meeting, Los Angeles, CA
99. Stauffer WR, Cui XT (2006) Polypyrrole doped with 2 peptide sequences from laminin. Biomaterials 27:2405–2413
100. Richardson-Burns SM, Hendricks JL, Foster B et al (2007) Polymerization of the conducting polymer poly(3,4-ethylenedioxythiophene) (PEDOT) around living neural cells. Biomaterials 28:1539–1552
101. Richardson-Burns SM, Hendricks JL, Martin DC (2007) Electrochemical polymerization of conducting polymers in living neural tissue. J Neural Eng 4:L6–L13
102. Ito Y, Yagi T, Ohnishi Y et al (2002) A study on conductive polymer electrodes for stimulating the nervous system. Int J Appl Electromagnetics Mech 14:347–352
103. Yang JY, Martin DC (2006) Impedance spectroscopy and nanoindentation of conducting poly(3,4-ethylenedioxythiophene) coatings on microfabricated neural prosthetic devices. J Mater Res 21(5):1124–1132
104. Abidian MR, Martin DC (2008) Experimental and theoretical characterization of implantable neural microelectrodes modified with conducting polymer nanotubes. Biomaterials 29:1273–1283
105. Abidian MR, Martin DC (2009) Multifunctional nanobiomaterials for neural interfaces. Adv Funct Mater, 19:573–585
106. Tolstopyatova EG, Sazonova SN, Malev VV (2005) Electrochemical impedance spectroscopy of poly(3-methylthiophene) and poly(3-octylthiophene) film electrodes. Electrochim Acta 50:1565–1571
107. Bobacka J, Lewenstam A, Ivaska A (2000) Electrochemical impedance spectroscopy of oxidized poly(3,4-ethylenedioxythiophene) film electrodes in aqueous solutions. J Electroanal Chem 489:17–27
108. Macdonald JR (1987) Impedance Spectroscopy: Emphasizing Solid Materials and Systems. New York, Wiley
109. Macdonald JR (1992) Impedance spectroscopy. J Ann Biomed Eng 20:289–305
110. Otero TF, Lopez Cascales JJ, Vazquez Arenas G (2007) Mechanical characterization of free-standing polypyrrole film. Mater Sci Eng C27:18–22
111. Bloor D, Hercliffe RD, Galiotis CG et al (1985) Electronic Properties of Polymers and Related Compounds, Springer-Verlang, Berlin, p. 179
112. Wallace GG, et al (2003) Conductive Electroactive Polymer, Intelligent Materials Systems, CRC Press, Boca Raton, FL
113. Xu H, Wang C, Wang CL et al (2006) Polymer actuator valves toward controlled drug delivery application. Biosens Bioelectron. 21:2094–2099
114. Norlin A, Pan J, Leygrafa C (2005) Investigation of electrochemical behavior of stimulation/sensing materials for pacemaker electrode applications I. Pt, Ti, and TiN coated electrodes. J Electrochem Soc 152:J7–J15

115. Williams RL, Doherty PJ (1994) A preliminary assessment of poly(pyrrole) in nerve guide studies. J Mater Sci Mater Med 5:429–433
116. Kamalesh S, Tan P, Wang J et al (2000) Biocompatibility of electroactive polymers in tissues. J Biomed Mater Res 52:467–478
117. Wang X, Gu X, Yuan C et al (2004) Evaluation of PPy with biological tissues biocompatibility of polypyrrole in vitro and in vivo. J Biomed Mater Res 68A:411–422
118. Bidez PR III , Li S, Macdiarmid AG et al (2006) Polyaniline, an electroactive polymer, supports adhesion and proliferation of cardiac myoblasts. J Biomater Sci Polymer Edn 17:199–212
119. Wangnd KN, Wang CH, Dong YQ et al (1999) In vivo tissue response to polyaniline. Synth Met 102:1313–1314
120. Farrington AM, Slater JM (1997) Prediction and characterization of the charge/size exclusion properties of over-oxidized poly(pyrrole) films. Electroanalytical 9:843–847
121. Gao ZQ, Zi MX, chen BS (1994) the influence of overoxidation treatment on the permeability of polypyrrole film. J Electroanal chem 373:141–148
122. Shiigi H, Kishimoto M, Yakabe, H et al (2002) Highly selective molecularly imprinted overoxidized polypyrrole colloids: One step preparation technique. Anal Sci 18:41–44
123. Smela E, Inganas O, Lundstrom I (1995) Controlled folding of micrometer-size structures. Sci 268:1735–1738
124. Smela E (1998) Thiol-modified pyrrole monomers: 4. Electrochemical deposition of polypyrrole over 1-(2-thioethyl)pyrrole. Langmuir 14:2996–3002
125. Smela E (1999) Microfabrication of PPy microactuators and other conjugated polymer devices. J. Micromech. Microeng 9:1–18
126. Smela, E, Zuccarello G, Kariis H et al (1998) Thiol-modified pyrrole monomers: 1. Synthesis, characterization, and polymerization of 1-(2-thioethyl) pyrrole and 3-(2-thioethyl)pyrrole. Langmuir 14:2970–2975
127. Liu Y, Gan Q, Baig S et al (2007) Improving PPy adhesion by surface roughening. J. Phys. Chem. C 111:11329–11338

Microelectronics of Recording, Stimulation, and Wireless Telemetry for Neuroprosthetics: Design and Optimization

Moo Sung Chae, Zhi Yang, and Wentai Liu

Abstract This chapter focuses on how to interface biological systems with electronics so as to implement bio-instruments to obtain the in-depth understandings about the animal behavior and human brain activities, and complex neuroprosthetic devices to treat various neurological diseases. The interdisciplinary nature of the system requires a wide range of knowledge on both biology and electronics to build such systems. A unique environment where the system should operate imposes challenging design constraints and system issues, which can be solved only by considering both biology and electronics simultaneously. Fundamental building circuits including amplifiers, filters, analog-to-digital converters (ADCs) are addressed first and subsystems which consist of those basic circuits are explained with emphasis on trade-offs which should be considered carefully to achieve optimal design. Several state-of-art systems such as integrated wireless neural-recording systems and retinal prostheses are presented to explain how the fundamental knowledge and principles are used in the real applications.

Contents

M.S. Chae (✉)
Department of Electrical Engineering, University of California, Santa Cruz, CA, USA
e-mail: gomdori07@gmail.com

D.D. Zhou, E. Greenbaum (eds.), *Implantable Neural Prostheses 2*,
Biological and Medical Physics, Biomedical Engineering, DOI 10.1007/978-0-387-98120-8_9,

1 Introduction

Since Hodgkin-Huxley's pioneering work on the analysis of the action potential [1], enormous amount of knowledge about the structure and functions of the nervous system has been accumulated. Extensive understanding of the neural activities at the cellular level has been achieved through the techniques of extracellular recording and stimulation, and there are a lot of on-going studies about the behavior of a large network of neurons, which will eventually lead to the thorough understanding of human behavior. A lot of technical innovations in microelectronics have been made so far, and people began to try to use those techniques and knowledge to treat disease related to nervous system and to help disabled people. Remarkable successes of cochlear implant for the deaf and deep-brain stimulation (DBS) for Parkinson's disease revealed a possibility that implantable neuroprosthetic devices can be an efficient and practical method to treat a lot of neurological diseases and nervous-system related disabilities. Retinal prostheses for blind people [2, 3], brain-machine interface to help the patient with spinal cord injuries [4–6], epilepsy suppression [7], more advanced cochlear implant [8], and DBS of the subthalamic nucleus for Parkinson's disease [9] are actively under development these days and gaining more and more interest and support. As the range of applications is growing and the functionality is getting more complicated and sophisticated, neuroprosthetic devices are evolving to a closed-loop control system, which is composed of three main functional blocks of neural recording, neural-signal processing, and neuromuscular stimulation. The bio-signals monitored by the recording blocks are processed to generate a command signal to stimulate a particular neuron or muscles to evoke expected results in the biological objects. Such recording and stimulation are analogous to the read and write function of computer system.

However, modern neuroprosthetic devices are facing challenging requirements and constraints on the electronics design. More and more recording and stimulation channels are required to improve the performance of prosthetic devices. However, the mobile nature of the applications forces the entire system to operate on very limited power budget. Most power sources available for this specific kind of system do not have enough capacity unfortunately. Moreover, the implanted devices should not consume too much power even when they operate on reliable power source, because of the possible heat damage to the tissue surrounding the device. As a consequence, low-power operation is inevitable. In addition, the space to host the system in the biological object is usually very small and restricted. Therefore, it is necessary to integrate as many components as possible on a single chip to miniaturize the system. However, as the number of the channel and the complexity of the system increase, the power consumption and size of the entire system should increase as well, which is against the constraint set by the applications. All these make it very challenging to design the electronics for implantable devices, leading to a necessity for a systematical approach at the early design stages. Wireless capability is another critical issue. Power and data should be transmitted to the implanted system, which is inside the body of the animals or patients. Wire connections have a critical problem of infections and they often seriously hinder the free movements of the objects. Thus, wireless telemetries are essential for the design of systems targeting for free moving or behaving biological objects. Special necessity for the wide bandwidth data telemetry is emerging as the number of the channels is increasing. Flexibilities of the system are also critical because the properties of the bio-signals have a very wide range of variations from one object to another, and it even severely changes in the same object as time goes. Therefore, flexibilities to change critical system parameters such as gain and bandwidth of the amplifiers are necessary to maintain the performance of the system. This chapter will discuss such design issues mentioned above and present design methodologies to overcome the barriers.

This chapter is organized as follows. Section 2 describes the design of the most basic circuits frequently used in implantable microelectronics such as preamplifiers, filters, and analog-to-digital converts (ADCs). Section 3 discusses how to put basic building blocks together and form subsystems that have specific purposes of recording, signal processing, stimulation, and wireless transmitting of power and data. Emphasis is put on the trade-offs between various design parameters to achieve optimal solutions. Section 4 presents the real design examples of the entire systems. A 128-channel wireless neural-recording system and 256-pixel retinal-prosthesis ICs are explained in detail.

2 Basic Building Blocks

2.1 Amplifier

The signals generated by neurons are very small in amplitude and very noisy. For example, the amplitude of the extracellular action potentials induced at the recording microelectrode is not more than a couple of hundred μV in general. One of the major noise sources in neural-recording systems is the microelectrode. The noise contributed by the electrode is given by the well-known Nyquist's formula, and this value is typically tens of μV when the recording bandwidth is set to 10 KHz [10]. In addition, the biological background noise, which has various origins, seriously degrades the signal quality. Local field potentials (LFPs) and other far-field signals tend to have larger signal amplitudes than that of extracellular action potential, but the signal amplitude is still too small to be used directly by the microelectronics. Therefore, a low-noise preamplifier to amplify the small potential difference between the recording electrode and reference electrode is required at the first stage of the recording systems.

The different neural signals have different signal amplitude and bandwidth and their typical values are presented in the Table 1. The amplifier's gain should be set large enough to ensure the proper processing of the signal at later stages of the system. In addition, the bandwidth of the amplifier, mostly determined by the high-frequency cut-off of the amplifier, needs to match that of the target signal so that the signal-to-noise ratio (SNR) is maximized. Low-frequency cut-off of the amplifier also has to be low enough to guarantee that the low-frequency content of the bio-potentials is not lost. For example, to properly record the extracellular action potentials from neurons, total signal gain of more than 1000 to achieve 100 mV signal is necessary and the high frequency cut-off of the amplifier needs to be set to 10 KHz while low-frequency cut-off should be smaller than 200 Hz.

Table 1 Amplitude and bandwidth of various bio-potentials

Signal	Bandwidth (Hz)	Signal range (mV_{pp})
ECG	0.05 ~ 256	0.1 ~ 10
EEG	0.05 ~ 128	0.02 ~ 0.4
ECoG	0.1 ~ 64	0.02 ~ 1
EMG-1	1 ~ 1 K	0.02 ~ 1
EMG-2	1 ~ 128	0.02 ~ 1
LFPs	0.1 ~ 100	0.1 ~ 1
EAPs	100 ~ 10 K	0.04 ~ 0.2

Usually, there is a large DC offset voltage at the electrode–electrolyte interface [11]. This DC offset voltage is called half-cell potential and has a critical effect on the bio-potential amplifier design. Because the half-cell potential has typically amplitude of more than a few hundred mV, the amplifier output would be saturated if these large DC offset voltages are not taken into consideration during the amplifier design. These unique constraints of both DC offset and the low-frequency nature of the signal make the bio-potential amplifier design distinguished from other types of amplifiers.

Input impedance is also important and should be considered at the design stage with care. Figure 1 shows the simplified circuit model for the neural recording.

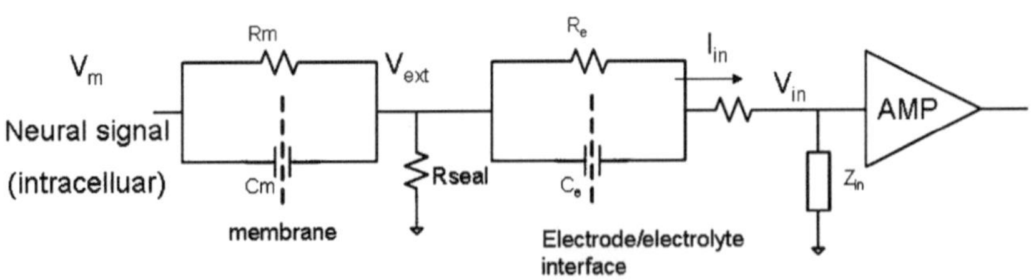

Fig. 1 Simplified circuit model for neural-recording environment

When a neuron fires an action potential, different parts of cell membrane become depolarized by the opening of voltage-controlled ion channels leading to a flow of ion current both inside and outside the neuron. Because the surrounding body fluid is the conducting media, a time-varying potential field is generated around the neurons. Since extracellular media is resistive, the extracellular potential is approximately proportional to the current across the neuron membrane. The membrane roughly behaves like an RC circuit and most current flows through the membrane capacitance. The extracellular action potential is sensed by the electrode and amplified by the neural amplifier. The actual input voltage of the amplifier is determined by the impedance ratio of the electrode and amplifier's input impedance. The impedance of the recording electrode varies according to the geometry and material. For microelectrode arrays, which are widely used in multichannel recordings, the impedance could be as high as 1 MΩ at 1 KHz. Therefore, to minimize the signal loss, the input impedance of the amplifier needs to be at least a few MΩ at 1 KHz.

Noise is another critical concern in recording system design. There are three major noise sources, which affect the signal quality in neural recording and depicted in Fig. 2. Among them, biological background noise has the largest contribution. The amplifier should be designed so that the input equivalent noise is smaller than those of electrode and biological background noise.

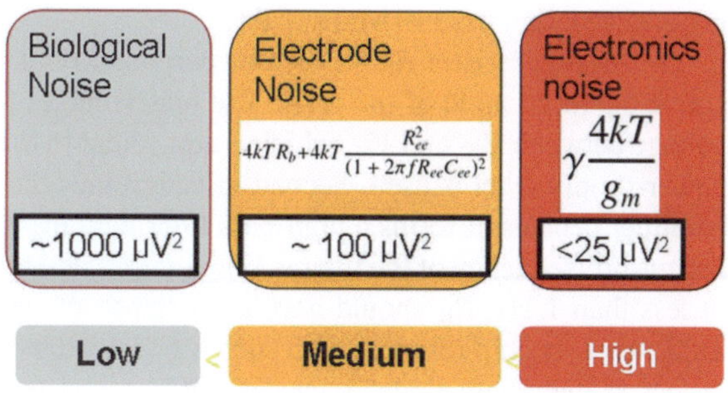

Fig. 2 Noise sources that affect the signal integrity of the neural recording

2.1.1 Negative-Feedback Amplifier

A capacitively coupled negative-feedback amplifier [12, 13] is most widely used in neural-recording systems and its circuit is shown in Fig. 3 [13]. It consists of an operational transconductance amplifier (OTA) as an amplification component, capacitors to form a feedback network, and biasing resistors. It has an advantage of automatically rejecting DC offset voltages by the input capacitance C_1 and this feature makes it very suitable for neural-recording applications [12–14]. The value of C_1 should be decided so that the input impedance of the amplifier is large enough compared to the electrode impedance. Since the input node of the OTA is virtual AC ground when both $C_L \gg C_2$ and $g_m/C_L \gg 1$ are satisfied, where the OTA can reasonably be regarded as an operational amplifier, the differential input impedance of the amplifier is simply $(j2\pi fC_1)^{-1}$. Because different neural signals have different

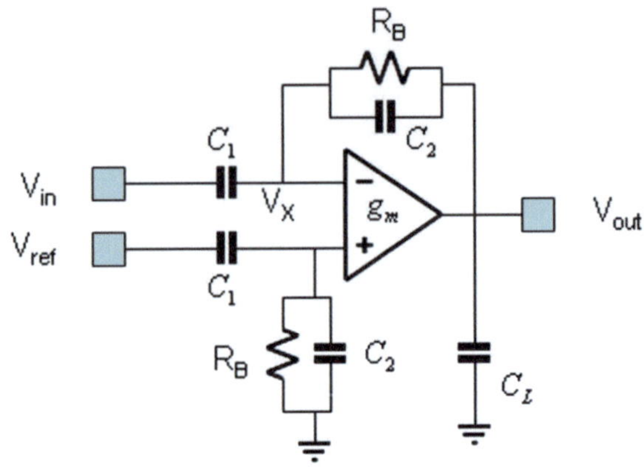

Fig. 3 A negative capacitive feedback amplifier

signal bandwidths and the impedance of the electrode is frequency dependent, C_1 should be chosen carefully according to the target signal. For instance, in the extracellular-action potential recording, because most of the signal energy is concentrated around 1 KHz and the impedance of the microelectrode ranges from several hundred KΩ to a few MΩ at 1 KHz, C_1 should be less than 16 pF to make the input impedance larger than 10 MΩ at 1 KHz.

The biasing resistor R_B serves two purposes. First, it sets a proper DC bias voltage at the input node of the OTA. The bias voltages are usually set to the midway voltage between V_{dd} and V_{ss}, which is equivalent to the circuit ground in case a dual power-supply level is used. Second, it determines the low-frequency cut-off of the amplifier because it forms a high-pass filter with the input capacitor C_1. Therefore, the cut-off frequency of the filter is $(2\pi R_B C_1)^{-1}$ and to achieve the cut-off frequency of less than 1 Hz, R_B should have a resistance of several GΩ. It is challenging to implement on-chip resistors with such high impedance. Several methods have been proposed including diode-connected MOS transistors [12], MOS-bipolar devices acting as pseudo resistors [13], and MOS transistors biased in subthreshold region [14].

When the condition $C_L \gg C_2$ is met, most output current of the OTA flows through C_L. Therefore, the output voltage V_{out} can be described as follows:

$$V_{out} = -g_m V_X \cdot \frac{1}{j\omega C_L} \tag{1}$$

Also, we can assume that the input impedance of the OTA is large enough that the following equation is valid:

$$C_1 (V_{in} - V_X) = C_2 (V_X - V_{OUT}) \tag{2}$$

By (1) and (2), the gain of the amplifier can be calculated as below:

$$A(\omega) = \frac{V_{\text{out}}}{V_{\text{in}}} = \frac{C_1/C_2}{1 + j\omega\frac{(C_1/C_2)C_L}{g_m}} \tag{3}$$

From (3), we can find out that the amplifier's mid-band gain A_M and high-frequency cut-off ω_{HF} are C_1/C_2 and $g_m(A_M C_L)^{-1}$, respectively.

OTA has critical effects on the overall performance of the amplifier such as input equivalent noise, offset voltages, and common-mode rejection ratio (CMRR). A current-mirror OTA with cascade output stage [13] is commonly used in conventional designs.

The width and length of the input transistors are chosen large enough to make the $1/f$ noise as small as possible. Subthreshold design is often employed for low-power operation [13]. However, the subthreshold design requires very careful simulations and layout techniques due to relatively large process variations and poor matching between critical devices.

A fully differential and self-biased OTA was also proposed [14] and its schematic is presented in Fig 4. A self-biased fully differential OTA enables the amplifier to operate at low voltage with an exceptionally large CMRR and small area. Fully differential output has a great advantage over single-ended output because it has larger CMRR. Because the OTA is self-biased, there is no need for common mode feedback (CMFB) circuitry to set the output common mode level and this enables the amplifier to achieve low power at low power-supply level and small area for miniaturization. When common-mode signal is applied to the input pair of OTA, the output signal is suppressed by the negative feedback loop formed by the self-biased current source pair M_{PSC1}, M_{PSC2}, M_{NSC1}, and M_{NSC2}. However, when differential signal is applied, the total source current supplied by M_{PSC1} and M_{PSC2} remains the same. Hence, the OTA amplifies only differential signals.

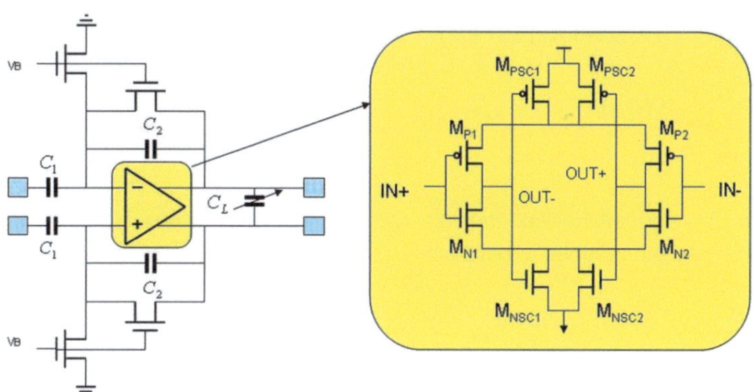

Fig. 4 An amplifier with a fully differential and self-biased OTA

By the symmetrical structure of the circuit, we can use the half-circuit model in Fig. 5 to find out the DC operating point from the following equation.

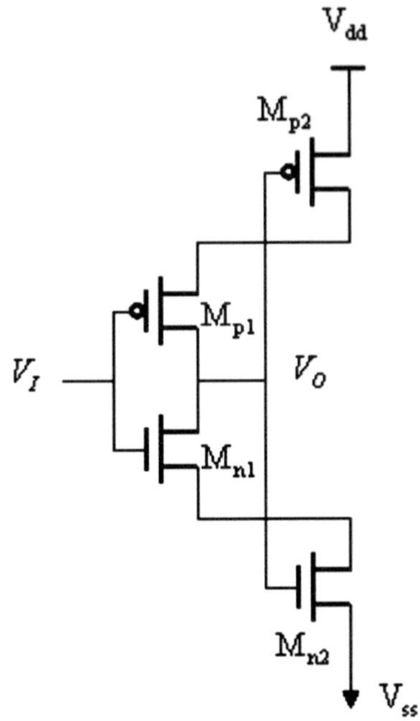

Fig. 5 The half-circuit model for the analysis of OTA

$$\frac{1}{2}\left(\frac{W}{L}\right)_{p2}\mu_p C_{\text{ox}}\left(V_{\text{SG,p2}}+V_{\text{th},}p\right)^2\cdot(1+\lambda_p V_{\text{SD,p2}})=\frac{1}{2}\left(\frac{W}{L}\right)_{n2}\mu_n C_{\text{ox}}\left(V_{\text{GS,n2}}-V_{\text{th},}n\right)^2\cdot(1+\lambda_n V_{\text{DS,n2}}) \tag{4}$$

Because longer channel length than minimum feature size of the process is chosen for the analog amplifier design in general, we can ignore Early effect, resulting in DC output common level as below:

$$V_{\text{O}}=\frac{V_{\text{DD}}+V_{\text{th},p}+\sqrt{\kappa}\cdot V_{\text{th},n}+\sqrt{\kappa}\cdot V_{\text{SS}}}{\sqrt{1+\kappa}}\text{,where }\kappa=\frac{\left(\frac{W}{L}\right)_{n2}\mu_n}{\left(\frac{W}{L}\right)_{p2}\mu_p} \tag{5}$$

From (5), we can understand that the DC output common level V_{O} is purely determined by the geometric parameter of M_{P2} and M_{N2}. In general, it is preferable to set V_{O} to (V_{dd} + V_{ss})/2, which is GND in this design, for maximum output range. Assuming symmetrical dual power-supply level, ($V_{\text{dd}}=-V_{\text{ss}}$) the desired geometric ratio of two source transistors, κ can be set as below:

$$V_{\text{O}}=\frac{V_{\text{DD}}+V_{\text{th},p}+\sqrt{\kappa}\cdot V_{\text{th},n}+\sqrt{\kappa}\cdot V_{\text{SS}}}{\sqrt{1+\kappa}}=0 \tag{6}$$

$$\kappa=\left(\frac{V_{\text{DD}}-\left|V_{\text{th},p}\right|}{V_{\text{DD}}-V_{\text{th},n}}\right)^2 \tag{7}$$

To find out the transconductance of the OTA, we again use the symmetry of the circuit for the analysis. As explained above, for the differential input, we can assume that M_{P2} and M_{N2} in Fig. 5 form a constant current source. Thus, the transconductance of the half circuit is simply the sum of transconductance of two input transistors, M_{N1} and M_{P1}, resulting in the overall transconductance as below:

$$g_m = 2g_{\mathrm{m,half}} = 2\left(g_{\mathrm{mn1}} + g_{\mathrm{m}p1}\right) \tag{8}$$

2.1.2 Chopper Amplifier

Several researchers have proposed using chopper-modulated amplifier for biomedical applications [15–16] because of the ability to suppress low-frequency noise such as 1/*f* noise and DC offset voltages. This technique is especially useful for the amplification of the far-field signal whose frequency is so low that the major noise source to affect the signal is 1/*f* noise rather than thermal noise. The fundamental idea of the chopper amplifier is to move the original signal to a higher frequency band and perform the amplification at that frequency band where the noise and offset of the amplifier do not exit. The block diagram to explain the concept and operation of the chopper amplifier is given in Fig. 6. The incoming neural signal is first chopper modulated so that the signal spectrum is moved to higher frequency band. After the modulation, an amplifier provides the required signal gain with the addition of the amplifier's intrinsic noise. It should be noted that the noise added by the amplifier is not chopper modulated with the neural signal. A low-pass filter then removes the low-frequency noise and the filtered signal is moved back to the original signal band by the chopper demodulation. Therefore, the 1/*f* noise added by the amplifier is effectively removed making it possible to achieve extremely small input equivalent noise of the amplifier. The chopper amplifier is inherently a DC-coupled amplifier and as a consequence, there is no need for DC-blocking capacitors at the input. The major disadvantage of the chopper amplifier is the clock noise, which can hurt the

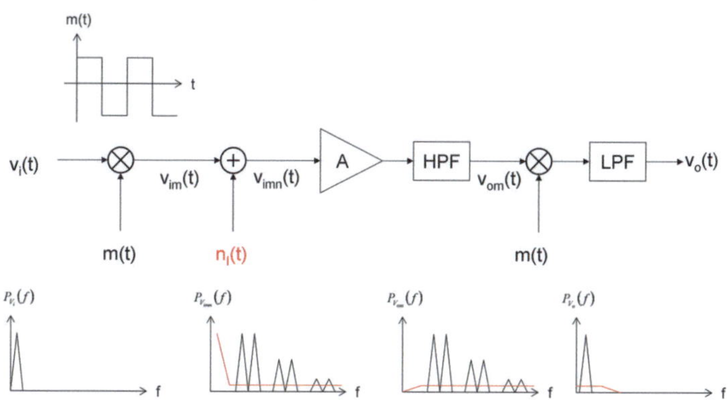

Fig. 6 Block diagram of basic chopper amplifier

weak neural signal unless care is not taken at the design stage of the microelectronics. In addition, the switch for the chopper modulation can add its intrinsic noise to the neural signal, making the signal noise of the original signal worse unless designed appropriately.

2.2 *Filter*

Although the amplifiers can serve as a first-order filter, it is usually not enough to obtain a reasonably good signal-to-noise ratio due to extremely weak power of neural signal. In poor-recording environments, the extracellular action potential could be as low as 50 μV. In addition, reducing noise power is always preferred because a larger number of neurons around the recording electrode can be monitored with the same number of recording electrodes. Therefore, most neural-recording systems employ separate band-pass or low-pass filters following the preamplifiers.

Those filters should have an appropriate cut-off frequency and filter order to maximize the signal-to-noise ratio. In most cases, first-order filter is used resulting in second-order overall filtering together with the amplifier. For extracellular action potential recording, a cut-off frequency of 10 KHz is typical, and this frequency should be even lowered for other far-field signals such as LFPs or EEG. Such low cut-off frequencies required for proper operation of the system impose challenging design issues in the integrated circuit design.

2.2.1 Passive R-C Filter

The most straightforward choice for the implementation of low-order filters such as first-order low-pass filters is passive R-C filters. Passive L-C filters are not usually used because the inductance for the required cut-off frequency is too large to be integrated on chip. For example, to achieve 10 KHz cut-off frequencies with 100 pF on-chip capacitor, on-chip inductor of 2.5 H is required, which is impossible to implement on chip. Passive filters have an advantage of not consuming any static power compared to other types of the filters, but have a disadvantage of large area even with R-C combinations. This limitation becomes more serious as the number of recording channel increases.

To overcome this problem, Miller capacitance technique was proposed at the expense of spending a little static power [14]. In Fig. 7(a), the equivalent impedance Z_{eq} can be calculated as below:

$$Z_{eq} = \frac{V_i}{I_i} = \frac{R_{out}}{1+A} + \frac{1}{j\omega(1+A)C} = R_{eq} + \frac{1}{j\omega C_{eq}} \tag{9}$$

where, R_{out} and A are the output impedance and voltage gain of the Miller amplifier, respectively. From (9), we can find out that the equivalent capacitance and resistance are modified by the factor (1+A).

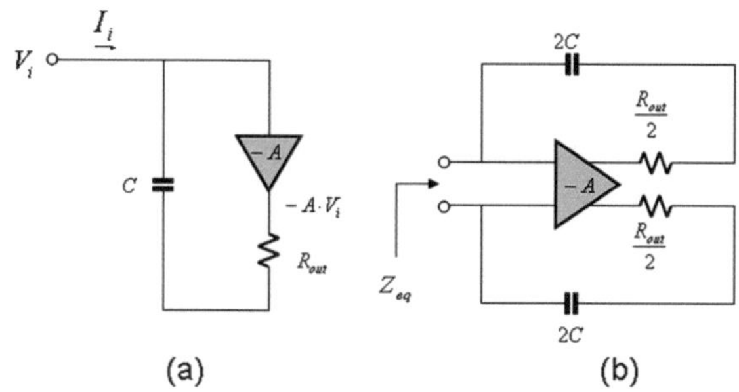

Fig. 7 (**a**) Increased capacitance using Miller effect and (**b**) for fully differential-signaling scheme

$$R_{\text{eq}} = \frac{R_{\text{out}}}{1+A}, \qquad C_{\text{eq}} = (1+A)\,C \tag{10}$$

For a fully differential-signaling scheme, the circuit should be modified as in Fig. 7(b). In this circuit, the gain of the amplifier should be greater than 3 to achieve the area reduction of the capacitance. The circuit of the proposed filter with Miller capacitance technique is presented in Fig. 8.

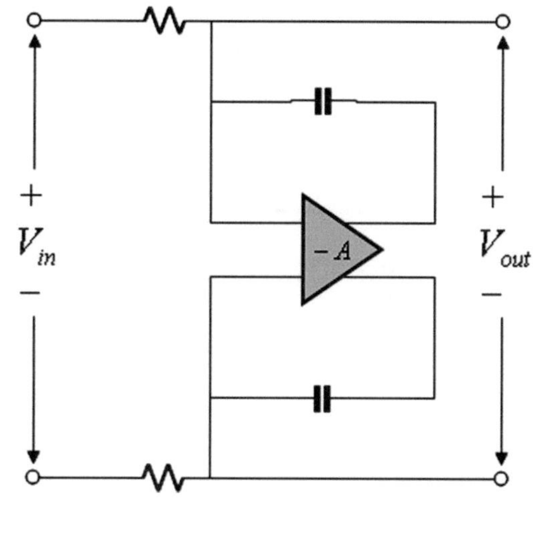

Fig. 8 Low-pass filter with Miller capacitance technique

2.2.2 Active Filter

Active filters are widely used for the implementation of high-order filters. A high-complexity fourth-order band-pass filter tuned for a one-octave passband of 20–40 Hz [17] was demonstrated using OTAs and capacitors. A technique of stagger tuning where two second-order band-pass filters are cascaded and tuned to slightly different frequencies resulted in a wider, flatter pass band than in either signal filter. All the transistors were operated in subthreshold region to save power consumption by the active circuits. The large resistors were realized by the OTAs with a very small transconductance. The current when test voltage source, V_{test}, is applied between two ports of the circuit in Fig. 9 is determined by the transconductance of the OTA. Therefore, the equivalent resistance between two ports is given as below:

$$R = \frac{V_{\text{test}}}{I_{\text{test}}} = \frac{1}{g_{\text{m}}} \tag{11}$$

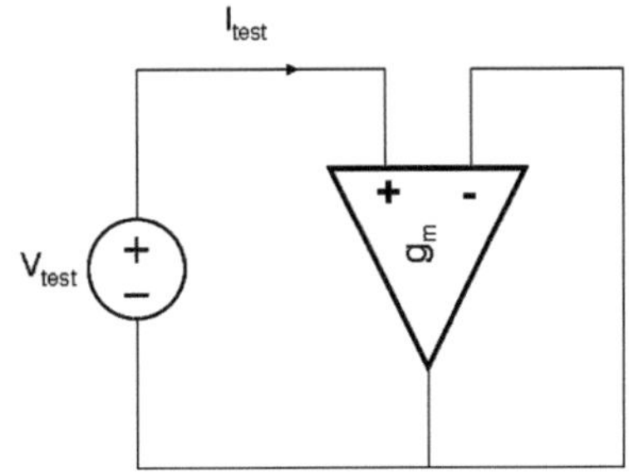

Fig. 9 A large resistor realized by an OTA

As shown in (11), a large resistance can be achieved by reducing the transconductance of the OTA.

2.2.3 Switched-Capacitor Filter

Switched-capacitor filters can also be used. In the switched-capacitor filter, resistors are realized with switched capacitors rather. To understand how switched capacitors can be used to replace resistors, consider the circuit shown in Fig. 10 where a capacitor is connected to two switches and two different voltages. When S_1 closes with S_2 open, and then S_2 closes with S_1 open, the amount of charge transferred from V_1 to V_2 is as below:

$$\Delta q = C \cdot (V_1 - V_2) \tag{12}$$

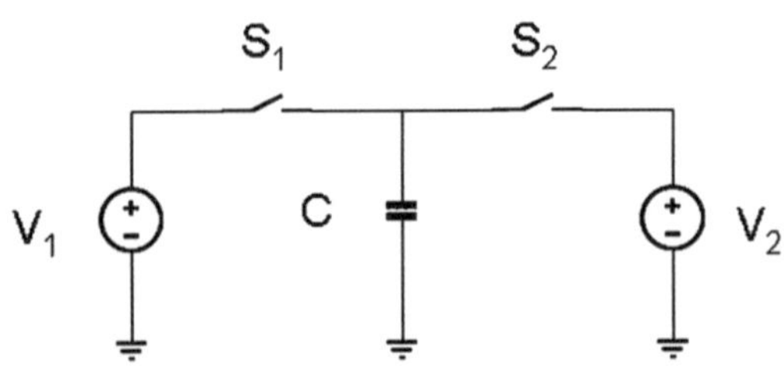

Fig. 10 A switched capacitor

If the switching frequency of the switches is f_S, then the total amount of transferred charge per second, the current from V_1 to V_2, can simply be written as below:

$$i_S = C \cdot (V_1 - V_2) \cdot f_S = \frac{(V_1 - V_2)}{(C \cdot f_S)^{-1}} \tag{13}$$

The above equation (13) states that the switched capacitor is equivalent to a resistor whose resistance is $(Cf_S)^{-1}$. Therefore, by lowering switching frequency f_S with given C, it is possible to achieve very large resistance. A major advantage of switched capacitor is that its value can be precisely controlled by adjusting switching frequency, making it less susceptible to process variations compared to other types of resistors available in integrated circuits. However, great care must be taken at the design stages to minimize the charge-injection noise and switching noise such as clock feedthrough [18].

2.3 ADC

Digital data are preferred to analog data in most of the on-chip signal processing as the feature size of the fabrication process gets smaller. Modern technologies enable a great amount of computation power with remarkably small power consumption and chip area by realizing complex signal-processing algorithms in digital domain. Therefore, it is necessary to convert analog neural signal amplified and filtered by the analog circuits into digital data so that it can be processed in digital domain. An analog-to-digital converter (ADC) is a circuit that digitizes analog signal into digital bits with a certain resolution and sampling rate. The resolution and sampling rate are the two most important circuit parameters of ADC in implantable electronics. These two parameters are chosen carefully based on the properties of the signal to be processed so that there is no overdesign, which leads to unnecessary power consumption and chip area [10]. One example of how to determine these parameters will be given in the later section.

Many different types of ADCs have been proposed so far, and ADC design itself is one of the important on-going research areas these days. Among several kinds of ADCs, the successive approximation register (SAR) ADC is the most appropriate for our application in terms of effective number of bits and required bandwidth [19]. The SAR ADC is also well suited for low-power and small-area applications since they usually require a minimal amount of analog circuitry [20]. Since SAR ADCs can use variable resolution easily, the user also has more flexibility in varying the resolution depending on the type of application. However, a buffer is necessary to overcome the relatively large input capacitance caused by the sampling capacitor of SAR ADC. Figure 11 is the general architecture of SAR ADC widely used in low-power mobile applications [20] and the detailed explanation on the operation of SAR ADC can be found in many literatures [21, 22].

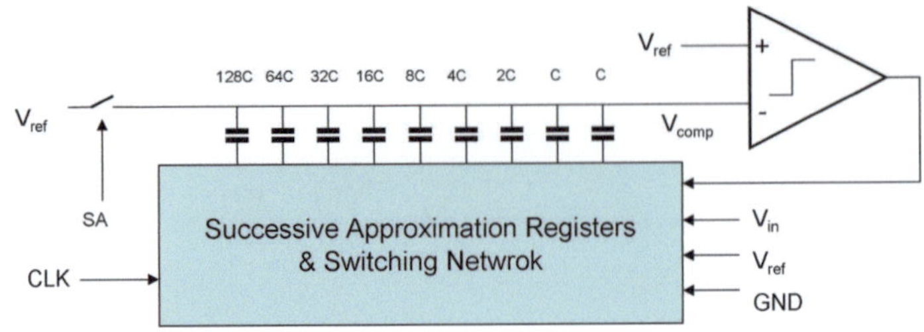

Fig. 11 A general architecture of SAR ADC widely used in low-power mobile applications [20]

3 Subsystem Design

3.1 Front-End Blocks for Neural Recording

Advances in microelectrode arrays (MEAs) have enabled neuroscientists and researchers in biomedical engineering to take advantage of a large number of channels [22], and this has made it possible to pursue a variety of neuroprosthetic applications such as brain-controlled limb prostheses to treat spinal cord injuries and paralysis. A brain-machine interface (BMI) is at the core of these applications to sense the signals from the brain.

These applications, by nature, impose serious limitations on power and area in the design of neural-recording systems. Researchers have developed several kinds of neural-recording systems [23–27], and those systems are generally composed of preamplifiers to amplify the small extracellular potentials, low-pass filters to reject the high frequency noise, multiplexers, and analog-to-digital converters (ADCs) followed by the wireless telemetry circuits to transmit data out of the body.

Much effort has been made to minimize the power of each individual circuit block to maximize the system's resolution, but almost no attention was paid to find the trade-offs among those circuit blocks to achieve an optimal design. For example, designers generally determine their system's resolution without considering electrode noise even though the noise contributed by the electrode is significant, leading to higher-than-required resolution, eventually consuming unnecessarily large power and chip area. The multiplexing ratio is also an important system design parameter because it is the key parameter in the trade-off between the system's power consumption and chip area. However, there has been no analysis of this critical parameter and is usually chosen in an arbitrary manner. An analysis of an optimal design would greatly help to optimally design a low-power neural-recording system and integrate more functions such as signal processing to overcome the interference among electrodes to reduce the data rate. We will investigate such a design methodology for a neural-recording system.

3.1.1 System Architecture and Circuit Modeling

Figure 12 shows the general architecture of the multichannel neural-recording system. The total number of channels is $N(=2^n)$ and the total number of ADCs is

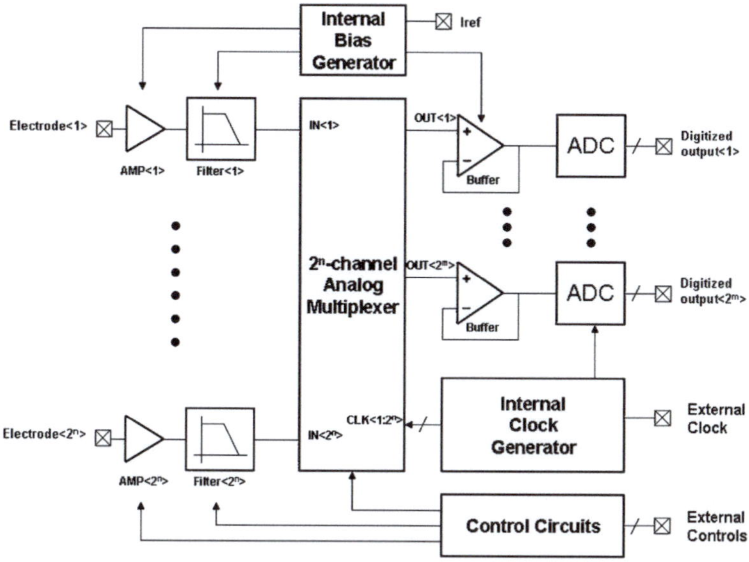

Fig. 12 General architecture of the multichannel neural-recording system

$M(=2^m)$, resulting in a multiplexing ratio of 2^{n-m}. A low-noise preamplifier is required to amplify the small potential difference between the recording microelectrode and a relatively large reference electrode when an action potential is generated inside the neuron. Typically, these extracellular potentials have amplitudes from 50 μV to 1 mV, and the signal energy is located between 300 Hz and 10 KHz.

Figure 13 shows the schematic of the preamplifier, which is composed of an operational transconductance amplifier (OTA) and a feedback network. A capacitive negative-feedback amplifier is widely used in neural-recording systems [28] because there is usually a DC offset of 1−2 V across the electrode-tissue interface [29]. The gain of the amplifier is determined by the ratio of the two capacitances in the feedback network.

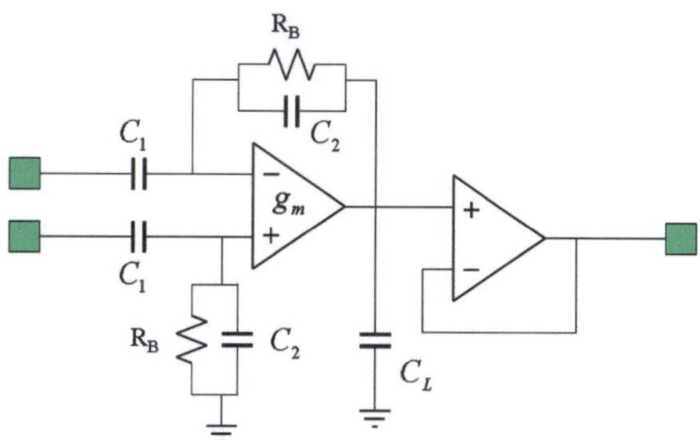

Fig. 13 Circuit diagram of preamplifier

$$G_{\text{amp}} = \frac{C_1}{C_2} \tag{14}$$

The input-referred noise voltage of this amplifier [29] is given by

$$\overline{v_{\text{ni,amp}}^2} = \left(1 + \frac{1}{G_{\text{amp}}}\left(1 + \frac{C_{\text{in}}}{C_2}\right)\right)^2 \cdot \frac{16k_\text{B}T}{3g_\text{m}} \cdot f_{\text{Neuron}} \tag{15}$$

where C_{in} is the input capacitance of the OTA, g_m is the transconductance of the OTA, and f_{Neuron} is the signal bandwidth of the action potential. The size of the amplifier is mostly determined by the feedback capacitances, which means that the size is approximately proportional to the amplifier gain.

Although the preamplifier can provide first-order low-pass filtering, dedicated low-pass filters are used to further minimize high-frequency noise. The cut-off frequency of low-pass filters is usually set to f_{Neuron}, 10 KHz. Several kinds of low-pass filters, such as gm-C filters and passive RC filters, can be used for this purpose. Among those, passive RC filters are the most straightforward and easy to implement. However, the area of a passive element is relatively huge for on-chip implementation due to the low cut-off frequency. To overcome this disadvantage, Miller effect can be used at the expense of increased power consumption [30]. Because the cut-off frequency of filters is independent of the multiplexing ratio, we can represent the power consumption for filtering as a constant, P_{LPF1}. Filters also need buffers to drive the analog multiplexer and these buffers must meet slew rate and output impedance constraints. Figure 14 shows a simplified model of low-pass filters and analog multiplexers. In most cases, the slew rate is the limiting factor while output impedance is easily achievable. Since the slew rate of the buffer is proportional to the biasing current, the following inequality holds:

$$\frac{2G_{\text{amp}}V_{\text{Neuron}}}{(1/\lambda)\,(1/f_\text{s})} \leq \frac{I_{\text{BF}}}{C_{\text{pmux}}} \tag{16}$$

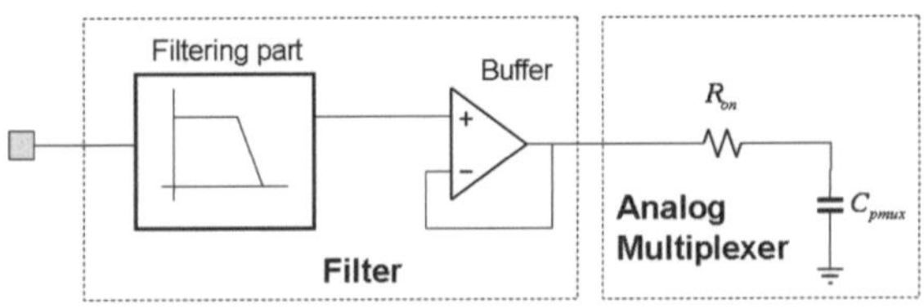

Fig. 14 Simplified modeling of filter and analog multiplexer

Here, V_{Neuron} is the maximum amplitude of the extracellular potential, which means that the maximum signal change between two consecutive channels can be 2 V_{Neuron}. The sampling rate of the ADCs, f_s, increases as the multiplexing ratio does. This can be expressed as $f_\text{s}=2^{n-m}2f_{\text{Neuron}}$, if we use the Nyquist sampling ratio. C_{pmux} is the parasitic loading capacitance, which is also proportional to the multiplexing ratio of the analog multiplexers and is equal to $C_{p\text{mux}1}2^{n-m}$ where

C_{pmux1} is the parasitic loading of just one channel. The timing margin ratio λ is determined by the sampling clock's jitter, skew, and the minimum timing window of the ADC. Therefore, (16) can be rewritten as:

$$I_{\text{BF}} \geq 4\lambda G_{\text{amp}} V_{\text{Neuron}} f_{\text{Neuron}} C_{\text{pmux1}} 2^{2(n-m)} \tag{17}$$

The mathematical model of the total power consumption of filters is, therefore:

$$\begin{aligned} P_{\text{FILTER,total}} &= 2^n \cdot (2V_{\text{dd}} \cdot I_{\text{BF}} + P_{\text{LFP1}}) \\ &= 2^n \left(2V_{\text{dd}} \cdot 4\lambda G_{\text{amp}} V_{\text{Neuron}} f_{\text{Neuron}} C_{\text{pmux1}} 2^{2(n-m)} + P_{\text{LPF1}}\right) \\ &= 8\lambda V_{\text{dd}} G_{\text{amp}} V_{\text{Neuron}} f_{\text{Neuron}} C_{\text{pmux1}} 2^{3n-2m} + 2^n P_{\text{LPF1}} \end{aligned} \tag{18}$$

However, if we use one ADC per one channel, there is no need for buffers and the above modeling simplifies to

$$P_{\text{FILTER,total}} = 2^n P_{\text{LPF1}} \ldots \tag{19}$$

Usually, the area of the filtering part is independent of the multiplexing ratio and buffers' area is negligible compared to filtering part. Thus, the total filter area can be described as:

$$\begin{aligned} A_{\text{FILTER,total}} &= 2^n \cdot (A_{\text{BF1}} + A_{\text{LPF1}}) && \text{when } n \geq m, \\ &= 2^n A_{\text{LPF1}} && \text{when } n = m, \end{aligned} \tag{20}$$

where A_{BF1} is the area of one buffer and A_{LPF1} is that of a filter.

Crosstalk noise can cause a serious problem in multichannel-recording systems. There are three possible crosstalk noise sources in an integrated neural-recording system. One is capacitive coupling between metal interconnection, second is the finite turn-off resistance of the analog switches due to subthreshold conduction, and the third is the capacitive coupling through the parasitic capacitance of the transistor. As for capacitive coupling between metal lines, those coupling capacitances can be eliminated by the careful shielding of each signal line at the expense of increased bussing area [31]. In general, the signal common mode level is set to the middle of the power-supply level and the amplified signal rarely exceeds a few hundred mV. Therefore, there is a large back-bias effect making the subthreshold conduction negligible. In general, the turn-off resistance is more than several GΩ, which is extremely larger than the turn-on resistance of the analog switches. As a result, the only remaining crosstalk noise source is the parasitic coupling in the MOS transistors through source-to-substrate junction capacitance and drain-to-substrate junction capacitance. We can model crosstalk noise from one channel using Thevenin's equivalent circuit as Fig. 15. The equivalent voltage amplitude and source impedance can be calculated as below:

$$Z_{\text{OFF}}(s) = \frac{1}{sC_{\text{JSB}}} + \left(R_{\text{Bulk}} \left\| R_{\text{OUT}} + \frac{1}{sC_{\text{JDB}}} \right.\right) \cong \frac{1}{sC_{\text{JSB}}} + \left(R_{\text{Bulk}} \left\| \frac{1}{sC_{\text{JDB}}} \right.\right) = \frac{1}{sC_{\text{JSB}}} + \frac{R_{\text{Bulk}}}{1 + sR_{\text{Bulk}}C_{\text{JDB}}} \tag{21}$$

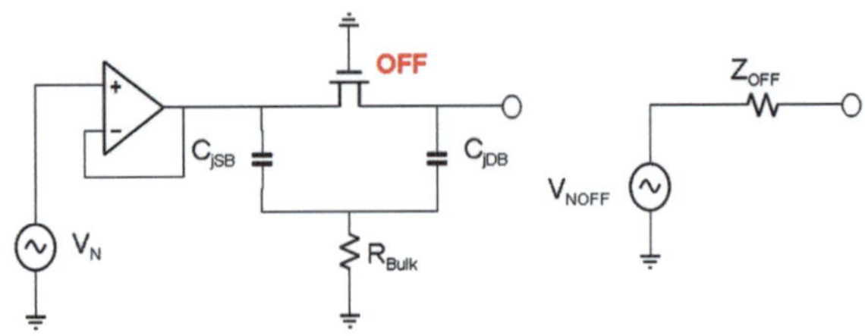

Fig. 15 Crosstalk noise due to parasitic of the switch and its Thevenin's equivalent circuit

$$V_{\mathrm{NOFF}}(s) = \frac{R_{\mathrm{Bulk}}}{R_{\mathrm{OUT}} + \frac{1}{sC_{\mathrm{JSB}}} + R_{\mathrm{Bulk}}} \cdot V_{\mathrm{N}} = \frac{sR_{\mathrm{Bulk}}C_{\mathrm{JSB}}}{1 + s\left(R_{\mathrm{OUT}} + R_{\mathrm{Bulk}}\right)C_{\mathrm{JSB}}} \tag{22}$$

where C_{JSB} is the source-to-substrate junction capacitance, C_{JDB} is the drain-to-substrate junction capacitance, R_{Bulk} is the substrate resistance, and R_{OUT} is the output impedance of the buffers in the filter. Using this equivalent circuit, we can model the worst-case crosstalk noise in an N-channel neural-recording system as shown in Fig. 16. From this, the crosstalk noise can be calculated as

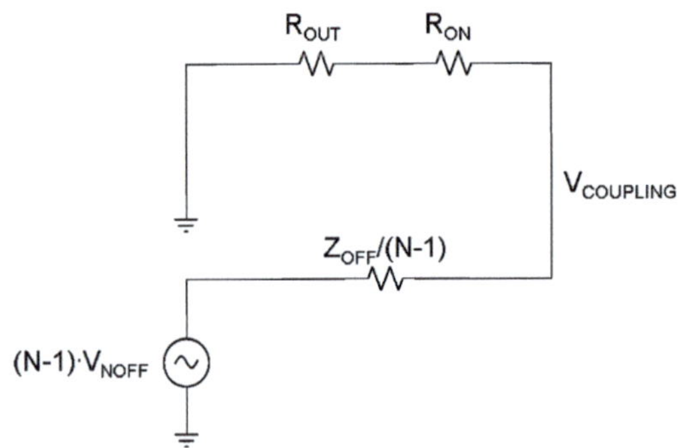

Fig. 16 Modeling of crosstalk noise in *N*-channel analog multiplexer

$$V_{\mathrm{COUPLING}} = \frac{R_{\mathrm{OUT}}\_R_{\mathrm{ON}}}{R_{\mathrm{OUT}}+R_{\mathrm{ON}}+\frac{Z_{\mathrm{OFF}}}{N-1}} \cdot (N-1)\,V_{\mathrm{NOFF}} = \frac{R_{\mathrm{OUT}}+R_{\mathrm{ON}}}{R_{\mathrm{OUT}}+R_{\mathrm{ON}}\,\frac{Z_{\mathrm{OFF}}}{N-1}} \cdot \frac{s(N-1)R_{\mathrm{Bulk}}C_{\mathrm{JSB}}}{1+s(R_{\mathrm{OUT}}+R_{\mathrm{Bulk}})C_{\mathrm{JSB}}} \cdot V_{\mathrm{N}}$$

$$\therefore \frac{V_{\mathrm{COUPLING}}}{V_{\mathrm{N}}}(j\omega) = \frac{R_{\mathrm{OUT}}+R_{\mathrm{ON}}}{R_{\mathrm{OUT}}+R_{\mathrm{ON}}+\frac{1}{j\omega(N-1)C_{\mathrm{JSB}}}+\frac{R_{\mathrm{Bulk}}/(N-1)}{1+j\omega R_{\mathrm{Bulk}}C_{\mathrm{JDB}}}} \cdot \frac{j\omega(N-1)R_{\mathrm{Bulk}}C_{\mathrm{JSB}}}{1+j\omega(R_{\mathrm{OUT}}+R_{\mathrm{Bulk}})C_{\mathrm{JSB}}} \tag{23}$$

where R_{ON} is the turn-on resistance of the switches and is set to 5 KΩ. C_{JSB} and C_{JDB} are set to 7.1 *f*F, which is typical for 0.35 μm CMOS process. R_{OUT} and R_{Bulk} are set to 4 KΩ and 10 Ω, respectively. With these values, we can find out that the worst-case crosstalk noise by the parasitic capacitances in the analog multiplexer is around –120 dB at 10 KHz for the 128 channels.

Similar to the buffers that drive the analog multiplexer presented before, the power consumption of the buffers to drive ADCs are also determined by the slew rate, which means the biasing current of one buffer is proportional to the multiplexing ratio as shown below:

$$\begin{aligned} I_{\mathrm{BA}} &\geq 2G_{\mathrm{amp}}V_{\mathrm{Neuron}}f_{\mathrm{s}}C_{\mathrm{in,ADC}} \\ &\geq 2G_{\mathrm{amp}}V_{\mathrm{Neuron}}2^{n-m+1}f_{\mathrm{Neuron}}C_{\mathrm{in,ADC}} \end{aligned} \tag{24}$$

Here, $C_{\mathrm{in,ADC}}$ is the input capacitance of the each ADC. Therefore, the total power consumption of the 2^m buffers is independent of the multiplexing ratio as below:

$$\begin{aligned} P_{\mathrm{BUF,total}} =& M \cdot 2V_{\mathrm{dd}}I_{\mathrm{BA}} \\ =& M \cdot V_{\mathrm{dd}} \cdot 2^{n-m+3}G_{\mathrm{amp}}V_{\mathrm{Neuron}}f_{\mathrm{Neuron}}C_{\mathrm{in,ADC}} \\ =& V_{\mathrm{dd}} \cdot 2^{n+2}G_{\mathrm{amp}}V_{\mathrm{euron}}f_{\mathrm{Neuron}}C_{\mathrm{in,ADC}} \end{aligned} \tag{25}$$

where V_{dd} is the power supply voltage. Also, the total power consumption of 2^m ADCs is constant regardless of the multiplexing ratio because the power consumption of one ADC is roughly proportional to the sampling rate and the sampling rate is proportional to the multiplexing ratio. In other words, the power consumed by the one ADC increases as the multiplexing ratio does, but the total power consumption of the ADCs remains the same because the number of ADCs decreases proportionately. However, the total area consumed by ADCs is directly proportional to the number of ADCs multiplied by the area of one ADC, A_{ADC1}.

$$A_{\mathrm{ADC,total}} = M \cdot A_{\mathrm{ADC1}} = 2^n \cdot A_{\mathrm{ADC1}} \propto 2^m \cdot C_{\mathrm{in,ADC}} \tag{26}$$

The last relationship is generally valid because sampling capacitors take most of chip area in SAR ADC.

3.1.2 System Resolution

The dynamic range to noise ratio (DNR) of a neural-recording system cannot exceed that of an input signal at the preamplifier's input regardless of the amplifier's input equivalent noise voltage and the number of bits of the ADC. Generally, the area of SAR ADC increases exponentially with respect to the number of bits. Also, the worst-case power consumption of the ADC is proportional to the sampling capacitance, which increases exponentially with respect to the number of bits as well [33]. In addition, the amplifier's input equivalent noise voltage decreases as the gain of the amplifier increases, which can be seen in (15). This means that there is a trade-off between the amplifier's area and noise. The amplifier gain is not the only design parameter that affects the amplifier's input equivalent noise. Noise can also be reduced by increasing the OTA's transconductance, which increases the power and area of the OTA. Therefore, to avoid wasting power and chip area by increasing the system's resolution more than what is actually needed, it is critical to know the fundamental limit of the system's dynamic range determined by the input signal to avoid overdesign.

The amount of thermal noise contributed by the electrode is given by Nyquist's formula:

$$V_{\mathrm{RMS}}\sqrt{4k_{\mathrm{B}}TR\Delta f} \tag{27}$$

where k_{B} is Boltzmann's constant ($k_{\mathrm{B}} = 1.38 \times 10^{-23}$ J/K), T is the temperature (in degrees Kelvin), R is the resistance of the electrode (in Ohms), and Δf is the recording bandwidth (in Hz). The resistance of the electrode includes the resistance of the metallic portion of the microelectrode and the seal resistance, sometimes called the spreading resistance, which is the resistance of the saline bath between the metallic interface and ground electrode [34]. For an electrode with a resistance of 1 MΩ operating at 27°C with a 10-KHz recording bandwidth, the RMS noise voltage is 12.6 μV. And there is as much background noise, which is a sum of many smaller spikes, as thermal noise [35]. In general, the maximum amplitude of extracellular spike potentials is smaller than 1 mV and the noise is around 20 μV for 1 MΩ. Therefore, the DNR of input signal is about 34 dB. This implies that regardless of how large the number of bits of the ADC is and how small the amplifier noise is, the system's effective number of bits cannot exceed 6 bits. Therefore, it is reasonable to make the ADC's DNR the same as that of input signal and make the amplifier's input equivalent noise negligible to achieve as large DNR as possible (in this case, 6 bits). Thus, we can set the number of bits in the ADC, N_{B}, to satisfy following relationship:

$$V^2_{\mathrm{noise,eltd}} = \left(\frac{1}{\sqrt{2}}\frac{V_{\mathrm{Neuron}}}{2^{N_{\mathrm{B}}-1}}\right)^2 \tag{28}$$

In this work, we set the amplifier's noise power to be 1/10th of the electrode noise power. This results in the small 4.8% increase in the total equivalent RMS noise voltage. From (15) and (28), we can express the relationship mentioned above as

$$\left(1+\frac{1}{G_{\mathrm{amp}}}\left(1+\frac{C_{\mathrm{in}}}{C_2}\right)\right)^2\cdot\frac{16k_{\mathrm{B}}T}{3g_{\mathrm{m}}}\cdot f_{\mathrm{Neuron}} = V^2_{\mathrm{noise,eltd}}\times 0.1 = \left(\frac{\frac{V_{\mathrm{Neuron}}}{2^{N_{\mathrm{B}}-1}}}{}\right)^2\times\frac{1}{20} \tag{29}$$

This equation can be rewritten as:

$$G_{\mathrm{amp}} = \frac{\left(1+\frac{C_{\mathrm{in}}}{C_2}\right)}{-1+\sqrt{\frac{g_{\mathrm{m}}}{g_{\mathrm{m,\,min}}}}}\left(\text{,where } g_{\mathrm{m,\,min}} = \frac{5\cdot 4^{N_{\mathrm{B}}+2}K_{\mathrm{B}}Tf_{\mathrm{Neuron}}}{3V^2_{\mathrm{Neuron}}}\right) \tag{30}$$

This equation sets the lower bound for g_{m}, the OTA's transconductance. If we choose a small g_{m} for low power, the chip area of the amplifier increases due to the large amplifier gain determined by (30). Therefore, there exists a trade-off where the optimal design can be found by setting the g_{m} and G_{amp} such that the power and area product of the amplifier is minimal. This power and area product of the amplifier is proportional to the product of ${g_m}^2$ and G_{amp}, if the transistor operates in saturation region and width of transistors is unchanged.

$$g_m^2 G_{amp} = \frac{b \cdot g_m^2}{\sqrt{g_m} - a} (,\, where\; a = \sqrt{g_{m,\,min}},\; b = \left(1 + \frac{c_{in}}{C_2}\right) \cdot \sqrt{g_{m,\,min}}\,) \tag{31}$$

To find out the minimum, we differentiate the above equation with respect to g_m.

$$\partial \frac{\left(g_m^2 G_{amp}\right)}{\partial g_m} = b \cdot \frac{2g_m \cdot \left(\sqrt{g_m} - a\right) \cdot \frac{1}{2\sqrt{g_m}}}{\left(\sqrt{g_m} - a\right)^2} = b \cdot \frac{g_m \cdot \left(\frac{3}{2}\sqrt{g_m} - 2a\right)}{\left(\sqrt{g_m} - a\right)^2} \tag{32}$$

Setting this equation to zero, we find the $g_{m,opt}$ that makes (31) minimum:

$$g_{m,opt} = \left(\frac{4}{3}\right)^2 \cdot a^2 = \frac{16}{9} \cdot g_{m,\,min} = 1.78 \cdot g_{m,\,min} \tag{33}$$

From (30), the optimal gain of the amplifier is given by

$$G_{amp,opt} = 3\left(1 + \frac{C_{in}}{C_2}\right) \tag{34}$$

For values of $C_{in}/C_2 = 7$, the optimal value of the amplifier's gain is 24. With conservative margins, it is proper to choose 50 as preamplifier's gain. However, because this preamplifier alone cannot provide enough signal gain, it is typical to employ additional second amplifier after the analog multiplexer. The second amplifier is shared by all the recording channels and does not need to be a low-noise amplifier; the power consumption and chip area required are usually negligible compared to other circuit blocks.

3.1.3 Trade-Off Between System Power and Chip Area

In this section, using real-circuit examples [30], we will discuss the optimal number of ADCs when the total number of channels is given. Simply, the total power consumption of the system is the sum of each block's power consumption, which is given by:

$$P_{system} = P_{AMP,total} + P_{FILTER,total} + P_{MUX,total} + P_{BUF,total} + P_{ADC,total} \tag{35}$$

In real implementations, the power consumption of one amplifier with the optimal transconductance and gain given by (33) and (34) respectively was 170 μW. Therefore, the total power consumption of 2^n amplifiers is simply:

$$PA_{MP,total} = 2^n \cdot 170\,(\mu W) \tag{36}$$

The power consumption for pure filtering was 99 μW. Using (18) and (19), the total power consumption of the filters was described as:

$$\begin{aligned} P_{\text{FILTR,total}} &= 99 + 0.13 \cdot 2^{2(n-m)}\,(\mu W)\,, && \text{When n} \geq \text{m} \\ &= 99\,(\mu W)\,, && \text{When n} = \text{m} \end{aligned} \tag{37}$$

The power consumed by the analog multiplexer was so small that it can be neglected here. Also, the power consumption of buffers and ADCs were respectively

$$P_{\text{BUF,total}} = 2^{n+2} \cdot 3.62\,(\mu W) \tag{38}$$

$$P_{\text{ADC,total}} = 2^{n} \cdot 0.89\,(\mu W) \tag{39}$$

Table 2 is the calculation result with various numbers of channels and ADCs. Here it can be seen that when it comes to power consumption, a one ADC per one channel system is the best, which is a very natural result because it does not need buffers to drive the analog multiplexer.

Table 2 System power [mW] according to various numbers of channels and ADCs

Unit [mW]	m=0 (1 ADC)	m=1 (2 ADC)	m=2 (4 ADC)	m=3 (8 ADC)	m=4 (16 ADC)	m=5 (32 ADC)	m=6 (64 ADC)	m=7 (128 ADC)
n=4 (16ch)	5.08	4.68	4.58	4.56	*4.5*	N.A	N.A	N.A
n=5 (32ch)	13.36	10.16	9.36	9.17	9.11	*9.1*	N.A	N.A
n=6 (64ch)	52.28	26.27	20.33	18.73	18.33	18.23	*18.2*	N.A
n=7 (128ch)	309	105	53.4	40.7	37.5	36.7	36.5	*36.4*

Also, the total chip is approximately the sum of areas of each block, which is

$$A_{\text{system}} = A_{\text{AMP,total}} + A_{\text{FILTER,total}} + A_{\text{MUX,total}} + A_{\text{BUF,total}} + A_{\text{ADC,total}} \tag{40}$$

However, the area of the multiplexer is very small when compared to those of other circuit blocks. Therefore, (40) can be rewritten as

$$A_{\text{system}} \cong A_{\text{AMP,total}} + A_{\text{FILTER,total}} + A_{\text{BUF.total}} + A_{\text{ADC,total}} \tag{41}$$

The area occupied by the preamplifiers is determined by the number of channels and not by the multiplexing ratio because the area of the amplifier is mostly determined by the size of the feedback capacitances, which is independent of the multiplexing ratio. Therefore, the area of the amplifier can be treated as a constant. The area of one amplifier whose gain is determined by the (34) was 120,000 μm^2

when implemented with 0.35 μm CMOS process. Therefore, we can model the area of the $2n$ amplifiers simply as below:

$$A_{\text{AMP,total}} = 2^n \cdot 12000\mu\text{m}^2 \tag{42}$$

Also, the total area of the filters was modeled according to (20) and it is given as:

$$\begin{aligned} A_{\text{FILTER,total}} &= 2^n \cdot (A_{\text{BF1}} + A_{\text{LPF1}}) = 2^n \times 82950\mu\text{m}^2, \quad \text{when}\, n \geq m \\ &= 2^n A_{\text{LPF1}} = 2^n \times 78330\mu\text{m}^2, \quad \text{when}\, n = m \end{aligned} \tag{43}$$

The area of one buffer to drive ADCs was 14100 μm^2 making the total area occupied by 2^m buffers as

$$A_{\text{BUF,total}} = 2^m \cdot 14100\mu\text{m}^2 \tag{44}$$

The optimal number of bits of ADCs determined was 6. With a conservative margin, we set the number of bits of the ADC to be 9. This resulted in a size of 848000 μm^2 per ADC. Therefore, the total area of 2 m ADCs is simply given as:

$$A_{\text{ADC,total}} = 2^m \cdot 84800\mu\text{m}^2 \tag{45}$$

Using (42), (43), (44), and (45), the total chip area described as (41) is:

$$\begin{aligned} A_{\text{system}} &= (2^n \times 0.203 + 2^m \times 0.862)\, mm^2, \quad \text{when}\, n \geq m \\ &= 2^n \times 1.06\text{mm}^2 \quad \text{when}\, n = m \end{aligned} \tag{46}$$

Table 3 is the calculated area of the neural-recording system with various channel numbers and ADCs. As shown, the total chip area increases as the number of ADCs does, which is a straightforward conclusion. Therefore, there must be an optimal multiplexing ratio that makes the system's power-area product minimum. Table 4 is the power-area product based on Tables 2 and 3. From Table 4, we can deter-

Table 3 Chip area [mm^2] according to various numbers of channels and ADCs

Unit [mm^2]	m=0 (1 ADC)	m=1 (2 ADC)	m=2 (4 ADC)	m=3 (8 ADC)	m=4 (16 ADC)	m=5 (32 ADC)	m=6 (64 ADC)	m=7 (128 ADC)
N=4 (16ch)	*4.1*	5.0	6.7	10.1	17.0	N.A	N.A	N.A
N=5 (32ch)	*7.4*	8.2	9.9	13.4	20.3	33.9	N.A	N.A
N=6 (64ch)	*13.9*	14.7	16.4	19.9	26.8	40.6	67.8	N.A
N=7 (128ch)	*26.8*	27.7	29.4	32.9	39.8	53.6	81.2	135

Table 4 Power and chip area product [mW•mm^2] with respect to various numbers of channels and ADCs

Unit [mW•mm^2]	m=0 (1 ADC)	m=1 (2 ADC)	m=2 (4 ADC)	m=3 (8 ADC)	m=4 (16 ADC)	m=5 (32 ADC)	m=6 (64 ADC)	m=7 (128 ADC)
n=4 (16ch)	*20.82*	23.40	30.69	46.06	76.50	N.A	N.A	N.A
n=5 (32ch)	98.86	*83.31*	92.66	122.8	184.9	308.5	N.A	N.A
n=6 (64ch)	726.7	392.8	*333.4*	372.7	491.2	740.1	1234	N.A
n=7 (128ch)	8281	2909	1570	*1339*	1493	1967	2964	4914

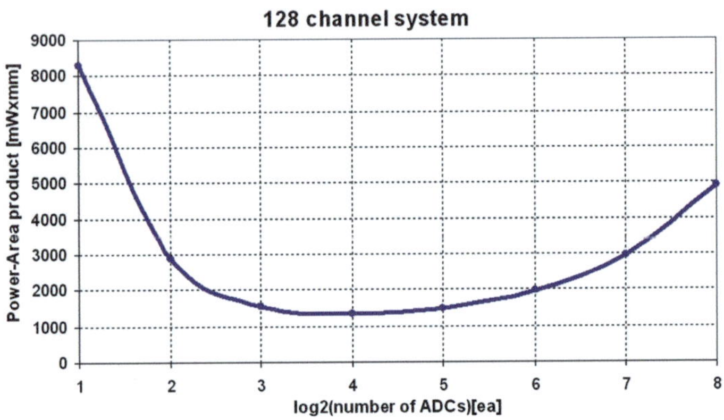

Fig. 17 Power-area product of a 128-channels system as a function of the multiplexing ratio

mine that the optimal number of channels per one ADC is 16. Figure 17 shows the power-area product of the 128-channels system as a function of the multiplexing ratio. As expected, at the extremes of the multiplexing ratio, the power-area product is dominated by either power or area. Though this example uses the area and power values for specific circuit topologies and process technology, the optimization methodology is applicable to all topologies and technologies. As a result, note that an optimized design is not necessarily the same for different applications even with the same specifications. It depends on the designer to assign priorities to different performance metrics (power and area of individual blocks and system) and take the additional constraints into account (for example, one might be reusing a circuit block that is part of the overall system, in which case, it is a given and the optimization has to be based on that). But it is necessary for the designer to ensure that a proper design methodology is adopted that takes into account all the parameters to ensure the right choice of system architecture.

3.2 Neural-Signal Processing Unit

Neurons communicate by firing action potentials, which induce transient voltage fluctuations in the surrounding neural tissue. The voltage fluctuations can be recorded with extracellular electrodes, where signals take the shape of spikes. A recording electrode is often surrounded by multiple firing neurons, causing the recorded activities become superimposed. To extract and understand the signal processing in the biological neural network, it is required to resolve spikes to individual neuronal sources [36–41]. This labeling process is called spike sorting.

Sorting neural spikes is challenging due to several factors, such as the presence of unresolved neuronal activity, similarity in recorded shapes and magnitudes of firing neurons, overlapping of spikes from simultaneously firing neurons, as well as changes in signal amplitude obtained from the same neuron at different times [36, 42–44]. If the sorting algorithm fails to differentiate signals from similar neurons, it will inaccurately report a "frequent" firing neuron, which compromises the reliability and accuracy of the neural information decoder.

Reported spike feature extraction algorithms include principal component analysis (PCA) [45, 46], Bayesian algorithm [47], template matching [48–51] and wavelets [52–54], independent component analysis (ICA) [55–59], interspike intervals-based algorithms [53, 60], and sample selection-based feature extraction algorithms [61, 62]. The performance of the feature extraction algorithms to differentiate similar neurons can be improved by advanced training algorithms [48, 63–65], which are usually computationally intensive, thus slowing down the training process. Retraining is required when the electrodes move, thus could be frequent in certain applications [66]. Although an early study has shown that computation of the training algorithm is affordable, it was assumed that retraining is performed daily [45]. For frequent retraining in a shorter period as suggested in [66, 67], the implementation of a complex training algorithm for large number channels becomes more difficult in terms of computation.

This work reports a method of using the first-order derivative of spikes for the sorting of similar neurons. The theoretical framework that includes neuronal geometry signatures and noise shaping is derived. By evaluating neuronal geometry signatures with compartment model, we find that emphasizing high-frequency signal spectrum helps differentiating similar neurons. Because the associated spike noise is dominant at lower frequency spectrum, a frequency-shaping filter could be used to reduce the noise. The operation of the derivative is a simple frequency-shaping filter almost linearly emphasizing signal spectrum according to the frequency. Combining spike derivative and a conventional spike feature extraction algorithm, improved differentiation of similar neurons is achieved without using complex training procedures.

3.2.1 Theory

Geometry Signatures for Similar Neurons

This section focuses on neuronal geometry signatures [67], which are used for sorting similar neurons. If the transmembrane current profiles from all membrane segments are obtained, the potential around the target neuron can be modeled by Laplace's equation [68] as below:

$$\nabla^2 V = 0, \tag{47}$$

with the boundary condition $V = 0$ at infinity and

$$\sigma_e \nabla \cdot \vec{n} = j_m \tag{48}$$

where j_m is the transmembrane current, σ_e is the conductivity, and $\vec{n}$ is the normal vector to the cell membrane. By further assuming that both the intra- and extra-fluids are neutral, the induced voltage waveform is:

$$V\left(\vec{r_0}\right) = \int \frac{j_m\left(\vec{r},t\right) dr}{4\pi\sigma_e \left|\vec{r} - \vec{r_0}\right|}, \tag{49}$$

where $\vec{r_0}$ denotes the location of the electrode and $\vec{r}$ describes the locations of the active membranes.

Since action potentials propagate slowly along the axonal branches of the cortical neurons (averaged 0.5 ~ 2 m/s [69]), the recorded active membranes usually do not fire simultaneously. As a result, the detailed geometry of the underlying neuron influences the shape of spikes. Following the computational model described in [70–73], a neuron is modeled as compartment elements including soma, axon hillock, and axonal branches. Different compartment elements may have different ion-channel densities, resulting in different transmembrane current profiles. An extracellular electrode only records those membrane segments within the recording radius, which is measured to be tens of μm [74]. Meanwhile, many cortical neurons have axonal branches up to hundreds of μm or more [69, 75–77], and membranes within the recording radius are simply modeled as one or few compartments with uniform ion-channel densities. As a result, the spike waveform can be expressed as the convolution of the transmembrane current profile and an implicit geometry kernel function:

$$V(t) = \int j_m(\tau)\, W(t - \tau)\, d\tau, \tag{50}$$

where $W(t)$ is the geometry kernel function determined by geometry properties of the recorded membrane segments, and the integration range is from $-\infty$ to ∞. From here on, this integration range is used as the default if not specified.

It is possible that the recorded neurons have distinguished spikes, which can be easily captured by a sorting algorithm. It is also possible that the recorded neurons have similar shape spikes, due to similar ion-channel densities. However, if the magnitudes of such spikes are also similar, the differentiation becomes challenging. A general spike sorting algorithm without using complex training procedures may fail to resolve such ambiguity and will inaccurately report a single, large, spike cluster, which affects the reliability and accuracy of the decoder. An approach of differentiating the associated kernel functions could be used to sort the similar spikes. Assume $W_1(t)$and $W_2(t)$ as the geometry kernel functions of two neurons with the same ion-channel population, the difference between the two spikes is:

$$\Delta V(t) = \int j_{\mathrm{m}}(\tau)\left[W_1(t-\tau) - W_2(t-\tau)\right] d\tau. \tag{51}$$

Small waveform difference appears if$\int\left[W_1(t) - W_2(t)\right] dt \approx 0$. Intuitively, the condition means the waveforms are identical, ignoring the skew of the activation of membranes.

Directly extracting the kernel functions from waveforms is difficult because of the uncertainties of neurons' geometry and transmembrane current profile. To differentiate the waveforms, we rewrite (51) in the frequency domain as:

$$F(\Delta V) = F(j_{\mathrm{m}}) F(W_1 - W_2), \tag{52}$$

where $F()$ denotes the Fourier transform. The condition of $\int[W_1(t) - W_2(t)]dt \approx 0$ is equivalent to $F(W_1 - W_2) \approx 0|_{\mathrm{f=0\,Hz}}$, which implies that the waveform difference caused by the geometry kernel functions has small contribution at lower frequency spectrum. A more quantitative explanation can be given by studying the derivative of $F(\Delta V)$ with respect to the frequency using (52).

The second term in (53), on the other hand, exhibits a strong frequency dependency within the dominant spectrum of$F(j_{\mathrm{m}})$. It can be expanded as

$$\begin{aligned} &F(j_{\mathrm{m}}) \tfrac{\partial F(W_1 - W_2)}{\partial f} = -j2\pi F(j_{\mathrm{m}}) \textstyle\int [W_1(t) - W_2(t)] t e^{-2\pi j f t} dt, \\ &= 2\pi F(j_{\mathrm{m}}) \textstyle\int [W_1(t) - W_2(t)] t \sin(2\pi f t)\, dt \end{aligned} \tag{53}$$

where the approximation holds good when kernel functions W_{i} are symmetrical.

As a summary, the waveform difference between similar neurons caused by geometry functions satisfies the following conditions:

$$\begin{cases} F(\Delta V) \approx 0|_{\mathrm{f=0\,Hz}} \\ \frac{\partial F(\Delta V)}{\partial f} \approx 4\pi^2 f F(j_{\mathrm{m}}) \int [W_1(t) - W_2(t)]\, t \frac{\sin(2\pi f t)}{2\pi f} dt \propto f. \end{cases} \tag{54}$$

In (54), $\frac{\partial F(\Delta V)}{\partial f}$ is linear to frequency $f$at low-frequency region, as $\frac{\sin(2\pi f t)}{2\pi f}$. The strong emphasis on frequency shows that $F(\Delta V)$ exhibits a higher frequency spectrum. As a result, a frequency-shaping filter with an emphasis on high-frequency spectrum may help to differentiates kernel functions.

The frequency-shaping filter also modifies the spectrum of noise; a further improved waveform differentiation can be achieved if it reduces the noise. A detailed analysis on noise is shown in "Frequency-Shaping Filter and Spike Derivative".

Frequency-Shaping Filter and Spike Derivative

In the previous section, we have discussed the differentiation of similar neurons using a frequency-shaping filter with emphasis on high frequency signal spectrum without considering the noise. The noise power spectrum, however, is clearly modified by the frequency-shaping filter. Intuitively, low frequency noise is reduced and high-frequency thermal noise is amplified. In general, the power spectrum of the input referred noise at the first-stage amplifier exhibits a decaying profile [61, 78, 79] and approximates as

$$N(f) = N_{\mathrm{neu}} + N_{\mathrm{e.e}} + N_{1/\mathrm{f}} + N_{\mathrm{therm}} \approx N_{\mathrm{f_{c1}}} \left(\frac{f_{\mathrm{c1}}}{f}\right)^{\alpha} + N_{\mathrm{thern}}, \tag{55}$$

where N_{neu} is the neuronal noise, $N_{\mathrm{e.e}}$ is the electrode-electrolyte interface noise, $N_{1/\mathrm{f}}$ is the flicker noise, N_{therm} is the thermal noise contributed by tissue impedance and transistors, f_{c1} is the high-pass corner frequency of the digital filter, and $N_{\mathrm{f_{c1}}}$ is the low-frequency noise at frequency f_{c1}. Except thermal noise, the remaining noise is featured at low frquency and assumed to have profile following $f^{-\alpha}$. Noise profiles vary among both of objects and recording systems, however, low frequency noise is typically dominant. As shown in Fig. 18, the noise power spectrums recorded from two objects are plotted. Within the signal band (hundreds of Hz to several KHz), a $f^{-\alpha}$ noise profile is observed.

Among various frequency-shaping filters, taking derivative is a simple one, which almost linearly emphasizes signal spectrum according to frequency. For a discrete time spike sequence, taking the derivative after the analog-to-digital converter (ADC) has the frequency response

$$H(f) = 2e^{j\pi f/s} \sin(\pi f/f_2), \tag{56}$$

where f_s is the sampling frequency of the ADC.

The effect of a frequency-shaping filter on noise can be quantitatively evaluated by the expression:

$$k = \frac{1}{N_0 \left|H\left(f_{\mathrm{spike}}\right)\right|^2} \int_{\mathrm{f_{c1}}}^{f_{c2}} N(f) \left|H(f)\right|^2 df, \tag{57}$$

where f_{ci} are the corner frequencies (3-dB attenuation frequency points) of the digital filter before feature extraction, f_{spike} is the center frequency of the spike signal, $N(f)$ is the estimated power spectrum of the noise, and N_0 is the integrated noise over passing band. If k is less than 1, the SNR increases further, which improves waveform differentiation.

After taking derivative, the noise spectrum density changes to:

$$N(f) \left|H(f)\right|^2 = 2\left[N_{\mathrm{f_{c1}}} \left(\frac{f_{\mathrm{c1}}}{f}\right)^{\alpha} + N_{\mathrm{therm}}\right] \sin^2(\pi f/f_{\mathrm{s}}). \tag{58}$$

For integer α, a closed loop expression of the integrated noise after derivative (N_1) can be obtained from (58). With a further assumption that the sampling frequency is sufficiently higher than the signal spectrum, the expression of N_1 could be generalized to noninteger α as:

$$N_1 \approx \frac{2N_{\mathrm{fc1}} f_{\mathrm{c1}} \alpha \pi^2}{(3-\alpha) f_{\mathrm{s}}^2} \left[f_{\mathrm{c2}}^{3-\alpha} - f_{\mathrm{c1}}^{3-\alpha} \right] + \frac{2N_{\mathrm{therm}} \pi^2}{3 f_{\mathrm{s}}^2} \left[f_{\mathrm{c2}}^3 - f_{\mathrm{c1}}^3 \right] \tag{59}$$

Combine (57) and (59), the parameter k that is used to quantify the modification to SNR due to the frequency-shaping filter is

$$k = \frac{\frac{N_{\mathrm{fc1}} f_{\mathrm{c1}}^{\alpha}}{3-\alpha} \left(f_{\mathrm{c2}}^{3-\alpha} - f_{\mathrm{c1}}^{3-\alpha} \right) + \frac{N_{\mathrm{therm}}}{3} \left(f_{\mathrm{c2}}^3 - f_{\mathrm{c1}}^3 \right)}{N_{\mathrm{fc1}} \frac{f_{\mathrm{c1}}^{\alpha}}{1-\alpha} \left(f_{\mathrm{c2}}^{1-\alpha} - f_{\mathrm{c1}}^{1-\alpha} \right) + N_{\mathrm{therm}} \left(f_{\mathrm{c2}} - f_{\mathrm{c1}} \right)} \frac{1}{2 f_{\mathrm{spike}}^2}. \tag{60}$$

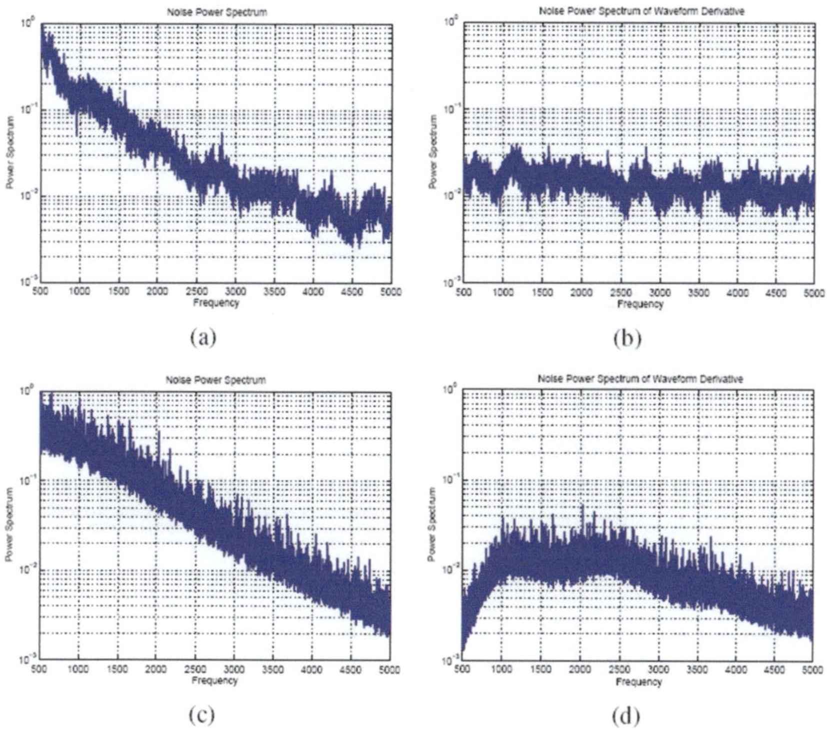

Fig. 18 Noise properties of recordings from two objects (500 Hz–5 KHz). (**a**) Noise power spectrum measured from a cat. (**b**) Noise power spectrum of the derivative (the same data in (**a**) are used). (**c**) Noise power spectrum measured from a monkey. (**d**) Noise power spectrum of the derivative (the same data in (**c**) are used)

The quantitative impact of frequency-shaping filter on noise is affected by the recording system and biological environment, and the typical values of _ we observe vary around 2 within the signal band as shown in Fig. 18. Here, we use $\alpha = 2$ to illustrate the analysis:

$$k = \frac{N_{\mathrm{fc1}} f_{\mathrm{c1}}^2 \left(f_{\mathrm{c2}} - f_{\mathrm{c1}}\right) + \frac{N_{\mathrm{therm}}}{3}\left(f_{\mathrm{c2}}^3 - f_{\mathrm{c1}}^3\right)}{N_{\mathrm{fc1}} f_{\mathrm{c1}}^2 \left(f_{\mathrm{c1}}^{-1} - f_{\mathrm{c2}}^{-1}\right) + N_{\mathrm{therm}}\left(f_{\mathrm{c2}} - f_{\mathrm{c1}}\right)} \frac{1}{2f_{\mathrm{spike}}^2} \approx \frac{\frac{f_{\mathrm{x}}^2}{f_{\mathrm{c2}}} + \frac{f_{\mathrm{c2}}}{3}}{\frac{f_{\mathrm{x}}^2}{f_{\mathrm{c1}}} + f_{\mathrm{c2}}} \frac{f_{\mathrm{c2}}^2}{2f_{\mathrm{spike}}^2} \tag{61}$$

with

$$f_{\mathrm{x}} = f_{\mathrm{c1}} \left(\frac{N_{\mathrm{fc1}}}{N_{\mathrm{therm}}}\right)^{1/2}, \tag{62}$$

where f_{x} is the frequency at which the noise spectrum approximately settles to the thermal noise floor. During the measurement, f_{x} varies according to the recording system and biological environment (varies from 6 to tens KHz in measurements).

In the case that the digital filter's low-pass corner frequency f_{c2} is designed smaller or comparable to f_{c2}, (61) can be simplified as:

$$k \approx \frac{f_{\mathrm{c1}} f_{\mathrm{c2}}}{2f_{\mathrm{spike}}^2} \approx \frac{2f_{\mathrm{c1}} f_{\mathrm{c2}}}{\left(f_{\mathrm{c1}} + f_{\mathrm{c2}}\right)^2} \le \frac{1}{2} \tag{63}$$

where the approximation holds well if the center frequency of the spike signal is close to the middle point of the filter's passing band.

As a summary, the spectrum of the recorded noise exhibits a decaying profile with respect to the frequency within the signal band. Therefore, an appropriate frequency-shaping filter could be used to further improve the SNR.

3.2.2 Spike Sorting Methods and Results

Quantitative and comparative sorting experiments are carried out to demonstrate the usefulness of a frequency-shaping filter. The sorting procedures and results are described in this section.

Spike Detection

In this work, spikes are detected with the nonlinear energy operator (NEO) [80–83]. NEO was formulated by Kaiser and is used for the amplitude and frequency demodulation and speech analysis. It is also effective to detect spikes that have localized high-frequency and instantaneous energy. With discrete time signal, NEO is

$$\psi\left(x(n)\right) = x^2(n) - x^2(n) - x(n+1)\,x(n-1). \tag{64}$$

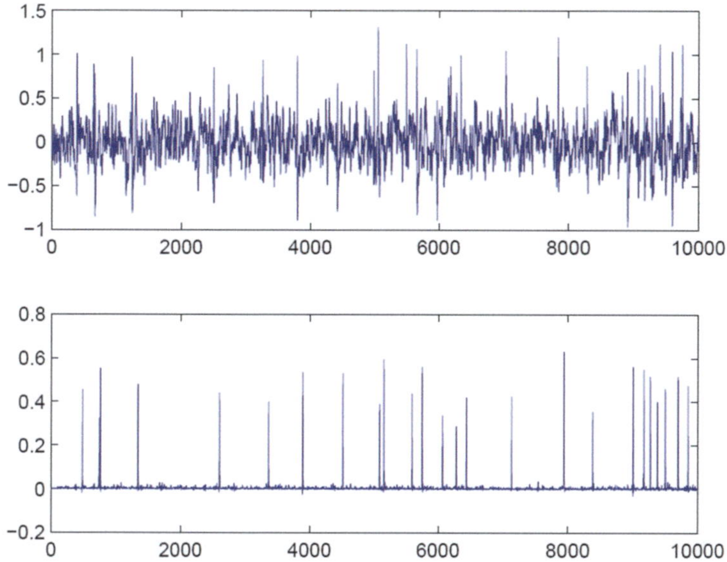

Fig. 19 The *upper trace* is the digitized neural signal. The *bottom trace* is the NEO's output

A typical processing result using NEO is illustrated in Fig. 19, where the upper trace is the raw neural signal and the bottom trace is the NEO's output. When high-frequency spike activity is present, the NEO generates a larger score.

Due to the improved differentiation of spike events (instantaneous high-energy events) and background activities (a mixture of low-frequency noise and thermal noise) by using

NEO, the detection threshold becomes a less sensitive parameter, which is shown by the ROC curves in Section 3.2.3. In this study, the threshold is set at three times of the averaged energy score corresponding to 1:4% error detection and identifying 99.5% spikes as a worst case in this data set. A more detailed description is shown in "Comparative Sorting Results Using Synthesized Spike Data".

Spike Feature Extraction

Two commonly used spike feature extraction algorithms, including PCA [45, 46] and spike peaks [84], are applied to demonstrate the improved sorting results by using the first derivative of spike waveforms.

All the algorithms are tested without performing interpolation. To quantify the performance of feature extraction algorithms alone, detected spikes with small interval are treated as overlapping events and ignored. In this design, the minimal spike interval is set to be 1.2 ms, which could generally represent well-isolated individual spike events. A Bessel type digital filter is applied to the identified spike waveforms before feature extraction. The corner frequencies (3-dB attenuation frequency point) of the filter are set to 250 Hz and 6 KHz with 60-dB out-of-band rejection. With

PCA-based feature extraction, the first two scores with the largest variance are used as the features. Compared with the feature extraction using spike peaks [84], a set with peaks of the spike derivative and spike height are used as counterpart features [67].

Spike Clustering

An example of extracted spike features using PCA is shown in Fig. 20. This example illustrates several challenges faced by a spike-clustering algorithm. First, the shapes of the clusters can be irregular and unpredictable. Second, the density and size of each cluster vary significantly. Third, the amount of data obtained is limited due to the acquisition and processing overhead. To overcome these challenges, we use a modified mean-shift clustering algorithm [62] to classify spikes.

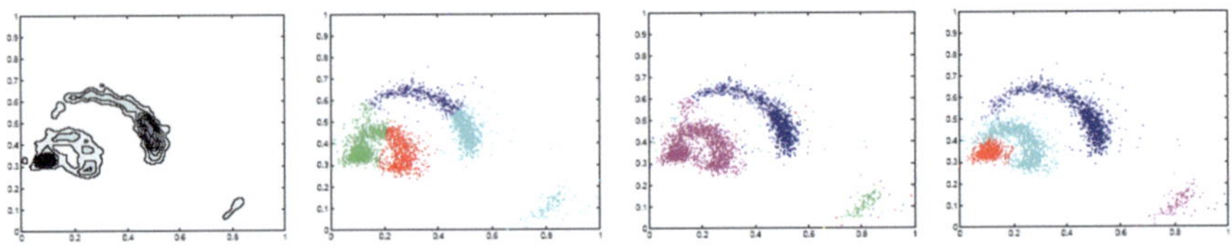

(a) (b) (c) (d)

Fig. 20 Comparative clustering results. (**a**) *Contour lines* of the spike feature map. (**b**) *k*-means clustering result with a four-cluster configuration. (**c**) Single linkage hierarchical clustering result with overspecified (50) partitions. (**d**) Modified mean-shift clustering result with manually specified kernel scope

The original mean shift algorithm is sensitive to the choice of the kernel radius [85–88]. Adaptive kernel radius estimation is reported in [89], which alleviates the problem to certain extent. In this work, the technique of adaptive kernel radius is applied, and the radius is chosen to be small. A side effect is that the algorithm tends to overpartition the data set into many subclusters, which is handled by a postmerging process. In the implementation, subclusters are merged based on two criteria. First, any modes with spike events less than 1% are forced to merge to the nearest mode. Second, subclusters are merged based on the density estimate at the boundary.

Comparative Sorting Results Using Synthesized Spike Data

Synthesized spike data from waveclusters (http://www.vis.caltech.edu/~rodri/) [53] are used as testing data set to compare the sorting results with both spikes and their derivative. The data are constructed using many averaged spike shapes compiled from recordings in the neocortex and basal ganglia. Simulated noise is further superimposed to the spike sequence to mimic background noise. Comparative results on challenging sequences, which contain three clusters, are presented in this section.

As described before, spikes are detected from the raw data using NEO. Receiver Operating Characteristic (ROC) curves are used to quantify the performance of detection algorithms, and comparative results are plotted in Fig. 21.

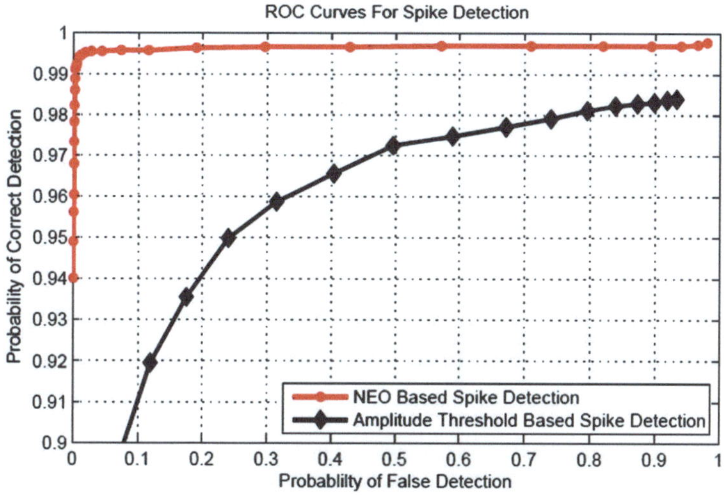

Fig. 21 ROC curves for spike-detection algorithms. The *red dotted curve* is the result from NEO-based spike detection. *The black diamond curve* is the result from amplitude-based spike detection

The "Probability of Correct Detection" used in Fig. 21 is defined as the ratio of the number of correctly detected spikes over the number of total neural spikes. The "Probability of False Detection" is defined as the ratio of the number of detected noise events over the number of total detected spikes. For example, if the number of neural spikes is 100 and the detector detects 120 spikes, among which 99 spikes are neural spikes and 21 events are noise, the "Probability of Correct Detection" is 99=100 and "Probability of False Detection" is 21=120. With NEO-based spike detection, the detection threshold hold is set to be three times of the RMS score, which corresponds to 1:4% error detection and detects 99:5% spikes as a worst case in this data set (detailed information is shown in Table 6). The detected spike with interval less than 1.2 ms are treated as overlapping events and ignored. It is worthy to clarify that a reduction of the minimal spike internal increases the error probability.

After detection, spikes are simply aligned according to the peaks before feature extraction. The alignment is performed without interpolation or waveform fitting. The results from PCA-based spike feature extraction algorithm are shown in Fig. 22(a)–(h) and (i)–(p). In Fig. 22(a)–(h), features are extracted from spikes. As a comparison, features extracted from the derivative of spikes are shown in Fig. 22(i)–(p), where a three-cluster configuration is clearly visible. The results of using waveform peaks-based feature extraction algorithms are shown in Fig. 22(q)–(x) and (y)–(af). In Fig. 22(q)–(x), the features are constructed from the peaks of

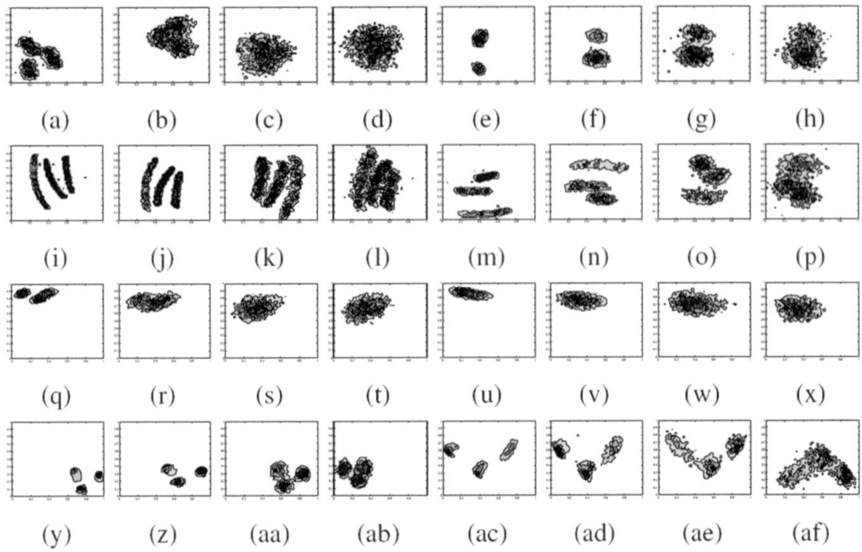

Fig. 22 Feature extraction results using spikes and their derivatives. (**a**)–(**h**) display PCA-based feature extraction result using spikes. (**i**)–(**p**) display PCA-based feature extraction result using the derivative of spikes (proposed). (**q**)–(**x**) display waveform peaks-based feature extraction using spikes. (**y**)–(**af**) display waveform peaks-based feature extraction using the derivative of spikes (proposed). All the algorithms are tested without performing interpolation. Overlapping spikes within 1.2 ms are ignored. Two-dimensional features are projected from a higher dimensional space

the spikes. Features using the peaks of spike derivative and spike height are shown in Fig. 22(q)–(x). Again, improved cluster isolation is observed when extracting features from the spike derivative.

Following [52], the Classification Matrix is used to quantify the performance of spike sorting algorithms. It is defined as

$$CM = \begin{vmatrix} & N_1 & N_2 & N_3 \\ C_1 & d_1 & r_4 & r_5 \\ C_2 & r_1 & d_2 & r_6 \\ C_3 & d_1 & r_2 & d_3 \end{vmatrix}$$

where N(1, 2, 3) represent spikes belonging to each neuron, while C(1, 2, 3) represent the clusters distinguished by the sorting algorithms. The ideal performance should have the sorting results perfectly matching the "ground truth," which corresponds to $r_k = 0, \forall k$. The sorting accuracy is intuitively defined as the ratio of the number of correctly sorted spikes over the total number of correctly detected spikes:

$$SA = \frac{\sum d_1}{\sum d_1 + \sum r_1}. \tag{65}$$

Quantitative sorting results of the sequences are displayed in Fig. 22. The sorting accuracies defined by (65) are listed in Table 5 as a performance measure to compare the results. In those sequences, improved performance is observed by sorting the derivative of spikes rather the original waveforms.

Table 5 Spike sorting accuracy comparison

Sequence number	1	2	3	4	5	6	7	8
Total neural spikes	3383	3448	3472	3414	3364	3462	3440	3493
Detected spikes	3382	3448	3470	3413	3361	3460	3438	3476
Noise events	0	0	0	0	0	0	2	52
Nonoverlapping spikes	3067	3179	3172	3415	3052	3203	3104	3139
PCA (SA) (%)	98	89	60	55	98	78	80	69
PCA + derivative (SA) (%)	98	98	98	96	98	98	94	90
Spike peaks (SA) (%)	34	34	35	34	36	38	36	36
Spike peaks + derivative (SA) (%)	98	97	96	95	98	98	93	92

Sorting Results Using Animal Data

Animal sequences are collected to test the performance of the proposed algorithm. An example with overlapped spike clusters is selected for demonstration. The sequence is recorded from the cat cerebral cortex. The sorting results are displayed in Fig. 23. In Fig. 23(a), the detected 1210 spikes are superimposed. Extracted spike features using the prespecified subset of samples implemented on chip are shown in Fig. 23(b). The discrete points in feature space are grouped into 8 clusters with different colors using off-line clustering. Less than 10% of noisy spikes and overlapping spikes are discarded, the rest are classified and plotted in Fig. 23(c). To

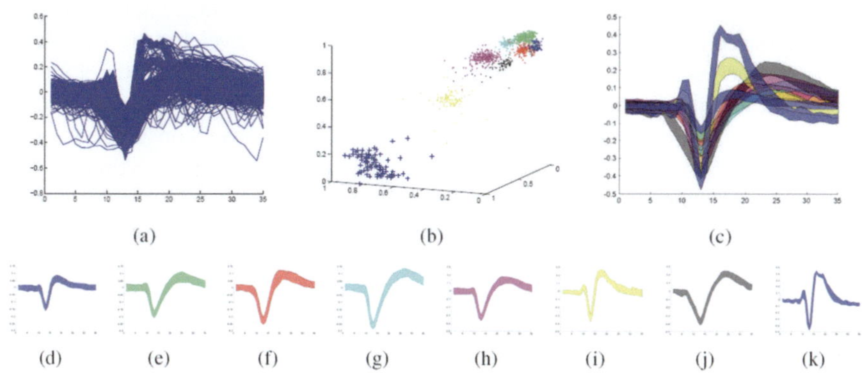

Fig. 23 (**a**) Recorded spikes from cat cerebral cortex are superimposed, (**b**) the extracted spike features using a subset of samples are plotted and grouped with a clustering algorithm implemented on PC, (**c**) the classified spike clusters are superimposed, (**d**)–(**k**) individual spike clusters superimposed in (**c**) are displayed. Spike clusters in (**d**)–(**g**) are plotted in a smaller vertical scale (–0.3, 0.15) compared with (**h**)–(**j**) in (–0.5, 0.3) and (**k**) in (–0.5, 0.5)

further quantify the validity of the classified spike clusters, superimposed clusters in Fig. 23(c) are individually plotted in Fig. 23(d)–(k).

3.3 Neuromuscular Current Stimulator with High-Compliance Voltage

These days, stimulus devices are being used in various biomedical applications such as cochlear implant, deep-brain stimulation, muscle stimulation, and retinal prosthesis. These devices are connected between two electrodes and deliver stimulation pulses to induce a contraction in muscle tissues or action potentials in neurons.

There are two different types of stimulators widely used in the applications. One is current stimulator and the other is voltage stimulator. The voltage stimulator produces voltage waveforms at the output resulting in a stimulus current determined by the load impedances, which consists of impedances of electrodes, electrode-electrolyte interface, and tissues. That load impedance is typically nonlinear and has a very wide range of variations, which implies it is very difficult to control the exact amount of charge delivered by the stimulators. This can cause a serious safety problem unless careful and proper control and protection methods are employed. On the contrary, the current stimulator directly supplies stimulation currents generated by a digital-to-analog converter (DAC). Therefore, the current does not change according to the variation of the load impedance and thus a precise control of the injected charge is achieved easily.

However, there is a main shortcoming for the current stimulators. Due to the high impedance of the electrode, which is getting more serious as the density of the electrode increases, high-compliance voltage is unavoidable around the output of the stimulus device. That requirement for high-compliance voltage naturally brings up a power-consumption issue because the power-supply level for the stimulator has to be increased if single power-supply level is used. Therefore, the stimulator should employ a mixed-voltage design to meet both high-compliance voltage requirement at the output and to minimize the overall power consumption. The power consumption of the entire stimulator can be reduced by making all the circuit blocks operate at the low-supply voltage levels, while the current output stage operates at high-supply voltage level for the high-compliance voltage. High-voltage process is inevitable for the transistors in output stages to make sure that the breakdown voltage of the output transistors is larger than the power-supply level. For the power-supply level of ± 12 V, the breakdown voltage of at least 30 V is necessary.

Figure 24 shows one example of circuit diagram of a current stimulator with a high-compliance voltage. The 4-bit DAC generates reference current according to the control signal assigned and this current is multiplied by the variable gain current mirror. The protection transistor interfaces circuits implemented with low-voltage transistors and the output stage implemented with high-voltage transistors. Cascode

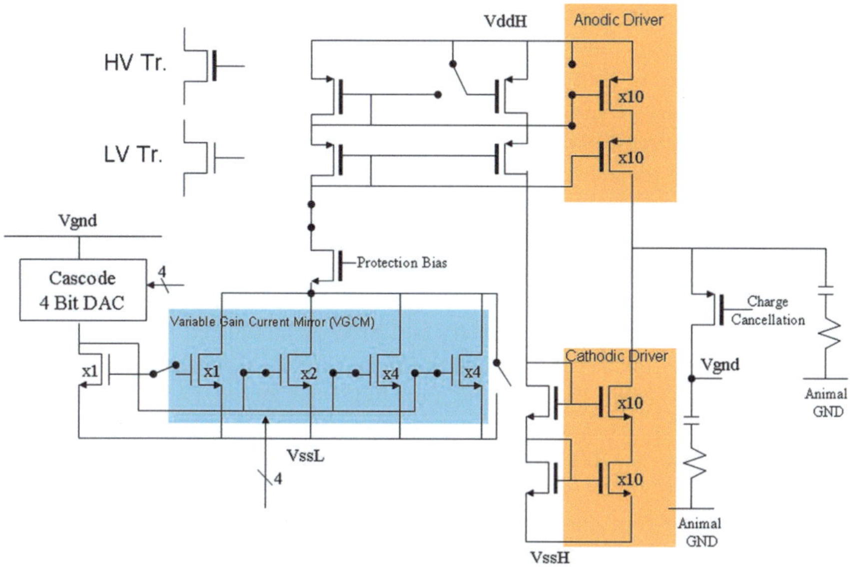

Fig. 24 The circuit diagram of exemplary current stimulator with high-compliance voltage

output structure is usually used to maximize the output impedance, which results in minimized anodic-cathodic tracking error. Charge cancellation transistor removes any residual charge after the stimulation is complete by shorting the stimulation electrode to the reference (counter) electrode.

3.4 Wireless Telemetry

Since 1960s, biomedical researchers have tried to use telemetry technology for collecting information such as organ temperature, blood flow, blood pressure, acceleration, EEG, EKG, etc from animals and human bodies in noninvasive method [90–93]. Such efforts have been accelerated with the invention of the integrated circuit that helped to minimize telemetry components and enabled bio-implantable devices. These days, the telemetry system is being more complicated and widely used in different neural prostheses such as artificial pacemakers, muscles stimulation, cochlear implants, and retinal prostheses.

Power and data should be simultaneously transferred to the implant, which is completely isolated from the external world after surgery. Instead of a tethering wire penetrating the skin that limits free running movement of the subject and inducing infection, transcutaneous telemetry using wireless inductive coupling has been preferred. At first, a single inductive coil was used to transfer data and power simultaneously in applications of artificial pacemaker, neuromuscular prostheses, cochlear implants, and retinal prostheses.

In the case of pacemaker, power delivery is not a problem because the device itself is of low power consumption and lithium iodine batteries can support a

10+ year lifetime. Therefore, the introduced programmable pacemaker focused on more data transfer than power delivery [94]. The data signal is wirelessly transmitted in the form of series of short RF pulses with a 175-KHz sinusoidal signal, a kind of amplitude shift keying (ASK), and it is demodulated in simple digital circuits after a passive front-end circuit [94]. The data rate of the device supports less than 1 kbps.

Neuromuscular prostheses and cochlear implants require higher data rate and power than artificial pacemakers. For neuromuscular prostheses, P. Troyk introduced frequency shift keying (FSK) [95], where two different frequencies of the carrier represent the binary data. The power signal serves as the data carrier, and an external Class-E amplifier sends the signal to increase the efficiency of power transfer. The data rate of this telemetry is 120 kbps modulated on 480 kHz that is power carrier frequency. For cochlear implants, several works have been discussing telemetry design [96–99], and amplitude shift keying (ASK) modulation, as an example, was used to transfer power and data to the implant in the ear [98]. The data rate is 400 kbps and it is modulated on 10-MHz power carrier frequency.

Since the research of the retinal prostheses actively began in the early of 1990s, various telemetry schemes to transfer power and data have been introduced. Unlike the above neuromuscular prostheses and cochlear implants, the retinal prostheses, which aim to support high-density electrodes (+1000), require high data rate and high-efficiency power transfer. At first, North Carolina State University (NCSC) designed the ASK modulation scheme that was coded using pulse width modulation (PWM) to control 100 electrodes [100]. This telemetry supports 250 kbps as a maximum data rate and the power carrier, which modulated with data signal, varies from 1 to 10 MHz. Frequency shift keying (FSK) was used to transfer power and data in [101]. 5 and 10 MHz used for two different carriers support a 2.5-Mbps data rate. This design, however, suffers from low power transfer efficiency due to the large frequency spacing and wide signal bandwidth.

Using single inductive coil is not enough to transfer high data rate and achieve high-efficiency power simultaneously. If the data rate is increased in a single-band telemetry scheme, the power efficiency will be reduced due to the quality factor (Q) of the power amplifier. In addition to this, it increases skin absorption of electromagnetic energy as the power carrier frequency goes higher. Thus, a dual-band approach, which requires dual inductive coils, was developed. Due to separate frequencies for power and data transfer, the trade-off between the power efficiency and the data rate is decoupled.

The MIT group developed ASK telemetry using the dual inductive coils [102]. Their targeted data rate is 700 kbps and the signal is modulated on 13.56-MHz data carrier. Also, 125 kHz is used for the power carrier frequency. Although this data rate is enough to control 15 electrodes, this data rate is low to support +1000 electrodes and the power interference, caused by inductive coupling, on the data coil still remains after filtering. Therefore, differential phase shift keying (DPSK) modulation scheme was introduced to eliminate the power interference on the data coil and to achieve high data rate [103]. This telemetry supports 1 Mbps and it is modulated on a 20 MHz carrier frequency. In addition, the 1-MHz power carrier is separately transmitted to the coil inside the eye and it is also used for the data clock in the receiver.

3.4.1 Power Telemetry

It has been demonstrated that the retina of patients affected with retinitis pigmentosa (RP) and age-related macular degradation (AMD) can be stimulated by charge-balanced biphasic current pulses to elicit visual perception [104]. The maximum amplitude of these current pulses can be as large as 1 mA at a pulse rate of 50–60 Hz to create strong visual perception, and the load on the stimulus varies around 10 K ohm at the stimulation frequency [105]. To sustain proper operation under the loading of a large number of stimulators, the power consumption of the implanted electronics is estimated to be more than 100 mW, which is transmitted wirelessly by an inductive link.

Power efficiency, defined as the ratio of the power delivered to load over the total power dissipation, is the most important parameter in designing power telemetry system. To optimize power efficiency under physical constraints, an accurate solution for power efficiency in terms of design parameters is desired. Earlier work in the literature provides power efficiency formulae based on linear models [106, 107]. Although such linear models provide design insights, they are oversimplified for the purpose of optimization. In this section, an expression of the power efficiency is derived. Based on derivations, the optimal component values are obtained and the fundamental upper limit of the power telemetry is concisely represented as a function of the quality factor (Q) and coupling coefficient.

The other important parameter of the inductive power link is the device size. Due to surgical and safety reasons, the size of the implanted coil is preferred to be small. An obvious trade-off is the low coupling coefficient, typically between 0.01 and 0.05, thereby reducing the power efficiency. To achieve comparatively high power efficiency with an implanted coil of small size, this section provides equations to calculate the Q using the geometry parameters. Based on calculation results, high Qcan be achieved by properly designing the coils, thus increasing power efficiency.

Power Telemetry Model

Power efficiency is a critical parameter of biomedical power telemetry circuits. Several studies have been conducted to estimate the power efficiency based on simplified linear models [106, 107]. Such linear models provide certain design insights; however, they are oversimplified for the purpose of optimization. In this section, an analytical approach is used to calculate the power efficiency and optimal component values to maximize the power efficiency are obtained analytically, followed by simulation results.

A general equivalent model of the power telemetry is illustrated in Fig. 25. Due to the rectifier and the time variant current loads, the equivalent model is nonlinear. The power efficiency of the telemetry system is defined as

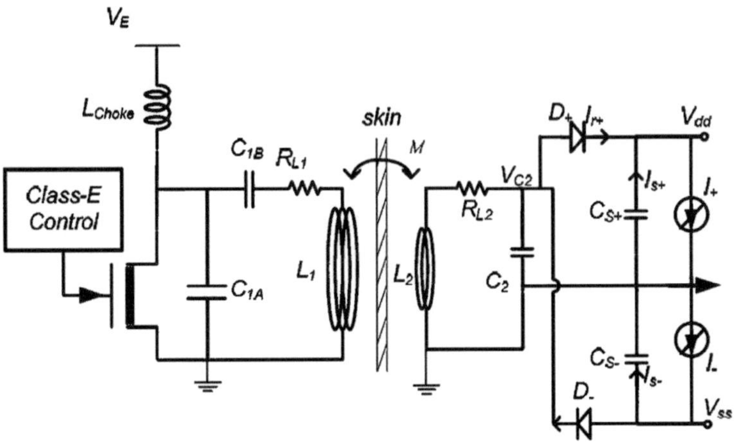

Fig. 25 A nonlinear equivalent circuit model for epi-retinal power telemetry. It consists of an external class E power amplifier, an inductive coil pair, a resonant amplifier, a rectifier, and time variant current loads

$$\eta = \frac{p_d}{P_d + P_1 + P_2}, \tag{66}$$

where η is the power efficiency, P_d is the power delivered to the implants, and P_1 and P_2 represent the external and internal power dissipation.

External power dissipation can be expressed as a function of the induced rms voltage on L_2 as below:

$$P_1 = \frac{V_{ind}^2}{\mu^2 \omega L_2 Q_1}, \tag{67}$$

where V_{ind} is the induced voltage on the secondary coil, μ is the coupling coefficient, L_2 is the inductance of the implanted coil, and Q_1 is the Q of the power amplifier.

Internal power dissipation due to diodes and coil's ESR is given by:

$$P_2 = \frac{1}{T}\int_0^T \left(i_{L2}^2 R_{L2} + 2 i_{r+} V_T\right) dt, \tag{68}$$

where i_{L2} is the current going through the implanted coil, R_{L2} is the implanted coil's ESR, i_{r+} and V_T are the current and voltage across the diode D_+.

During steady states, the rectified voltage should have constant DC component with small ripples and the current going through the diodes satisfies the following condition:

$$\frac{1}{T}\int_0^T (i_{r+} - i_+) dt \approx 0. \tag{69}$$

Therefore, the power dissipated by the two diodes can be calculated as

$$\frac{1}{T}\int_0^T 2i_{r+}V_T dt = P_d \frac{V_T}{V_{dd}}. \tag{70}$$

Since the diode turn-on voltage V_T (0.3~0.7 V) is one order less than the power-supply level V_{dd} (>10 V), the two diodes should have very short turn-on time per cycle. Thereby, it is reasonable to ignore the higher order harmonics of V_{C2} (voltage on the resonant capacitor C_2), and the amplitude of V_{C2} is approximately $V_{dd} + V_T$. Assuming V_{C2} with zero-phase, the relative phase information of V_{ind}, i_{C2}, and i_{L2} is illustrated in Fig. 26.

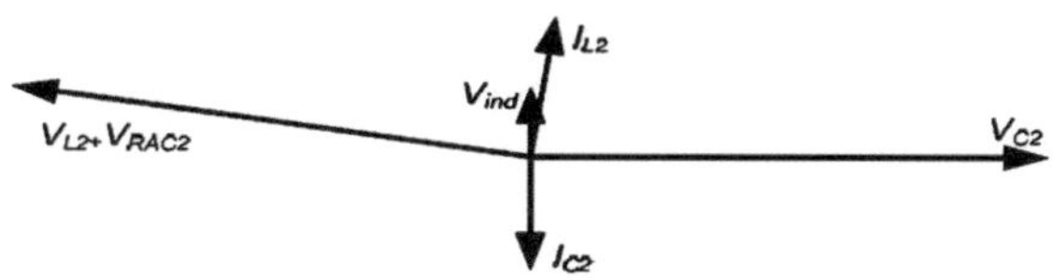

Fig. 26 Phase diagram for the resonant amplifier

Since the phase difference between V_{ind} and V_{C2} is close to 90°, we obtain:

$$\left|\frac{V_{L2} + V_{RL2}}{V_{C2}}\right| = \frac{\sqrt{V_{C2}^2 + V_{ind}^2 + 2V_{C2}V_{ind}\cos\theta}}{V_{C2}} \approx 1. \tag{71}$$

where V_{RL2} is the voltage on the implanted coil's ESR and θ is the phase difference between V_{ind} and V_{C2}. Therefore, the amplitude of i_{L2} is approximated as:

$$|i_{L2}| = \left|\frac{V_{L2} + V_{RL2}}{\omega L_2}\right| \frac{Q_2}{\sqrt{1+Q_2^2}} \approx \frac{V_{dd} + V_T}{\omega L_2}. \tag{72}$$

Combining (68), (70) and (72), $P_2$is computed as

$$P_2 = \left(\frac{V_{dd} + V_T}{\omega L_2}\right)^2 \frac{\omega L_2}{2Q_2} + P_d \frac{V_T}{V_{dd}}. \tag{73}$$

In (73), diodes are assumed to be ideal switches with turn-on voltages of V_T. To calculate the power efficiency, one has to find the relationship between the induced voltage and the delivered power to the implant, which is

$$P_d = i_{L2}V_{ind}\cos(\alpha_{L2} - \alpha_{ind}) - P_2 \approx \frac{V_{dd} + V_T}{\sqrt{2}\omega L_2} V_{ind} - P_2, \tag{74}$$

where P_d is the power delivered to the implant and α_{L2} and α_{ind} are the phases of i_{L2} and V_{ind}. When the resonant capacitor C_2 is properly chosen, the phase difference between i_{L2} and V_{ind} is close to zero such that the approximation in (74) holds.

Combining (66), (67), (73), and (74), the power efficiency is expressed as a function of P_d and V_{dd}:

$$\begin{aligned} \eta &= \frac{P_d}{P_d+P_1+P_2} \\ P_1 &= \frac{1}{\mu^2 Q_1}\left[\frac{(V_{dd}+V_T)}{2\omega Q_2^2 L_2} + \frac{2P_d^2 \omega L_2}{V_{dd}^2} + \frac{2P_d(V_{dd}+V_T)}{Q_2 V_{dd}}\right]. \\ P_2 &= \frac{(V_{dd}+V_T)^2}{2\omega Q_2^2 L_2} + \frac{P_d V_T}{V_{dd}} \end{aligned} \quad (75)$$

In (75), there is an optimal inductance for L_2 which minimizes the sum of P_1 and P_2 and therefore maximizes the power efficiency. The optimal L_2 is given by:

$$L_{2,\mathrm{OPT}} = \frac{(V_{dd} + V_T)}{2\omega P_d Q_2}\sqrt{1 + \mu^2 Q_1 Q_2}, \quad (76)$$

where $L_{2,\mathrm{OPT}}$ is the optimal inductance value in terms of maximum power efficiency. In applications where μ is fixed, (76) suggests a single optimal value for L_2. However, μ can be time-variant due to coil movements, where L_2 should be calculated with a chosen μ that characterizes real condition.

The corresponding C_2 is given by:

$$C_{2,\mathrm{OPT}} = \frac{2P_d Q_2}{\omega\,(V_{dd} + V_T)^2\sqrt{1 + \mu^2 Q_1 Q_2}} \quad (77)$$

where $C_{2,\mathrm{OPT}}$ is the optimal resonant capacitance as shown in Fig. 25. One noticeable fact is that V_{C2} has higher order harmonics. Therefore, the optimal value for C_2 slightly deviates from (77) and requires simulations to obtain the exact value. In this study, we assume that C_2 is properly chosen to maximize the gain of the resonant amplifier.

Applying (66), (67), (68), (69), (70), (71), (72), (73), (74), and(75), the maximum achievable power efficiency is obtained as

$$\eta = \frac{\mu^2 Q_1 Q_2}{2 + \mu^2 Q_1 Q_2 + 2\sqrt{1 + \mu^2 Q_1 Q_2}}, \quad (78)$$

Equation (78) sets the upper limit of the power efficiency, which is achieved once L_2 and C_2 are optimally chosen. Besides the high power efficiency, it is desirable to have low internal heat dissipation. As demonstrated in (75), P_2 is a decreasing function of L_2. If $L_{2,\mathrm{OPT}}$ is very small, the designer may trade the overall power efficiency with $L_{2,\mathrm{OPT}}$ by increasing L_2.

To give a numeric example, simulations of several key parameters, such as V_{ind}, P_2, and η, are compared with theoretic predictions, assuming the following specifications: $P_d = 150mW$, $V_{dd} = 12V = -V_{ss}$, $Q_1 = 100$, $Q_2 = 75$, $V_T = 0.7\,V$, $f = 2.5\,MHz$ (Figs. 27 and 28).

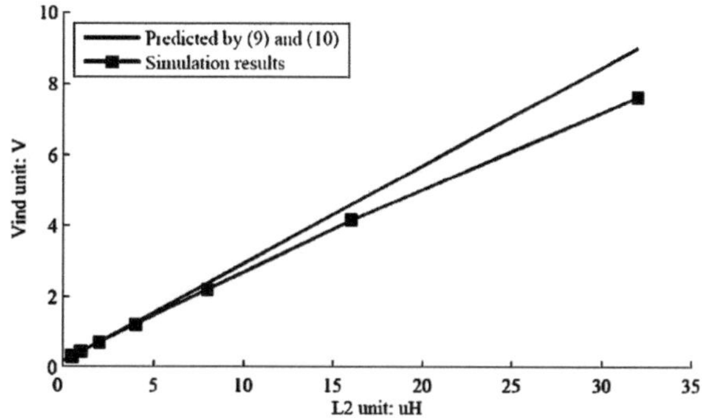

Fig. 27 Required induced voltage V_{ind} vs. implanted inductance L_2

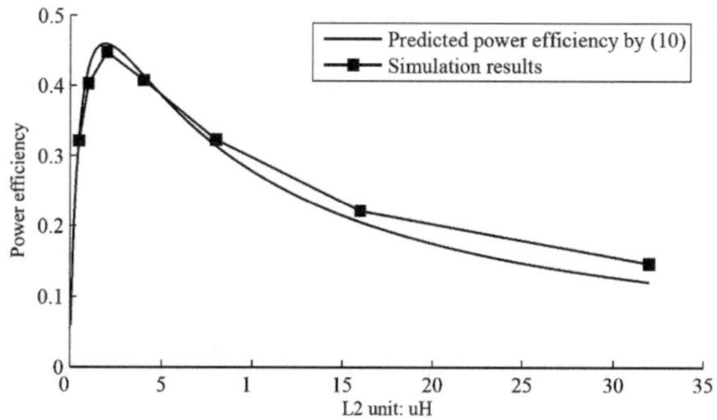

Fig. 28 Power efficiency η vs implanted inductance L_2

High Q Coil Design

In this section, an analytical model is presented to calculate the q of a coil with given geometry parameters. In order to reduce the AC resistance, and thereby the Q at high frequencies, litz wire is widely used in biomedical power telemetry to transmit both power and data, as shown in Fig. 29.

According to (78), it is necessary to have high Q coils to boost the power efficiency. In a distributed circuit model as shown in Fig. 29, the coil's Q, inductance, and ESR are approximately given by:

$$Q \approx \frac{2\pi f L_{AC}}{ESR}, \tag{79}$$

$$L_{AC} = \frac{L_{DC} f_{self}^2}{f_{self}^2 - f^2}, \tag{80}$$

$$ESR = \frac{R_{AC} f_{self}^4}{\left(f_{self}^2 - f^2\right)^2}, \tag{81}$$

where L_{AC} and L_{DC} are the AC inductance and DC inductance of the coil, R_{AC} is the series AC resistance of the coil without considering any parasitic capacitance, and f_{self} is the coil's self resonant frequency. As frequency approaches f_{self}, the Q quickly decreases to zero. Therefore, it is crucial to make sure that f_{self} is much larger than the carrier frequency. As derived in [108], f_{self} can be dramatically increased by applying a separation layer (h) between turns. In this study, we assume that $h$is properly chosen such that f_{self} is at least 3 times higher than the carrier frequency. In this case, the assumption of infinite f_{self} brings in less than 10% error.

In (79), (80), and (81), it is necessary to have L_{DC} and R_{AC} to analytically calculate the Q. Accurate expression for L_{DC} is readily found in the literature [109]. R_{AC} can be obtained analytically by computing power losses from skin effect and proximity effect. For litz coils, where each turn has multiple strands, the ratio of power losses by proximity effect and skin effect is approximately given by

$$\frac{P_{\mathrm{PRO}}}{P_{\mathrm{SKIN}}} \approx 6N_{\mathrm{t}}N_{\mathrm{s}}\beta\chi \tag{82}$$

where P_{PRO} and P_{SKIN} are power losses due to proximity effect and skin effect, respectively, N_{t} and N_{s} are the numbers of turns and strands, β is the ratio of the total conducting area over the cross section, and χ is a parameter to characterize a coil's geometry properties [110].

From (80), one can easily see that proximity effect induces much more power dissipation than skin effect in a litz coil. Power dissipation by proximity effect without considering any parasitic capacitance can be accurately described by Bessel Functions [111]. Under the condition that the radius of single strand is less than or comparable to the skin depth, the first-order harmonics of Bessel Functions are sufficient to provide an accurate result. In such cases, the power dissipation caused by proximity effect is given by

$$P_{\mathrm{PRO.Wind}} = \frac{\pi^2 d_{\mathrm{s}}^4 \mu_{\mathrm{r}}^2 \sigma N_{\mathrm{t}}^3 N_{\mathrm{s}} I^2 f}{128 A_{\mathrm{Wind}}} \chi\,(b/t)\,, \tag{83}$$

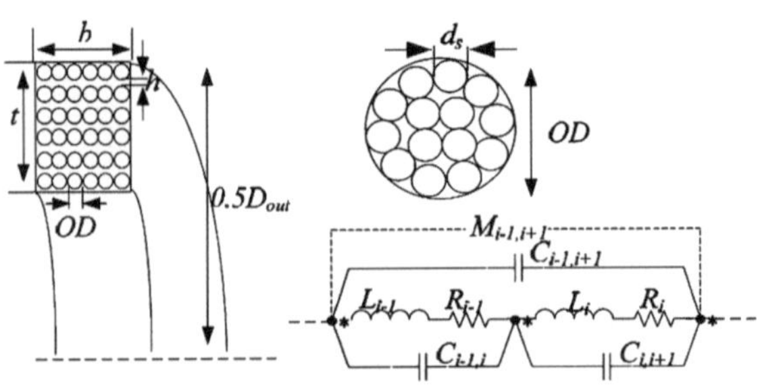

Fig. 29 t, b, D_{in} and D_{out} represent a coil's thickness, width, inner diameter, and outer diameter, respectively. OD and D_{s} are the diameters of single turn and single strand, respectively and h is the distance between layers. In the equivalent distributed model, L_{i} and R_{i} represent the self inductance and series resistance of one turn, and $C_{\mathrm{I,j}}$ denotes the parasitic capacitance between turn i and turn j

Table 6 Numeric values of X

b/t	0.125	0.25	0.5	1	2	4	8
X	0.41	0.62	0.86	0.96	0.86	0.62	0.41

where μ_r is magnetic permeability, I is the current passing through the coil, f is the current frequency, A_{Wind} is cross-sectional area of the coil, and χ is the parameter shown in Table 6. Therefore, the coil's ESR at RF is obtained as

$$R_{\mathrm{AC}} = R_{\mathrm{DC}} \frac{P_{\mathrm{DC}} + P_{\mathrm{PRO}}}{P_{\mathrm{DC}}} = \frac{N_t \left(D_{\mathrm{out}} + D_{\mathrm{in}}\right)}{2\sigma r_s^2 N_s} \tag{84}$$

where f_h is the frequency at which the AC power dissipation is twice the DC power dissipation. This f_h can be expressed as:

$$f_h = \frac{4\sqrt{2}}{\beta d_s \mu_0 \sigma \sqrt{\pi \chi A_{\mathrm{Wind}}}} = \frac{8\sqrt{2}}{\pi d_d^2 \mu_0 \sigma \sqrt{N_t N_s \chi \beta}}. \tag{85}$$

Combining Equations (79), (80), (81), (84), and (85), one can conclude that the coil has the highest Q when the operating frequency is equal to f_h. As shown in (78), the power efficiency is strongly dependent on the Q of coils. Therefore, it is desirable to have the operating frequency close to f_h for the purpose of high power efficiency. In real designs, however, the carrier frequency is always prechosen to meet other design specifications. As a useful guideline, the designer can move f_h to the operating frequency according to (85) for high Q. One preferable procedure is to change d_s such that f_h is close to the operating frequency with the A_{Wind} well fitting the geometry limitations. Once A_{Wind} and β are fixed, the Q is comparatively flat with different numbers of turns. Thereby, N_t, N_s, and even A_{Wind} can serve as additional parameters for designing appropriate inductance to achieve the maximum power efficiency as given by (78).

Design Example

As a design example to illustrate the high-efficiency power telemetry design, we assume the following specifications: $V_{dd} = 12V$, $V_{ss} = -12V$, $P_d = 150mW$, $V_T = 0.7V$, $D_{\mathrm{out}} = 6\,mm$, $D_{\mathrm{in}} = 4\,mm$, $b = 1\,mm$, $A_{\mathrm{Wind}} = 1\,mm^2$. The separation between two coils is $h = 10\,mm$ and the carrier frequency $f = 2.5\,MHz$. To guarantee a high self-resonant frequency, area efficiency is assumed to be $\beta = 0.4$.

To design highly efficient power telemetry, firstly, we maximize the coupling coefficient by sweeping the loop diameter of the external coil. The maximal coupling coefficient is calculated to be around 3%. Secondly, we design high Q coils at the operating frequency by choosing proper strand diameter d_s. For the implanted coil, since the cross-sectional area $A_{\mathrm{Wind}} = 1\,mm^2$ is given, the optimal strand diameter $d_s = 45\mu m$ is chosen according to (85). Q_2 is calculated to be around

50. Thirdly, we find the optimal inductance L_2 for the implanted coil using (76). With the assumption that the primary Class E power amplifier has $Q_1 \sim 50$(switch power loss is included), the optimal inductance L_2 is calculated to be $1.91 \mu H$ and achieved by 19 turns with 12 strands. The power efficiency η is computed to be 28% using (78).

3.4.2 Data Telemetry

DPSK Data Telemetry

For transcutaneous neural implants, it is necessary to transfer power and wireless data simultaneously to the implant inside body from the external world. For this purpose, a single-band approach, in which wireless data are modulated on power carrier and sent through an inductive coupling coil, has been applied to most existing neural implants [112–115].

However, some neural implants such as retinal prosthesis with high electrode count (hundreds to thousands) require high power efficiency as well as high data rate transfer [116]. If the data rate is increased in a single-band telemetry scheme, the power efficiency will be reduced due to the quality factor (Q) of the power amplifier. In addition to this, it increases skin absorption of electromagnetic energy as the power carrier frequency goes higher [117]. Thus, a single-band approach is not suitable to increase both data rate and power efficiency simultaneously.

In order to overcome these shortcomings of a single-band, a dual-band approach was introduced in [118]. Due to separate frequencies for power and data transfer, the trade-off between the power efficiency and the data rate is decoupled [119]. For the data receiver design, there are three main issues that should be taken into account. First, the interference caused by low power carrier frequency should be eliminated. Since the V_{pp} of power carrier is much larger than that of data carrier, the power spectral density (PSD) of the power carrier still remains high and interferes with the data signal at the input of the receiver even if a high-pass filter is used. Second, all functional blocks for the data demodulator should be fully integrated in as small an area as possible. External components should be minimal for neural implants because of implantable packaging reliability. Third, high data rate and high power efficiency should be achieved simultaneously through two inductive coupling coils. In the case of the retinal prosthesis with over 1000 channels, for instance, 2-Mbps data rate and 100-mW power delivery are required [118].

Noncoherent Differential Phase Shift Keying (DPSK) receiver, which is based on maximum 2 Mbps with 16 sampled and 1-MHz power carrier, was introduced for the purpose of eliminating the power interference [120]. The data demodulator, however, still has two outstanding issues; full integration and high data rate along with high power efficiency. Therefore, this design presents a fully integrated DPSK demodulator with increasing power efficiency twice compared to [120] while providing max 4 Mbps.

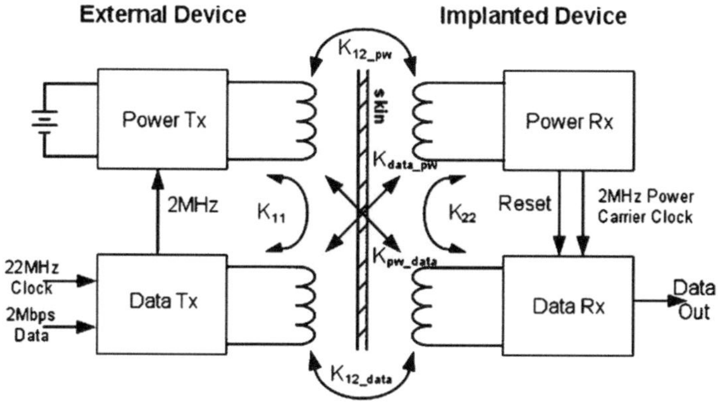

Fig. 30 A simplified diagram for dual-band telemetry

Receiver Architecture

Figure 30 shows a simplified dual-band wireless power and data telemetry. There are six different coupling coefficients, named K_{12_pw}, K_{12_data}, K_{11}, K_{22}, K_{data_pw}, and $K_{pw-data}$, between 4 coils. Among these six coefficients, the K_{12_pw} and K_{12_data} affect power and data transfer efficiencies, and the others, K_{11}, K_{22}, K_{data_pw}, and $K_{pw-data}$ determine the power interference to the data transmitter and receiver [117].

In order to eliminate the interferences caused by K_{pw_data} and K_{22}, differential phase shift keying (DPSK) scheme is used in this work. The main idea of DPSK is to detect the phase shift by comparing the current symbol with the previous symbol. If the 180° phase shift is observed, it indicates "1," and "0," if there is no phase shift. In more detail, the specific sample in the current symbol, $S_{n,m}$, where n is symbol state and m is sample order, is compared with $S_{n-1,m}$, and the result appears in the next sampling state, $S_{n,m+1}$. At that moment, the next sample, $S_{n,m+1}$, is compared with $S_{n-1,m+1}$, and the new result comes out on the state of $S_{n,m+2}$.

In order to sample each symbol regularly, the band-pass sampling theory, which samples the band-pass signal without down-mixing, is employed for the receiver system, and the equation is as follows:

$$f_s = \frac{4f_c}{(2n+1)} = 4kf_d \quad n = 0,1,2,\ldots \quad k = 1,2,3,..$$

where f_s, f_c, and f_d present sampling rate, data carrier frequency, and data rate, respectively [121].

Since the proposed receiver uses power carrier as data clock, higher the data rate is, higher is the power carrier frequency. In this design, 2-MHz power carrier is chosen to produce 2-Mbps data rate, which the retinal implants with 1000 electrodes require [118]. By the band-pass sampling theory, the sampling rate can be 8, 16, 24 MHz, and so on. Since the sampling rate is proportional to the number of switched-capacitor units, which occupies significant chip area, the minimum

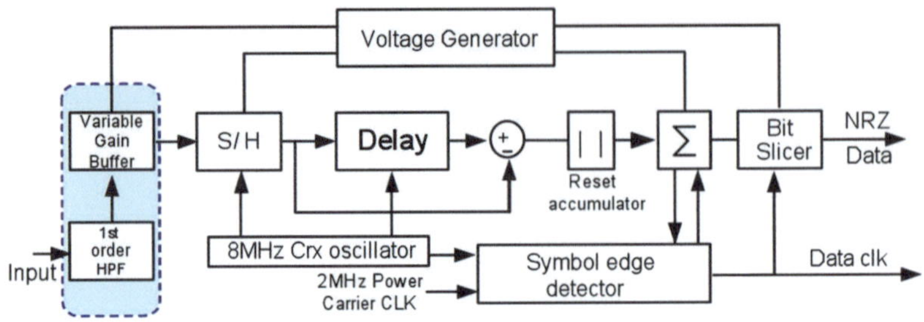

Fig. 31 The proposed data telemetry

8-MHz-sampling rate, which requires eight switched-capacitor units, is selected in this work. The data carrier frequency is 22 MHz, determined by 8-MHz sampling rate and 2-Mbps data rate.

Receiver Circuit Implementation

Figure 31 shows the block diagram of the proposed data receiver system. A 2-MHz power carrier, passing through a clock recovery buffer, goes into the symbol edge detector in order to generate the synchronized 2-MHz data clock, and the incoming modulated DPSK data pass through first-order high-pass filter (HPF) to eliminate the interference of 2-MHz power carrier. After that, the signal is amplified, sampled, compared, and integrated through the variable gain buffer, sample and holds, comparator, and integrator, respectively. The detailed circuit implementation, according to the signal sequence, is described in the following subsections.

The leftmost block in Fig. 32 shows the HPF and variable gain buffer. The capacitor and resistor connected to the noninverting input of the opamp functions as first-order HPF. The gain of the buffer, which is composed of a noninverting amplifier and a variable resistor, varies from 1 to 4. The operational amplifier of the buffer has 49.5 dB of gain and 41.2 degree of phase margin.

For sample and hold units, the switched capacitors are used whose circuit is described in Fig. 33. The digital switch control in the middle block of Fig. 32, which divides the 8-MHz oscillator clock, generates Ph1 and Ph2 with 50-ns pulse width and repeats every 125 ns. Figure 34 shows the timing diagram. Once Ph1 is high on $S_{n-1,4}$, the sampled charge is stored to the capacitor in the branch III of Array +. After one cycle, 500 ns, another sampled charge is stored to the capacitor in the branch II when Ph1 is high on $S_{n,4}$. The subtraction between the branch II and branch III occurs on Ph2 in $S_{n,4}$, and the comparator starts operating to decide whether the phase shift has occurred or not. If the 180° phase shift is detected, the subtraction result is finite, and the result should be zero unless there is a phase shift.

Quantitatively, the stored charge, Q_0, is $C \times V_0$ in the branch III on the Ph1 of $S_{n-1,4}$. After 500 ns, the stored charge, Q_1, is $C \times V_1$ on the Ph1 of $S_{n,4}$, in branch II, and then on the next Ph2 of $S_{n,4}$, the subtraction, $Q_0 - Q_1$, takes place. Since the

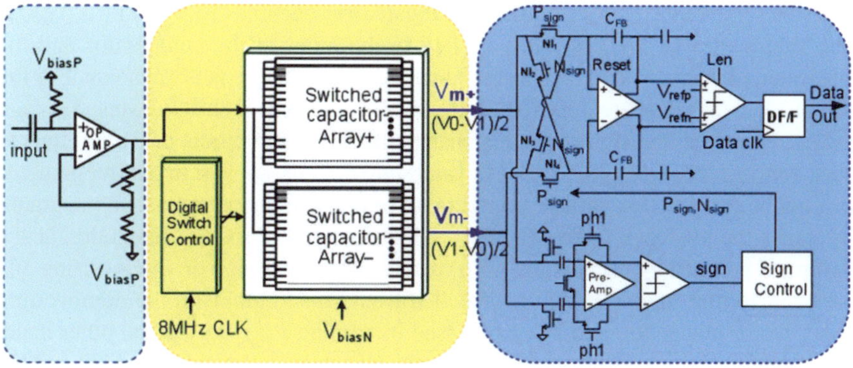

Fig. 32 The specific functional blocks in the data telemetry

Fig. 33 The switched-capacitor arrays

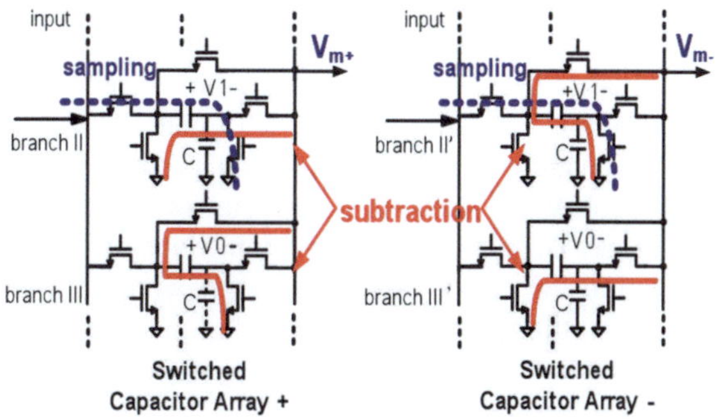

Fig. 34 Timing diagram for the switched capacitor

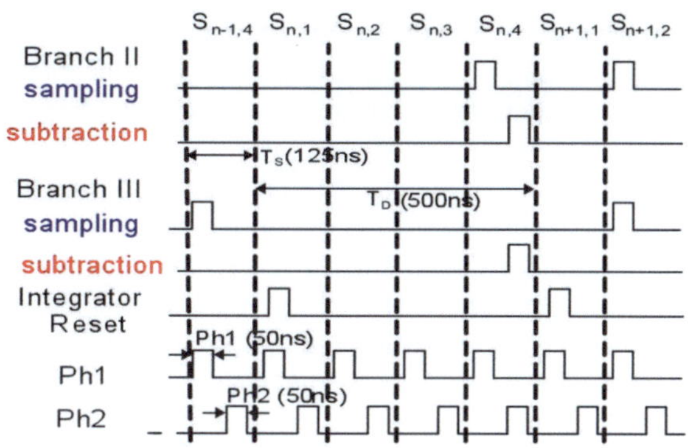

total capacitance is $2C$ at that moment:

$$2C \times V_{m+} = C \times (V_0 - V_1) \text{ and } V_{m+} = (V_0 - V_1)/2$$

In the same time, the V_{m-} of the counterpart array becomes $V_{\mathrm{m}-} = (V_1 - V_0)/2$. The values of V_{m+} and V_{m-} are applied to the input of the comparator and the integrator, simultaneously. While the absolute value of V_{m+} is always equal to the one of V_{m-}, the signs are opposite. In the case of 180° phase shift between $S_{n,\mathrm{m}}$ and $S_{n-1,m}$, therefore, the sign of V_{m+} is changed from plus to minus or from minus to plus, and vice versa for V_{m-}. If these values are directly applied to the integrator, it will not be possible to continue pumping up or down at the output of the integrator. By using the switches (NI_1, NI_2, NI_3, and NI_4) in front of the integrator in the rightmost box of Fig. 32, the integrator receives the same sign on the inverting and the noninverting inputs. If V_{m+} is switched from plus to minus in the noninverting input, for instance, NI_1 and NI_3 are closed by turning P_{sign} on. On the other hand, when the sign is changed from plus to minus, NI_1 and NI_3 are open, and N2 and N4 are closed by turning N_{sign} on, so that V_{m-}, which has plus sign, is connected to the noninverting input.

In order to control the switches, the comparator should observe the replacement of the sign and pass the result to the switches. As mentioned above, the V_{m+} and V_{m-}, which are smaller than the stored sampling V_0 and V_1, are sensitive to the offset of the comparator. Therefore, input offset storage technique, which consists of preamp and dynamic positive latch, is employed to reduce the offset [122]. If the sign is changed, the output of the comparator becomes low which causes N_{sign} to be on. Otherwise, the output keeps high which results in turning P_{sign} on.

Due to the comparator, the charge, which has consecutively the same sign, is stacked in the feedback capacitor C_{FB}. The integrator works for three out of four samplings and is reset for the rest of the sampling. The symbol edge detector generates the reset pulse.

Before the reset is high, the analog output of the integrator should be switched to a digital signal. Therefore, the bit slicer composed of dynamic positive latch is used. The latch compares the analog output with the reference voltages, 3.2 and 1.8 V, which are supplied by the DC generator on the chip, whenever L_{en} of the latch is high. NRZ data are recovered along with 2-MHz data clock through D flip-flop followed by the bit slicer.

The symbol edge detector that is proposed in [120] is applied in this work by reducing the number of D flip-flops in the frequency divider. 2- and 8-MHz clocks are fed into the digital control block that generates various control signals. While the preamble signal “010101” is detected, the symbol edge detector samples the magnitude of the integrator every 500 ns, stores the charge on a capacitor, and compares it with the previous charge on another capacitor. Once the magnitudes of these charges are the same, the pulse generator initiates the frequency divider to produce synchronized 2-MHz data clock.

Here, a beta-multiplier reference (BMR), which functions as a supply-independent bias, is employed instead of a bandgap reference as the human body temperature is relatively stable. The BMR consists of the peaking-current configuration [123] and is shown in Fig. 35. The 3.2-uA reference current is generated and V_{ref} produces independent 1.0 V on V_{DD}. The BMR has 2.8-mV/V and 103.7-dB static supply dependency and PSRR, respectively. The DC generator supplies

Fig. 35 The DC generator with BMR

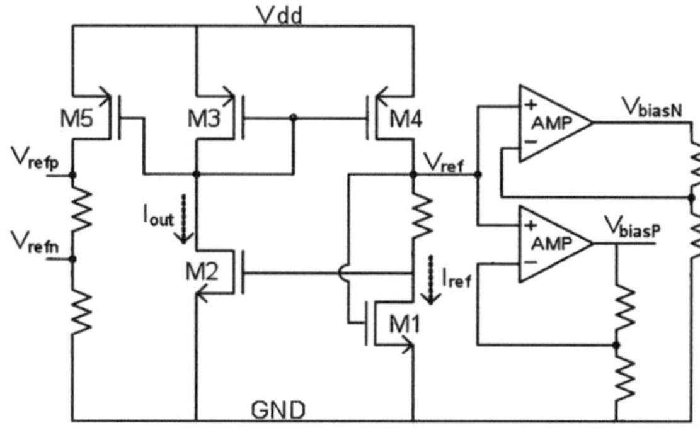

1.5 V_{biasN} and 1.8 V_{biasP} bias to the input buffer, the S/H units, and comparator, and 3.2 V_{refp} and 1.8 V_{refn} reference to the input of the bit slicer. Furthermore, the tail current of all amplifiers in the chip is copied from I_{ref}.

Ultra Wideband Telemetry

Recently, FCC assigned 3.1–10.6 GHz spectrums for unlicensed use of ultra wideband devices to support high data rates [124]. FCC has defined the UWB devices as one that has fractional bandwidth greater than 20% or absolute spectrum bandwidth greater than 500 MHz. Also, there are strict regulations on spectral shape and maximum power spectral density of UWB transmitter to prevent the interference issues. Impulse Radio UWB (IR-UWB) and multi-carrier UWB (MC-UWB) are the two general categories of UWB devices. The former is more appropriate for implantable neural-recording systems because the latter is generally suitable for multiple access communication systems and thus has rather complex transmitters consuming high power. IR-UWB uses simple short pulses for sending data and this makes the transmitter design very simple, small-area, and low-power. Also, it can provide enough data bandwidth due to its wideband nature. Theoretically, we can achieve data rate of 200 Mbps for 1028-channels neural-recording systems with IR-UWB. These features make IR-UWB the best candidate for the wireless telemetry of multichannel neural-recording systems.

Ultra-Short Pulse Generation

In this section, we will analyze pulse-generation schemes in both time and frequency domains. The method described here can be applied to different applications to meet the spectral mask of the UWB band. There are various methods to generate pulses. Among all methods using the delay-and-AND gate or delay-and-X-OR gate is the least complex way in CMOS-integrated circuit technology [125]. The delay unit can be realized using digital gates such as inverters, analog differential delay cells [126], and flip-flops or controllable capacitors [127]. A general scheme for such pulse generations is given in Fig. 36.

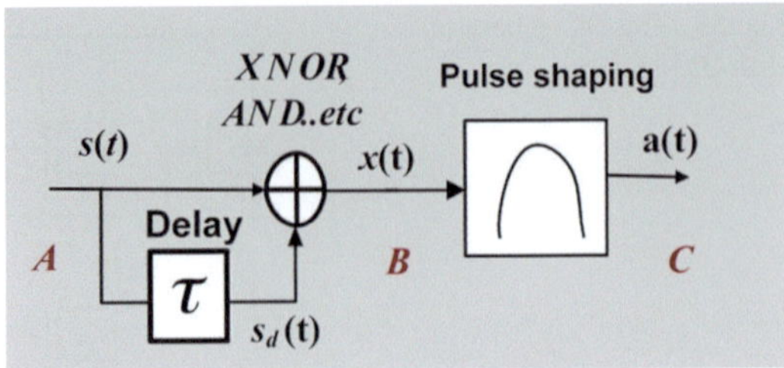

Fig. 36 An IR-UWB pulse-generation scheme

The data signal $s(t)$ and the delayed replica $sd(t)$ are passed through XOR gate or an AND gate to obtain a UWB narrow pulse $x(t)$ (e.g., $x(t) = S(t).S(t{-}\tau)$). A narrow band square wave can be represented by:

$$x(t) = \sum_{n=-\infty}^{\infty} g_{\mathrm{T}}(t - nT_{\mathrm{b}}) \tag{86}$$

where T_{b} is the bit period and

$$g(t - nT_{\mathrm{b}}) - \begin{Bmatrix} A & nT_{\mathrm{b}} < t \le (nT_{\mathrm{b}} + \tau) \\ 0 & (nT_{\mathrm{b}} + \tau) < t \le (n+1)T_{\mathrm{b}} \end{Bmatrix} \tag{87}$$

where A is the amplitude of the pulse and τ is the width of the UWB pulses obtained from the delay element as depicted in Fig. 37. Manchester NRZ data are preferred because they always have transition repeated every bid period T_{b}. Assuming there is a pulse repeated in every bit period, the Fourier series of the signal in (86) is given by [128]:

$$x(t) = \frac{A\tau}{T_{\mathrm{b}}} + \frac{2A\tau}{T_{\mathrm{b}}} \sum_{\mathrm{k}=1}^{+\infty} \frac{\sin(\pi k\tau/T_{\mathrm{b}})}{(\pi k\tau/T_{\mathrm{b}})} \cos(kwt) \tag{88}$$

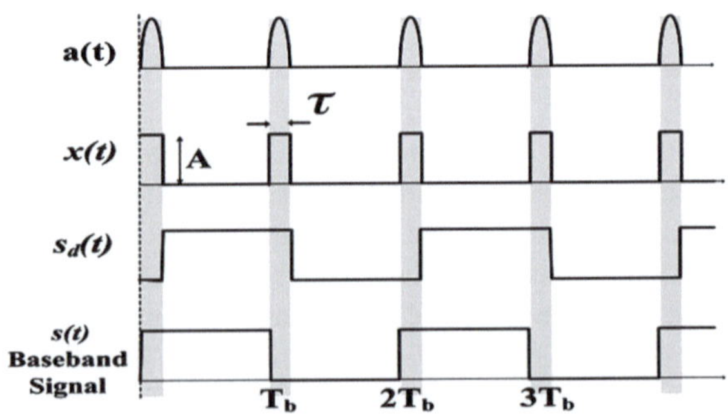

Fig. 37 Timing diagram for UWB pulse generation

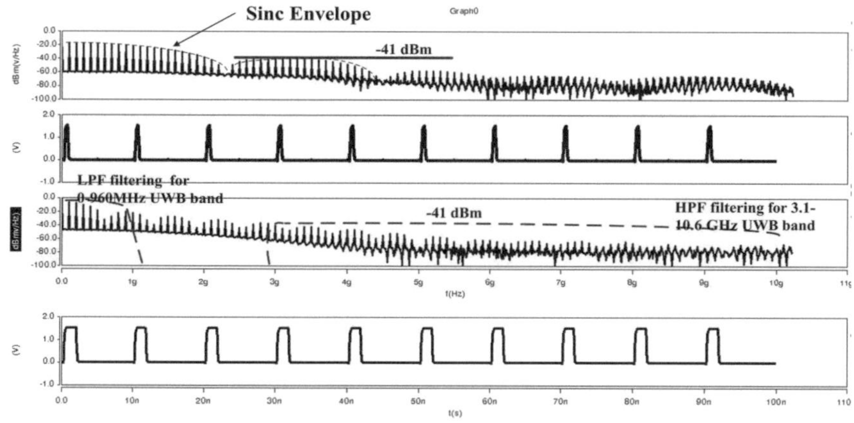

Fig. 38 Waveforms for UWB square waves

The signal includes a DC term and the fundamental frequency together with harmonic frequencies. As can been seen, a rectangular UWB pulse has a sinc envelope as a coefficient. The first zero of sinc(x) will occur at $x = \pi$, that is, $n = T_b/\tau$ in (88). It corresponds to a frequency of n/T_b or $1/\tau$. This first zero defines the distribution of the UWB signal in frequency domain as well as the number of the discrete spectral components. Figure 38 shows waveforms for UWB pulses with two different widths that are obtained from the circuit design. Spectrum plots are for the pulses that have the width of 2 ns (Fig 38a) and 500 ps (Fig 38c). Since the design targets a data rate of 100 Mbps, $T_b = 1/100$ MHz $= 10$ ns. If we select the pulse width $\tau = 2$ ns, the number of spectral components is therefore $T_b/\tau = 5$ (Fig. 38(b)). When $\tau = 0.5$ ns, the number of spectral lines is 20 (See Fig. 38). The distance between two spectral lines defines the data frequency (i.e., 100 MHz).

The best value for the delay is $\tau = T_b/2$. It results in the maximum power for the discrete spectral lines at the symbol rate frequency [128]. Using (88), the amplitude of the spectral lines become inversely proportional to their frequencies (amplitudes $= 2\,A/k\pi$). In the UWB transmission, the power spectrum of these discrete lines should be lower than that of the allowed spectral mask by the UWB regulations. That is why it makes it easy to have both A and the delay τ in (88) to control the power level of the signal such that it will fall within UWB spectral mask. Another observation is that once τ is arranged, increasing the bit period T_b yields a large number of spectral lines as illustrated in Fig. 38.

The square pulses $x(t)$ are passed through a pulse-shaping filter to decrease inter symbol interference (ISI) during transmission. Since square waves cause higher ISI, a pulse-shaping filter is used. A Gaussian pulse in UWB is generated after the pulse-shaping filter. In practice, one or more order high-pass filter is used to obtain such a shape [125]. If we look into (88), we see rectangular-shaped data extend over an unlimited frequency band. When a high-pass filter (HPF) is used, the UWB frequency from 3.1 to 10.6 GHz frequencies can be selected. However, for the UWB band of 0–960 MHz, a low-pass filter should be used. In the frequency domain, the distance between two discrete spectral lines is $1/T_b$=100 MHz. Note that the data information is contained in these discrete spectral lines at $1/T_b$ (Fig. 38.). When one of the UWB bands shown in Fig. 38 is transmitted, a narrow bandwidth band-pass filter (BPF) is used at the receiver site to obtain one of the spectral line for the symbol detection.

CMOS UWB Transmitter Design

There are two common transmitter categories used for UWB technology. The transmitters in the first category include a pulse generator and an up-converter that uses a mixer and a local oscillator (LO) to transfer the based band signal into UWB band. The transmitters in the second category consist of a pulse generator and a pulse-shaping circuitry only where the pulse directly falls in the UWB band. In those transmitters, there is no need for a mixer and LO that significantly reduces the complexity and power consumption of the transmitter [129]. Since the transmitter in this application does not require a multiaccess communication protocol and the power consumption is the most critical design specification, the second type of transmitter design techniques is used for our multichannel neural-recording system.

Figure 39 is the block diagram of the IR-UWB transmitter. The first stage of the transmitter is an encoder. The encoder enables the receiver to recover clocks directly from the encoded data and also to distinguish the data from different channels. The encoded data are then passed to a narrow pulse generator. The pulse generator circuit used is shown in Fig. 40. In this circuit, a pulse width is adjusted by control voltage *Vc*. Generated pulses are passed through the pulse-shaping filter to fit them into the FCC emission mask and to eliminate the transmission of unnecessary bands.

Unlike other UWB applications, power amplifiers are not necessary due to the low transmitted power and short distance range in neural-recording systems. Instead, a wideband-matching filter is used to regulate the transmitted power.

The transmitter can be configured to different pulse modulation schemes: on-off keying (OOK), pulse-position modulation (PPM), and binary phase shift keying (BPSK). A signal OOK_in is generated by passing the NRZ and Manchester NRZ baseband signals through an AND gate. As shown in Fig. 41, when the signal OOK_in is given to the pulse generator circuit depicted in Fig 6, $x(t)$ will be an OOK-modulated signal (Fig. 41). During the bit "1," a pulse is transmitted and

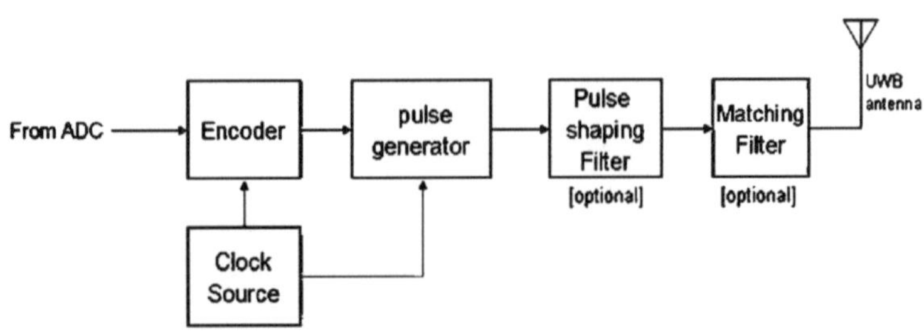

Fig. 39 Block diagram of IR-UWB wireless transmitter

meanwhile there is no pulse during the bit "0." The PPM signal is generated as follows. Manchester NRZ is passed through the pulse-generation circuit. The resulting narrow pulses are added (using XOR gate) with the OOK UWB to obtain PPM UWB in Fig. 41. As can be seen, the pulse position is different for bit "1" than that of bit "0." The bits are positioned such that to make the bit detection easier at the receiver site. To generate a BPSK signal, the pulse is inverted by 180° when the bit is "0."

4 System Design Examples

4.1 Recording: 128-Channel Wireless Neural-Recording System

A multichannel neural-recording system is used in neuroscience experiments to study complex neural networks of animals in their natural environments [130]. It is also a critical component in brain-computer interface used for cortical-controlled neural prosthetics, which has a wide range of applications such as upper and lower limb prostheses [131–134], bladder and bowel movement control for spinal cord injury (SCI) patients [135–136], respiration control for SCI patients [137], and hand grasping function restoration [138].

To support these applications, a neural-recording system has to meet challenging requirements imposed by the environment. First, it should be able to record a large number of channels simultaneously; and high-resolution recording can advance fundamental neuroscience studies and has the potential to improve the performance of neural prosthetic devices. Second, a wireless telemetry that transmits recorded neu-

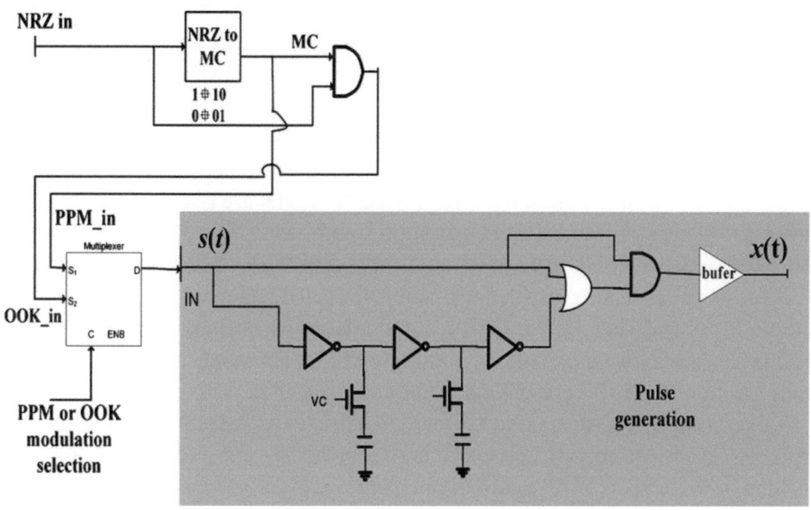

Fig. 40 Circuit used for pulse generation and modulation selection

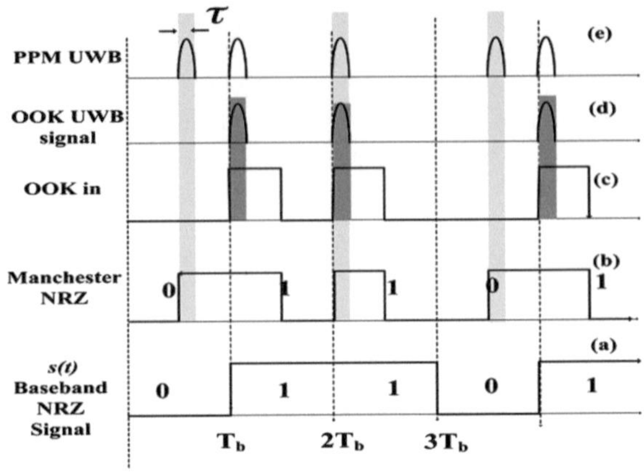

Fig. 41 Time diagram for PPM and OOK modulation

ral data is preferable, because tethering wires impose significant restrictions on the subjects and inhibit free movement in their natural environment. Third, on the fly processing of neural data is necessary to enable prosthetic devices functioning in a real-time. In addition, a fast processing capability removes the necessity of storing the large amount of raw data. Fourth, the specifications of the recording system should be designed with programmability and versatility that accommodate a wide range of bio-potentials in different applications. Finally, power consumption and chip area have to be optimized due to the limited space available at the recording site and the system should be powered wirelessly or operated on rechargeable battery.

Several neural-recording ICs previously reported in literature can support simultaneous multichannel recording [139–141], wireless data telemetry (spike information for 100 channels and raw data for one channel) [142], and on-chip spike detection [142, 143]. However, one of the major limitations of previous systems in the literature is that it allows recording from only a subset of the electrodes to be recorded simultaneously and transmitted to the outside mainly due to the limited bandwidth of the wireless telemetry. Many systems are not optimized and have unnecessary increase in power consumption and chip area. Although there have been several systems that have integrated functions of recording, processing (spike detection or spike feature extraction), and wireless telemetry [142, 144–145] with multiple ICs, a fully integrated IC with simultaneous recording, on-chip spike *detection and feature extraction,* and low-power wireless telemetry that can support raw data from more than 100 channels has not been reported.

To overcome the limitations of previous works, we present an integrated low-power IC that can record, process, and wirelessly transmit neural signals in real-time. This chip is able to simultaneously record and transmit raw data from 128 channels [146] wirelessly and serves as the core hardware for applications mentioned above when combined with additional components.

4.1.1 Chip Architecture

The chip has eight 16-channel front-end blocks. Each block consists of amplifiers, a multiplexer, and an ADC. The front-end blocks are followed by a data-serializing circuit, a DSP unit for spike detection and feature extraction, a digital MUX, an encoder, and a UWB transmitter (Fig. 42). On-chip bias generators provide DC bias voltages to the front-end blocks. At front-end, weak and noisy neural signals from electrodes are amplified by self-biased, fully differential preamplifiers and time-multiplexed by an analog multiplexer (Fig. 43). A second amplifier provides additional gain for the proper operation of the subsequent ADC. A successive-approximation-register (SAR) ADC is used to digitize the 16-to-1 time-multiplexed analog signal. The amplifiers are designed to have programmable gain and bandwidth to meet the requirements of various biological experiments. There is a trade-off between power and area because as we increase the number of channel per one ADC, the chip area decreases while the power consumption increases due to the increased multiplexer loading [147]. A careful analysis shows that the power and area product is minimized when 16:1 multiplexer is in the technology used [147]. A sequential turn-on method in the front-end blocks is utilized to save power. Two of the 16 channels are fully turned on at any given time, which leads to 71% additional power reduction when using this technique. For example, only the first and second channels are fully turned on when a sample from the first channel is being digitized (Fig. 43). At the next clock cycle, when the ADC accesses second channel, first channel is turned off and third one is turned on instead. In this technique, only the buffer to drive the analog multiplexer and ADC is sequentially turned on, and the preamplifiers, which draw very small current compared to the buffers, are always turned on.

For an electrode with resistance of 1 MΩ operating at 27°C with a 20-kHz recording bandwidth, the RMS noise voltage is 18.2 μV according to Nyquist's formula. In general, the magnitude of extracellular spikes is within 1 mV, resulting in a 35-dB signal-to-noise-ratio (SNR) of the input signal. With this estimated noise figure, we designed our ADC to have a 9-bit resolution after a conservative margin [147]. The sampling rate of one channel is chosen to be 40 k sample/s to avoid aggressive interpolation of spike samples, producing data rate as below.

$$40\text{ksample/s/channel} \times 128\text{channels} \times 9\text{bits/sample} = 46.08\text{Mbit/s} \qquad (89)$$

The 9-bit sampled data from eight front-end blocks are fed into digital data-serializing circuits or a DSP engine for spike feature extraction according to the mode setting. The chip can operate in one of the two modes. In streaming mode, all the sampled data from eight front-end blocks are fed into the digital data-serializing circuits and serialized by blocks resulting in 9-bit parallel digital data stream. This 9-bit data are expanded to 16-bit data to include 7-bit filler data for channel separation purpose at the receiver side and serialized again in the encoder. This final serialized data at the rate of 81.92 Mbps are then Manchester coded at the UWB transmitter to generate UWB pulses, which are transmitted through an off-chip

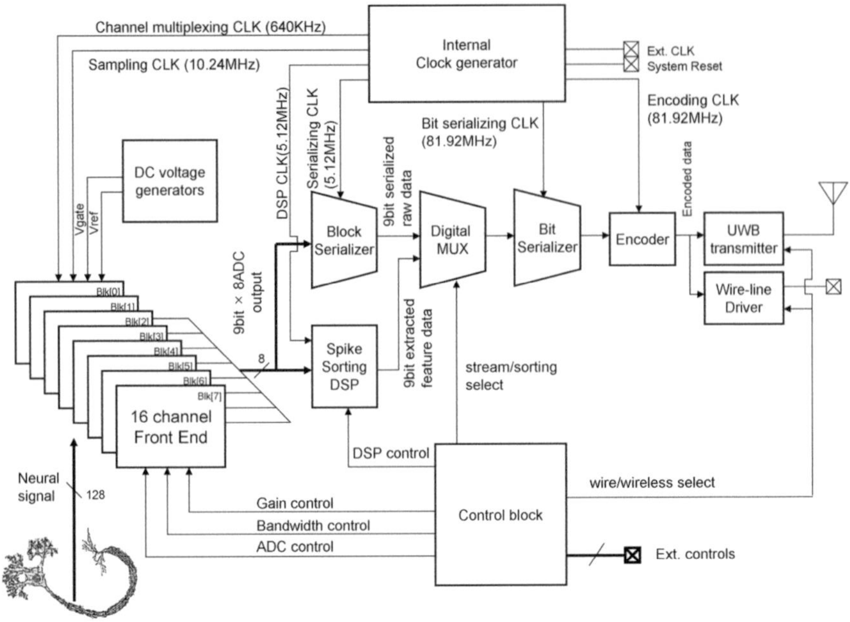

Fig. 42 Block diagram of the integrated neural-recording system

UWB antenna. A pulse-shaping filter is used to ensure that the emitted power spectrum of the UWB pulses is under the FCC regulation mask. When operating in DSP mode, a selected channel is connected to the on-the-fly spike feature extraction block and the features are transmitted wirelessly for further processing.

The clock signal of 81.92 MHz is applied to the chip externally and the on-chip internal clock generation circuit provides appropriate clock signals for each circuit block as shown in Fig. 42. In this design, we chose to use an off-chip crystal oscillator that can supply the required clock signal with a sufficient accuracy. The physical dimension of the commercial available crystal oscillator is small enough to be integrated with this chip in a hermetic sealing to form a higher-level system.

4.1.2 Front-End Block Design

The preamplifier (Fig. 44) uses AC-coupling at the input to reject the large DC offset occurring at the electrode-tissue interface [142]. The gain of the preamplifier is 40 dB, which is set by the ratio of feedback capacitances C_1/C_2. The second amplifier with a noninverting resistive negative feedback provides an additional gain of 17–20 dB according to external controls. C_1 and C_2 were chosen to be 20 pF and 200 fF resulting in differential input impedance of 16 MΩ at 1 kHz. The high-frequency roll-off of the preamplifier is configurable from 2 to 20 kHz in 16 steps by varying the load capacitance C_L. The low frequency roll-off is tunable from 0.1 to 200 Hz by changing the gate voltage V_B of the NMOS used as bias resistors. The

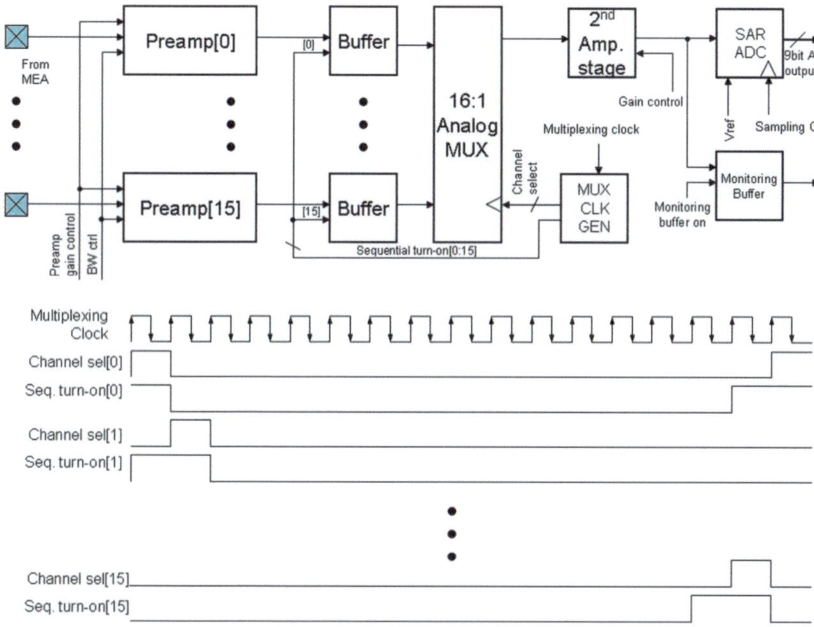

Fig. 43 Schematic of 16-channel front-end blocks (Sequential turn-on scheme is used to reduce the power consumption)

gate voltage provided by the bias generator is adjustable from 600 mV to 1 V by 50-mV steps. The adjusting range and the step size of V_B are designed to achieve the target programmability of low frequency roll-off in the presence of the process variations.

Operational transconductance amplifiers (OTA) have critical effects on the overall performance of the preamplifier. A high common-mode rejection ratio (CMRR) is preferred to suppress the 60-Hz power interference in the neural recordings [148], which makes the preamplifiers saturated and disables signal processing at the later stages of the system. A proposed fully differential self-biased OTA [149] shown in Fig. 44 enables a 90-dB CMRR and a 80-dB power supply rejection ratio (PSRR) with 4.9-μV_{rms} input referred noise integrated from 0.1 Hz to 20 kHz. To improve the common mode noise rejection, a fully differential output signaling was chosen. A common mode feedback (CMFB) circuit is not required because the output common voltage level is self-biased by the negative feedback leading to low power and small area of the preamplifier.

Each preamplifier and buffer draws 2 and 20.3 μA respectively, to drive the analog multiplexer when it is turned on by the sequential turn-on control signals, leading to an average current of 2.54 μA. The second amplifier draws 40.6 μA to drive 10-pF input sampling capacitance of the SAR ADC.

Care was taken to minimize the coupling noise from adjacent channels. All input signal lines that connect the inputs of the preamplifiers to bonding pads were

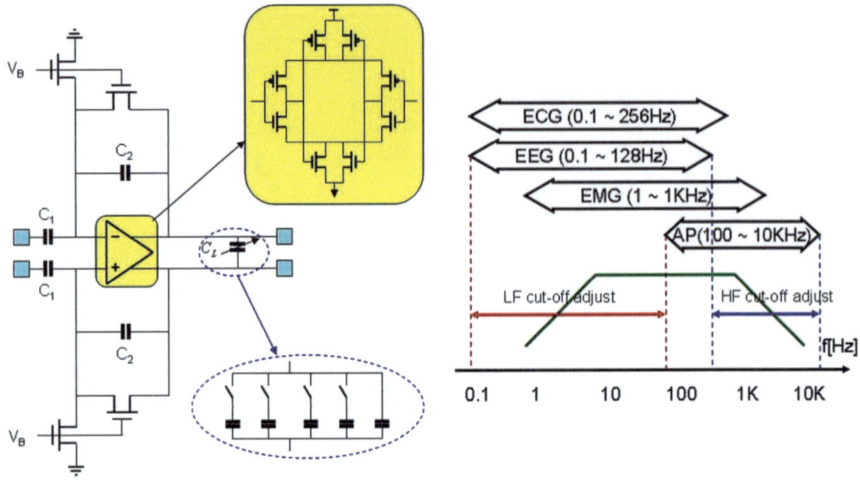

Fig. 44 Schematic of preamplifier and OTA used

shielded by quiet ground lines. In addition, the space between input signal line and shielding line was determined so that the metal parasitic capacitance is small enough not to affect the input impedance of the preamplifiers.

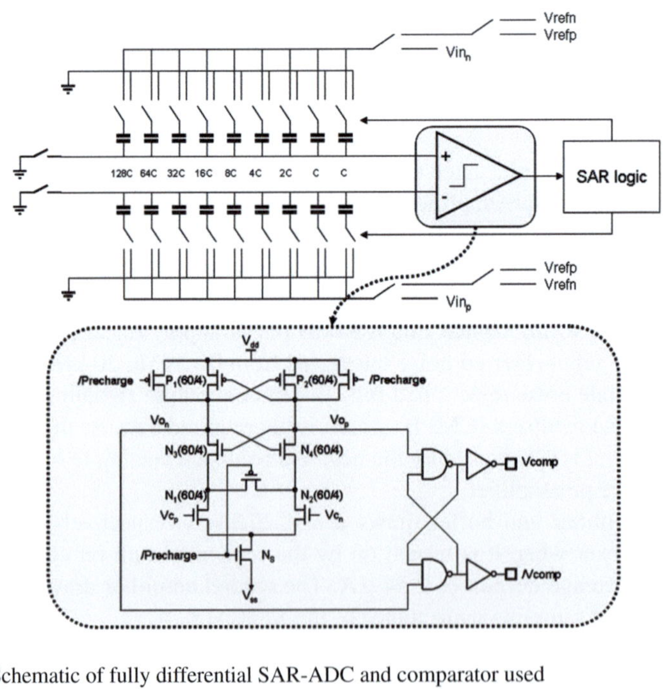

Fig. 45 Schematic of fully differential SAR-ADC and comparator used

SAR ADC is well suited for low-power and small-area applications because it requires minimal amount of analog circuitry [150]. The resolution of the ADC can be adjusted from 6 to 9 bits by external control signals. The total sampling capacitance of the SAR ADC is 10 pF and the power consumption is 14 μW, which is very small compared to other analog circuits. To reject common mode noise, the ADC is designed to digitize differential signal directly. The block diagram of the SAR ADC is given in Fig. 45. The on-chip DC voltage generators supply the reference voltages, and the voltage levels controlled by the external control signals are variable from 100 mV to 500 mV by 50 mV steps. The comparator of the ADC is based on track-and-latch comparator and its schematic is also shown in Fig. 45. The sizes of the transistors N1, N2, N3, N4, P1, and P2 can directly affect the offset of the ADC and therefore are chosen large enough to guarantee that offset is only a few least significant bits (LSBs) when minimum reference voltage is applied.

4.1.3 Neural-Signal Processing Engine

The recorded neural signals by a single electrode may contain spikes from multiple-firing neurons. Spike sorting is the procedure to attribute spikes to individual neurons and recover the encoded information in the neural signals. Spike sorting is challenging due to several factors, such as the presence of unresolved neuronal activity and similarity in recorded shapes. Training may be required when the electrodes move, and this could be frequent in certain applications [151]. The frequent training over a large amount of recording channels significantly increases the computational power and memory required [152]. Thus, for an implantable neural-recording system where the hardware needs to be integrated and miniaturized, power and area efficient algorithms are advantageous.

Figure 46 shows the overview of the algorithm that enables on-chip spike feature extraction. The 9-bit ADC output containing the time-multiplexed neural signals is fed to the spike detector, where spikes are detected by using a nonlinear energy operator (NEO) [153, 154]. This method is particularly effective when detecting a spike with localized high frequency and instantaneous energy. According to [155], a derivative-based frequency-shaping filter significantly attenuates the low-frequency noise and helps differentiating similar spikes from different neurons. As a complementary approach to principle component analysis, spike feature extraction algorithm based on informative sample set was first reported in [156] to identify uncorrelated local features. This concept requires only a subset of samples containing the necessary information to cluster the data. Intuitively, a sample is considered to be informative if the superimposed spikes can be classified into multiple clusters by evaluating the sample alone. Combining derivative operation and sample selection, improved sorting results were demonstrated in [156]. As a preliminary implementation to our feature extraction algorithm [157], the peaks of the original spike waveforms and maximum and minimum values of their first derivatives are used as the features to classify spikes. The choice of this simplified sample set for implementation is based on three reasons. First, it requires small computation and little memory [157]. Second, samples during the fast transition period frequently exhibit high information score [156]. Third, obtaining these three features requires no training.

Figure 47 shows the block diagram of the proposed on-the-fly spike-detection and feature extraction engine. The feature extraction engine is operated in event basis. If a spike event is recognized, the feature generation engine is reset and then extracts features. In this implementation, a subset of samples include the positive and negative peaks of the spike derivative and spike heights are chosen as the features for clustering. The final feature scores along with event-timing information are packed and output through the coder and packer. These procedures, including detection, filtering, feature extraction, and coding, are operated in parallel to meet the real-time requirement.

4.1.4 UWB Telemetry

Although there are several communication standards for biomedical applications, most of them cannot provide enough bandwidth for simultaneous recording from more than 100 channels. For example, MICS band, which is allocated for the unlicensed use of implantable devices, allows only 300 kHz for data transmission. Other wireless technologies such as Wi-Fi and Bluetooth cannot be used directly due to the power and area constraint of the implanted device. Therefore, there is a need for higher-bandwidth data transmission telemetry that consumes low power and occupies less physical area.

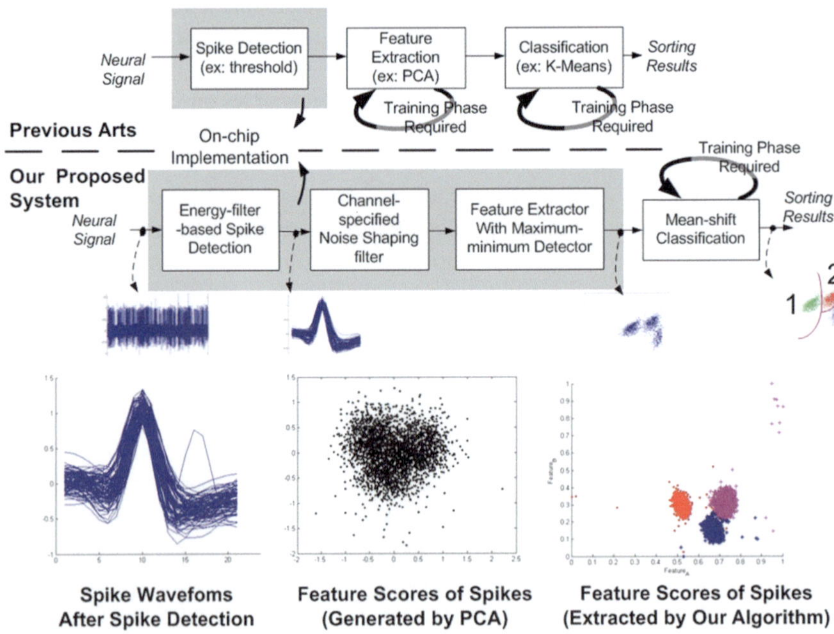

Fig. 46 Proposed spike sorting algorithm

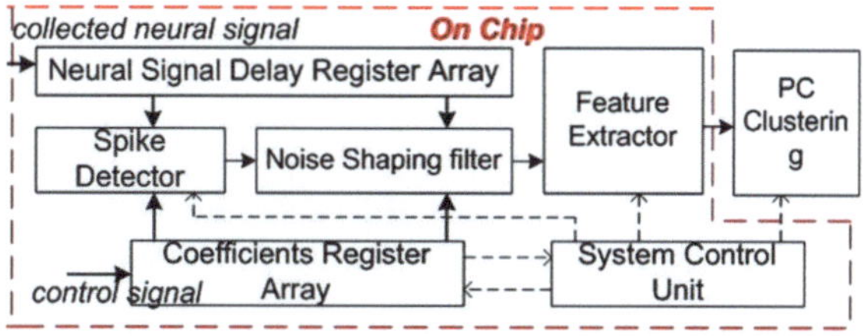

Fig. 47 Block diagram of an on-line spike sorting engine

UWB is a recently formed wireless technology that is used for low/high data rate wireless personal area network (WPAN) and short-range applications. It has the widest bandwidth (3.1–10.6 GHz) among all technologies and the smallest emission power density (–41.3 dBm/MHz), providing an opportunity for a wideband wireless telemetry for neural-recording systems. Although the reported receiver designs of UWB systems have been consuming more power than those of narrow band communications [158], in our neural-recording applications the implanted device requires only a transmitter and the receiver is outside the body, unlike the other short-range applications. Hence, the transmitter power consumption and complexity have been traded-off with that of the receiver as the receiver is located outside and its power consumption and size are not crucial. This greatly simplifies the complexity of the implanted telemetry design leading to reduction in power and area.

In this system, impulse radio-based UWB (IR-UWB) is employed as the targeted application does not require a multiple access communication, which requires complex transmitters consuming high power [159–160]. In IR-UWB, short pulses are generated for sending data, and both the center frequency and bandwidth of the pulse do not have to be very accurate. Such a process is simple to design in CMOS technology resulting in a very simple, small-area, and low-power transmitter design while providing enough data bandwidth because of its wideband nature.

Figure 48 shows the block diagram of the UWB transmitter. The sampled and serialized data are Manchester encoded first and then either on-off keying (OOK) or pulse-position modulation (PPM) is employed to generate short pulses. The redundant fillers are set to all "1" for channel, and all "0" for other channels in case of PPM, and the fillers for channel are set to 7-bit "logic low" of Manchester coding in case of OOK (Fig. 49). Therefore, the final maximum pulse repetition rate at the UWB transmitter output is twice the output data rate, 163.84 M pulses/s in case of using PPM and the same as the output data rate, 81.92 M pulses/s for OOK. A short pulse generator is based on a simple edge detector [161] and the circuit implementation is shown in Fig. 48. The pulse width can be controlled from 180 to 980 ps by an external control voltage V_C for adjusting the transmitting power. The generated short pulses are then passed through a high-pass filter to remove low-frequency component of the pulses and for the transmitting power to fit under the FCC emission mask. The filtered pulses are finally fed into the off-chip UWB

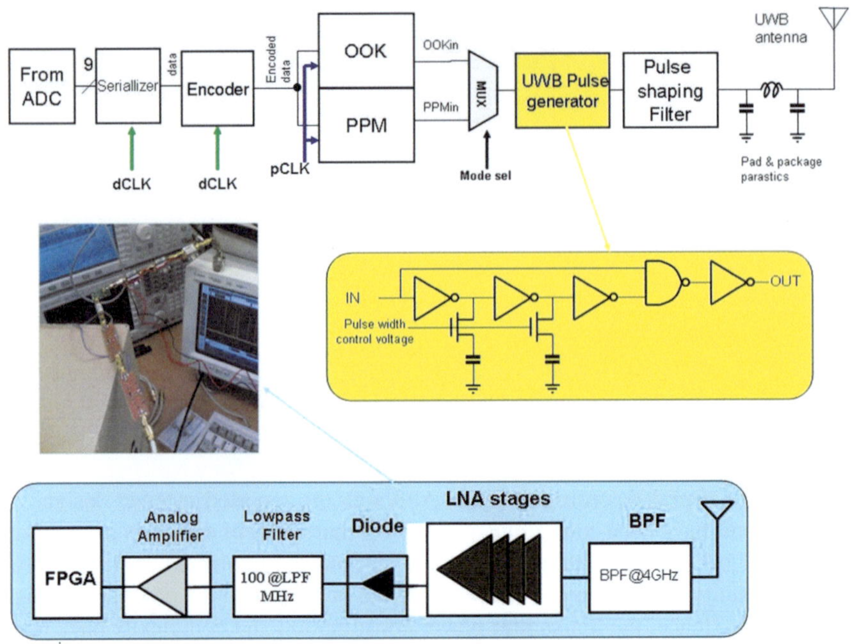

Fig. 48 Block diagram of on-chip UWB Tx, and UWB Rx with photo of its implementation using off-the-shelf components

antenna. Because of the high-frequency nature of the UWB, the size of the antenna is small enough to be implantable or to be carried by the animals.

The UWB receiver was built from off-the-shelf components (Fig. 48) since it is outside the biological objects. The signal received by an UWB antenna is passed through 1-GHz band-pass filter (BPF), whose center frequency is 4 GHz, to remove the interfering signals from other narrowband wireless devices. The signal is then amplified by the low-noise amplifier (LNA) stages, which plays a critical role in determining the maximum distance between the implanted transmitter and receiver outside the biological subject. An RF diode and a low-pass filter (LPF) down convert the UWB signal to a low-frequency one by performing envelope detection, and the digital data are finally recovered by FGPA (Fig. 48).

4.1.5 Test Results

The IC was fabricated in a 0.35-μm 4M2P CMOS process and the chip size is 8.8 × 7.2 mm^2 (Fig. 50). The measured power consumption of the entire chip is 6 mW from ±1.65 V when operating in streaming mode. The 8 front-end blocks consume 50% of the total power. The UWB TX consumes 1.6 mW with PPM modulation. The power consumption of the DSP block is only 0.1 mW. Waveforms at the output of amplifier, spike sorting, UWB TX, and UWB RX were monitored by the built-in test circuits (Fig. 51).

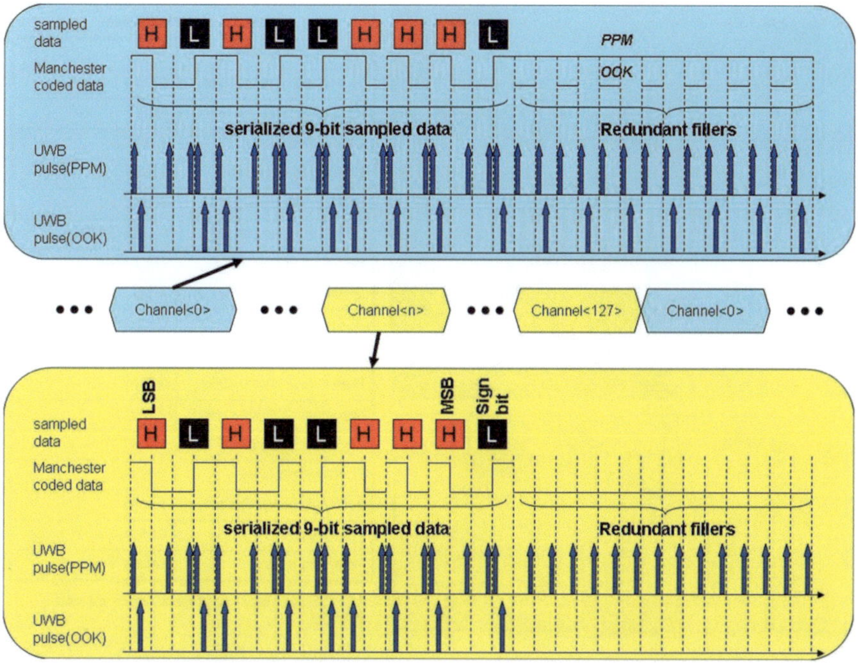

Fig. 49 Data format of the UWB transmitter

4.1.6 60-Hz Power Interference Issue

60-Hz power interference is a critical issue in biomedical instrumentation. It often makes the amplifier saturated that no further processing is possible. The power line interference occurs mainly due to capacitive coupling between the power line and any node of the recording system [162]. Power line interference caused by magnetic couplings is easily solved by reducing the size of the loop [162], which induces interference. A lot of studies have been made [148, 163–166] to obtain an equivalent circuit model to understand and eventually find out a practical solution for neural-recording applications. However, most of the work was focused on noninvasive recording system such as electrocardiogram (ECG), electroencephalography (EEG), and electromyography (EMG). The recording environment of implantable neuroprosthetic devices is quite different from those of noninvasive recordings and thus a new model to analyze the phenomena is required. It is important to devise practical solutions rather than to put a notch filter at every recording channel because the notch filter to eliminate the 60-Hz interference usually takes up large area and power consumption for implantable devices. The model also should consider the multichannel effect since most of the neuroprosthetic devices target high-density recording. A complete analytical model is developed [167] and can be used for that purpose.

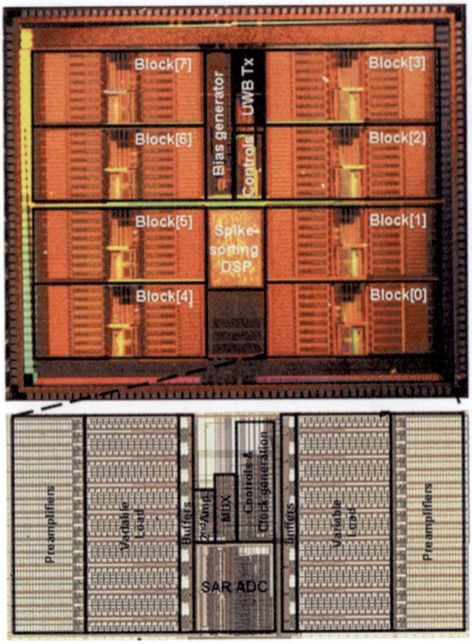

Measured performance	
Number of channels	128
Signal gain of the preamp	40dB
Input impedance (at 1KHz)	8MΩ
Input referred noise	4.9μV$_{rms}$
CMRR of the preamp	90dB
PSRR of the preamp	80dB
LF roll-off of the preamp	0.1Hz ~ 200Hz *
HF roll-off of the preamp	2KHz ~ 20KHz *
Signal gain of the 2nd amp	17dB ~ 20dB *
ADC resolution	6 ~ 9 bits *
ADC sampling rate	640Ksample/sec
Power dissipated by DSP	0.1mW
Maximum UWB data rate	90Mbps
Power dissipated by UWB	1.6mW
Power supply level	±1.65V
Total chip power dissipation	6.0mW
Technology	0.35μm 4M2P CMOS
Total chip area	8.8mm ×7.2mm

* specification is programmable through external controls

Fig. 50 Chip photo and performance summary

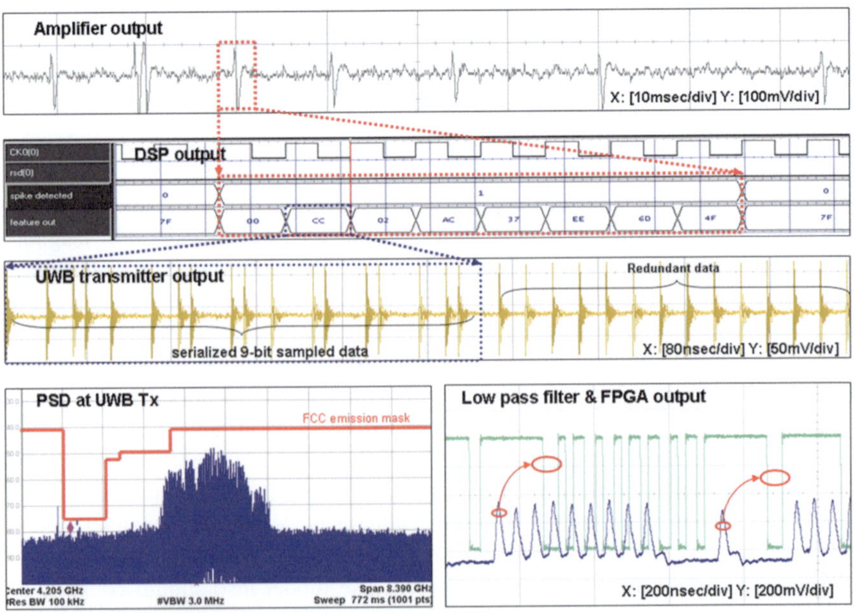

Fig. 51 Measured output waveforms of the chip

4.2 Stimulation: 256-Channel Retinal Prosthesis Chip

Each year over 70,000 new patients are reported legally blind in the United States due to the most common retinal degenerative diseases, retinitis pigmentosa and age-related macula degeneration [168]. Applying electrical pulses to retina made it possible for the patients to elicit visual perceptions [168]. Thus, prosthetic devices based on epiretinal [169–170] and subretinal [171] implants have been proposed and evaluated clinically. Experiments show the patients with 4 $\times$ 4 epiretinal prosthesis can perform simple visual tasks like identifying objects and detecting motions [172]. More complex visual functions such as finger count and mobility without a cane require at least 200 pixels; face recognition and reading require even more than 1000 pixels [168], imposing several challenging design issues.

As the number of stimulating electrodes increases, the physical dimension of those electrodes must be scaled down due to the limited size of the fovea area, leading to higher electrode impedance. Therefore, high-compliance voltages up to 30 V are inevitable [168] for the current stimulating output stages. Also the amount of the power and data to be delivered wirelessly to the implant increases with the number of pixels. Both 100 mW power and 2 Mbps data rate (19 bits/pixel $\times$1024 pixel/frame $\times$100 frame/s) are necessary for epiretinal prosthesis with more than 1000 pixels [169]. Since space constraint on the implant site severely limits the chip size and implant system package footprint, it is preferred to have a single-chip solution that integrates as many components as possible. In addition, the flexibility of the stimulator is desired for the proper operation of the implant in patients with different thresholds. The footprint limitation and available high-voltage process allow us to design a fully integrated 256-pixel epiretinal chip. This section presents the mixed-mode, mixed-voltage design of the chip fabricated in 32-V CMOS process.

4.2.1 System Architecture

The epiretinal prosthesis is composed of an external camera, a signal-processing unit, power and data Tx, intraocular Rx coils, a retinal IC, and a stimulating electrode array on inner retinal surface. The coils, IC, and electrode array are sealed in a hermetic package. The chip consists of a dual-band telemetry unit, a reverse telemetry, a global digital controller (GDC), and 256 stimulators. The dual-band telemetry employs two separate coils to optimize both power and data link for delivering 100-mW and 2-Mbps forward data [169]. A full-wave rectifier composed of on-chip diodes (D_1–D_4) and off-chip storage capacitors (C_S) generate unregulated voltages of $\pm$13 V and $\pm$2.8 V from a 2-MHz power carrier, and regulators generate 4 supply voltages of $\pm$12 V and $\pm$1.8 V. The 2-Mbps data are recovered by a first-order HPF and a demodulator, and then delivered to the GDC where CRC and flexible data packet parser are developed to trade-off between pixels drop rate and data bandwidth. A packet can carry as small as one pixel data in a noisy channel to reduce pixel drop rate, or as big as all 256-pixel data in a clear channel to reduce the error-detection code overhead. The stimulation parameter data are distributed

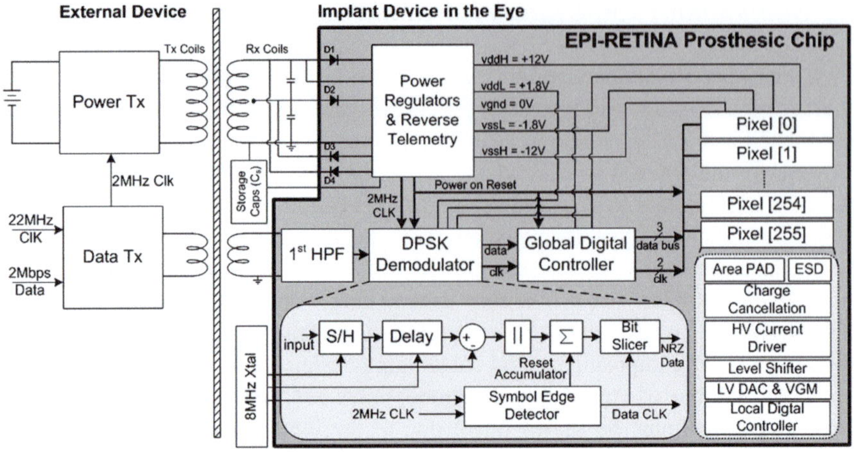

Fig. 52 System block diagram

to pixels via global data bus. Each pixel responds only to data on the bus with a corresponding ID address (Fig. 52).

4.2.2 Stimulator Pixel Design

The schematic of stimulator pixels is shown in Fig. 53. A pixel consists of a local digital controller (LDC), level shifters, a 4-bit current steering DAC, a variable gain current mirror (VGCM), a current output stage with a charge cancellation switch, an ESD protection unit, and an area pad. As the reference current of the DAC is 0.333 μA, the output current of the DAC ranges from 0 to 5 μA in 16 steps. VGCM has a variable gain of 1 to 10 and the output stage has an additional gain of 10, resulting in a maximum output current of 500 μA. A high-voltage (HV) PMOS transistor is employed as a charge-cancellation switch to remove any residual charge on

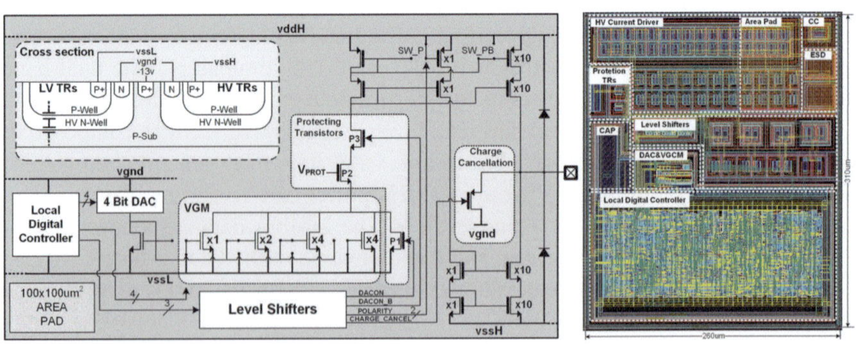

Fig. 53 Schematic of a stimulating pixel with high-compliance voltage

the electrodes and a pair of back-to-back diode is used as an ESD protection unit. To achieve high-compliance voltage at the output, HV transistors with breakdown voltage of 32 V are used, while DAC, VGCM, and LDC are built exclusively with low-voltage (LV) transistors. LV p-substrate biased to –1.8 V is isolated from the wafer substrate. This prevents LV transistors from malfunction caused by the noise in the wafer substrate due to significant ripples on unregulated voltage of −13 V. Also, the output stage using HV transistors is isolated from the wafer substrate and the HV p-well is biased to −12 V as shown in Fig. 53. Protecting transistors (Fig. 53) is connected to between the VGCM and HV transistors, making it possible to use LV transistors for DAC and VGCM and to reduce the overall size of the pixel by eliminating level shifters for DAC and VGCM control. By placing the ESD, the charge-cancellation switch, and parts of the output stage under the 100×100 μm^2area pad, which is composed of metal 4 through metal 6, the pixel size is reduced by 13.5% with this Circuit-Under-Pad (CUP). With a timing controller, the LDC is able to generate highly flexible stimulation waveforms. Each pixel has a timing unit and can be independently programmed to generate various stimulation pulse waveforms with different starting time. Programmable starting time is a mechanism to evenly distribute the stimulation of all pixels in the entire frame period. In addition, dual clock is employed in LDC to reduce static power consumption.

4.2.3 Dual-Band Power and Data Telemetry Design

To overcome interferences due to the coupling of data and power coils, the paper [173] proposed a noncoherent differential phase shift keying (DPSK) Rx at 16-MHz sampling rate to recover 2-Mbps forward data from a DPSK signal at 20-MHz carrier. In this work, the sampling rate is 8 MHz rather than 16 MHz, thus reducing the number of switched-capacitor arrays and accordingly the chip area and power consumption. Band-pass sampling theory allows the system to use a 22-MHz data carrier and a 2-MHz power carrier while providing 2-Mbps forward data. All rectified inputs of the regulators are directly derived from cascaded resonant tank, which avoids voltage conversion from ±13 V to ±2.8 V and the associated power loss. By tapping the power coil to generate the low voltage instead of using an additional one, the space occupied by the inductor coil at the implant side is reduced and more than 50 mW power is saved compared to the conventional method with a single resonant tank and voltage conversion.

5 Summary

The design methodology and example designs of microelectronics for neuroprosthetics have been presented. The device has quite a different requirement and constraint imposed by the unique operating environments than other pure electronics-oriented applications and thus requires the system designer to consider those constraints collectively and find out the optimal system parameters rather than just to focus on achieving best state-of-the-art performance in each circuit block. In this chapter, we investigated a way to achieve the goal systematically together with the design results. Basic circuits were explained first and how to organize those circuits to achieve a certain function was presented with emphasis on trade-off between various system parameters. Multichannel wireless recording system and state-of–the-art retinal prosthesis chip were discussed as examples of entire system design. Although there are technical issues and challenges to be solved for the complete neuroprosthetic device to treat various diseases and disabilities, the technical innovation occurring in various fields of electronics will eventually overcome those obstacles and will contribute to the human society.

Acknowledgment The authors greatly thank Hyunchul Kim for helping in editing, Jungsuk Kim for providing materials on DPSK data telemetry, Bin Liang for helping in chip layout, Linh Hoang for CAD tool supports, Kuanfu Chen for providing data on retinal prostheses, Lihsien Wu for providing data for power telemetry, Eric Basham for experimental setup, Tungchien Chen for works and data on signal processing, Dr. Mingcui Zhou for materials on data telemetry, and Dr. Mehmet R. Yuce for his works related with UWB telemetry. This study was partially supported by grants from NSF, NSF-ERC, and DOE.

References

1. Hodgkin AL, Huxley AF (1952) A quantitative description of membrane current and its application to conduction and excitation in nerve. J Physiol 117:500–544.
2. Liu W, Humayun MS (2004) Retinal prosthesis Solid-State Circuits Conference. Digest of Technical Papers. IEEE Int ISSCC, Vol. 1, pp. 218–219, 15–19 Feb. 2004.
3. Ortmanns M, Rocke A, Gehrke M, Tiedtke HJ (2007) A 232-Channel epiretinal stimulator ASIC. IEEE J Solid-State Circ 42(12): 2946–2959, Dec. 2007.
4. Taylor DM, Tillery SI, Schwartz AB (2002) Direct cortical control of 3-D neuroprosthetic devices. Science 296: 1829–1832, June 2002.
5. Schwartz AB (2004) Cortical neural prosthetics. Annu Rev Neurosci 27: 487–501, July 2004.
6. Jackson A, Moritz CT, Mavoori J, Lucas TH, Fetz EE (2006) The Neurochip BCI: Towards a neural prosthesis for upper limb function. IEEE Trans Neural Syst Rehabil Eng 14(2): 187–190, June 2006.
7. Anschel DJ, Ortega EL, Kraus AC, Fisher RS (2004) Focally injected adenosine prevents seizures in the rat. Exp Neurol 190: 544–457.
8. Bhatti PT, Lee S, Wise KD (2006) A 32-site 4-channel cochlear electrode array. IEEE J Solid-State Circ 41:2965–2973, Dec. 2006.
9. Limousin P, Krack P, Pollack P, Benazzouz A, Ardouin C, Hoffmann D, Benabid A (1998) Electrical stimulation of the subthalamic nucleus in advanced Parkinson's disease. New Eng J Med 339: 1105–1111.
10. Chae M, Liu W, Sivaprakasam M (2008) Design optimization for integrated neural recording systems. IEEE J Solid-State Circ 43: 1931–1939
11. Ferris CD (1978) Introduction to Bioinstrumentation. Humana, Clifton NJ.
12. Olsson RH, Gulari MN, Wise KD (2002) Silicon neural recording arrays with on-chip electronics for in-vivo data acquisition. IEEE EMBS Int Conf 237–240
13. Harrison RR, Charles C (2003) A low-power low-noise CMOS amplifier for neural recording applications. IEEE J Solid-State Circuits 38: 958–965
14. Chae M, Kim J, Liu W (2008) Fully-differential self-biased bio-potential amplifier. Electron Lett 44: 1390–1391.

15. Dagtekin M, Liu W, Bashirullah R (2001) A multichannel chopper modulated neural recording system. Proc IEEE EMBS Int Conf 757–760.
16. Denison T, Consoer K, Santa W, Avestruz AT, Cooley J, Kelly A (2007) A 2 μW 100 nV/rtHz chopper-stabilized instrumentation amplifier for chronic measurement of neural field potentials. IEEE J Solid-State Circ 42: 2934–2945
17. Harrison R (2008) The design of integrated circuits to observe brain activity. Proc IEEE 96:1203–1216.
18. Razavi B (2001) Design of Analog CMOS Integrated Circuits. McGraw-Hill, Boston, MA.
19. Vogels M, Gielen G (2003) Architectural selection of A/D converters. In Proc. Design Automation Conference, June 2–6, 2003, pp. 974–977
20. Scott M. D, Boser B. E, Pister S. J(2003) An ultralow-energy ADC for smart dust. IEEE J Solid-State Circ. 38(7): 1123–1129, July 2003.
21. Kugelstadt T (2000) The operation of the SAR-ADC based on charge redistribution. Tex Inst Analog Appl J 10–12, Feb. 2000.
22. Pang C, Cham JG, Nenadic Z, Musallam S, Tai YC, Burdick JW, Andersen RA (2005) A new multi-site probe array with monolithically integrated parylene flexible cable for neural prostheses. In Proc. 27th Annu. Int. Conf. Engineering in Medicine and Biology Soc., Shanghai, China, Sep. 1–4, 2005, pp. 7114–7117.
23. Demichele GA, Troyk PR (2003) Integrated multichannel wireless biotelemetry system. In Proc. 25th IEEE EMBS Int. Conf., Cancun, Mexico, Sep. 17–21, pp. 3372–3375.
24. Obeid IL, Nicolelis MA, Wolf PD (2004) A multichannel telemetry system for single unit neural recordings. J Neurosci Methods 133: 123–135, Feb. 2004.
25. Mohseni P, Najafi K (2005) A battery-powered 8-channel wireless FM IC for biopotential recording applications. In Dig. Tech. Papers 2005 IEEE Int Solid-State Circ Conf, San Francisco, CA, Feb. 6–10, 2005, pp. 560–561.
26. Harrison R, Watkins P, Kier R, Lovejoy R, Black D, Normann R, Solzbacher F (2006) A low-power integrated circuit for a wireless 100-electrode neural recording system. In Dig. Tech. Papers 2006 IEEE Int Solid-State Circuits Conf, San Francisco, CA, Feb. 6–9, 2006, pp. 2258–2267.
27. O'Driscoll S, Meng T, Shenoy K, Kemere C (2006) Neurons to silicon: Implantable prosthesis processor. In Dig. Tech. Papers 2006 IEEE Int Solid-State Circ Conf, San Francisco, CA, Feb. 6–9, 2006, pp. 2248–2257.
28. Harrison R, Charles C (2003) A low-power low-noise CMOS amplifier for neural recording applications. IEEE J Solid-State Circ 38(6): 958–965, June 2003.
29. Ferris CD (1978) Introduction to Bioinstrumentation. Humana, Clifton, NJ.
30. Liu W, Sivaprakasam M, Wang G, Chae MS (2006) A neural recording system for monitoring shark behavior. In Proc. IEEE Int. Symp. Circuits and Systems, May 21–24, 2006, pp. 4123–4236.
31. Razavi B (2001) Design of Analog CMOS Integrated Circuits, 2nd edn. McGraw-Hill, Boston, MA, Chapter 18, pp. 653–655.
32. Vogels M, Gielen G (2003) Architectural selection of A/D converters. In Proc. Design Automation Conference, June 2–6, pp. 974–977.
33. Scott MD, Boser BE, Pister SJ (2003) An ultralow-energy ADC for smart dust. IEEE J Solid-State Circ 38(7): 1123–1129, July 2003.
34. Robinson DA (1968) The electrical properties of metal microelectrodes. Proceedings of the IEEE, 56(6): 1065–1071, June 1968.
35. Fee MS, Mitra PP, Kleinfeld D (1996) Variability of extracellular spike waveforms of cortical neurons. J Neurophysiol 76:3823–3833.
36. Lewicki MS (1998) A review of methods for spike sorting: the detection and classification of neural action potentials. Network Comput Neural Syst 9:53–78.
37. Rutishauser U, Schuman EM, Mamelak AN (2006) Online detection and sorting of extracellularly recorded action potentials in human medial temporal lobe recordings in vivo. J Neurosci Methods. 154(1–2):204–224, June 2006.

38. Wood F, Black MJ, Vargas-Irwin C, Fellows M, Donoghue JP (2004) On the variability of manual spike sorting. IEEE Trans Biomed Eng 51(6): 912–918, June 2004.
39. Chapin JK (2004) Using multi-neuron population recordings for neural prosthetics. Nat Neurosci 7: 452–455, Apr. 2004.
40. Serruya MD, Hatsopoulos NG, Paninski L, Fellows MR, Donoghue JP (2002) Instant neural control of a movement signal. Nature. 416: 141–142.
41. Hochberg LR, Serruya MD, Friehs GM, Mukand JA, Saleh M, Caplan AH, et al. (2006) Neuronal ensemble control of prosthetic devices by a human with tetraplegia. Nature. 442: 164–171.
42. Quirk MC, Blum KI, Wilson MA (2001) Experience-dependent changes in extracellular spike amplitude may reflect regulation of dendritic action potential back-propagation in rat hippocampal pyramidal cells. J f Neurosci 21(1): 240–248, Jan. 2001.
43. Kaneko H, Tamura H, Suzuki SS (2007) Tracking spike-amplitude changes to improve the quality of multineuronal data analysis. IEEE Trans Biomed Eng 54(2): 262–272, Feb. 2007.
44. Bar-Hillel A, Spiro A, Stark E (2006) Spike sorting: Bayesian clustering of non-stationary data. J Neurosci 157(2): 303–316, Oct. 2006.
45. Zumsteg ZS, Kemere C, O'Driscoll S, Santhanam G, Ahmed RE, Shenoy KV, et al. (2005) Power feasibility of implantable digital spike sorting circuits for neural prosthetic systems. IEEE Trans Neural Syst Rehabil Eng 13(3):272–279, Sep. 2005.
46. Thakur PH, Lu H, Hsiao SS, Johnson KO (2007) Automated optimal detection and classification of neural action potentials in extra-cellular recordings. J Neurosci Methods 162(1): 364–376, May 2007.
47. Lewicki MS (1994) Bayesian modeling and classification of neural signals. Advances NIPS 590–597.
48. Vargas-IrwinC, Donoghue JP (2007) Automated spike sorting using density grid contour clustering and subtractive waveform decomposition. J Neurosci Methods; 164(1):1–18.
49. Lewicki MS (1994) Bayesian modeling and classification of neural signals. Neural Comput 6:1005–1030.
50. Vollgraf R, Obermayer K (2006) Improved optimal linear filters for the discrimination of multichannel waveform templates for spike-sorting applications. IEEE Signal Proc Lett. 13(3):121–124, Mar 2006.
51. Zhang P, Wu J, Zhou Y, Liang P, Yuan J (2004) Spike sorting based on automatic template reconstruction with a partial solution to the overlapping problem. J Neurosci Methods. 7(5):446–451, May 2004.
52. Letelier JC, Weber PP (2000) Spike sorting based on discrete wavelet transform coe_cients. J Neurosci Methods 101:93–106.
53. Quian Quiroga R, Nadasdy Z, Ben-Shaul Y (2004) Unsupervised spike detection and sorting with wavelets and superparamagnetic clustering. Neural Comput. 16(8):1661–1687, Aug. 2004.
54. Pavlov A, Makarov VA, Makarova I, Panetsos F (2007) Sorting of neural spikes: When wavelet based methods outperform principal component analysis. Natural Computing. 6(3):269–281, Sep. 2007.
55. Takahashi S, Sakurai Y (2007) Coding of spatial information by soma and dendrite of pyramidal cells in the hippocampal CA1 of behaving rats. Eur J Neurosci Methods 26(7): 2033–2045, Oct. 2007.
56. Sakurai Y, Tamura H, Takahashi S (2006) Dynamic synchrony of firing in the monkey prefrontal cortex during working-memory tasks. J Neurosci. 26(40):10141–10153, Oct. 2006.
57. Snellings A, Anderson D, Aldridge J (2006) Improved signal and reduced noise in neural recordings from closespaced electrode arrays using independent component analysis as a preprocessor. J Neurosci Methods 150(2):254–264, Jan. 2006.

58. Takahashi S, Sakurai Y (2005) Real-time and automatic sorting of multi-neuronal activity for sub-millisecond interactions in vivo. Neuroscience 134:301–315, Mar. 2005.
59. Mamlouk AM, Sharp H, Menne KML, Hofmann UG, Martinetz T (2005) Unsupervised spike sorting with ICA and its evaluation using GENESIS simulations. Neurocomputing 65:65–66.
60. Delescluse M, Pouzat C (2006) E_cient spike-sorting of multi-state neurons using inter-spike intervals information. J Neurosci Methods 150(1):16–29, Jan. 2006.
61. Yang Z, Zhao Q, Liu W (2009) Spike Feature Extraction Using Informative Samples. Spot light presentation Advances in Neural Information Processing Systems NIPS, 2009, MIT Press, Cambridge, MA.
62. Yang Z, Zhao Q, Liu W. Neural Signal Classification Using a Simplified Feature Set with Energy Based Non-parametric Clustering. To appear in Neurocomputing.
63. Jung HK, Choi JH, Kim T (2006) Solving alignment problems in neural spike sorting using frequency domain PCA. Neurocomputing 69(7–9):975–978, Mar. 2006.
64. Blanche TJ, Swindale NV (2006) Nyquist interpolation improves neuron yield in multiunit recordings. J Neurosci Methods 155(1):207–216, July 2006.
65. Sahani M (1999) Latent variable modelss for neural data analysis. PhD dissertation, California Institute of Technology.
66. Shenoy KV, Santhanam G, Ryu SI, Afshar A, Yu BM, Gilja V, et al. (2006) Increasing the performance of cortically-controlled prostheses. Proc 28th Ann Int Conf IEEE EMBS, p. 6652–6656.
67. Yang Z, Chen T, Liu W (2008) Neuron signature based spike feature extraction algorithm for on- chip implementation. Lecture Proc. 30th Ann. Int. Conf. IEEE EMBS, pp. 1716–1719, August 2008.
68. Holt GR, Koch C (1999) Electrical interactions via the extracellular potential near cell bodies. J Comp Neurosci 6(2):169–184, Mar. 1999.
69. Buzsaki G, Penttonen M, Nadasdy Z, Bragin A (1996) Pattern and inhibition-dependent invasion of pyramidal cell dendrites by fast spikes in the hippocampus in vivo. Proc Natl Acad Sci USA 93(18):9921–9925, Sep. 1996.
70. Tuckwell HC (1988) Introduction to Theoretical Neurobiology. Cambridge University Press, Cambridge.
71. Traub RD, Wong RKS, Miles R, Michelson H (1991) A model of a CA3 hippocampal pyramidal neuron incorporating voltage-clamp data on intrinsic conductances. J Neurophysiol 66(2):635–650.
72. Greenberg RJ, Velte TJ, Humanyun MS, Scarlatis GN, De Juan EJ (1996) A computational model of electrical stimulation of the retinal ganglion cell. IEEE Trans Biomed Eng 46(5):505–514, May 1999.
73. Rattay R (1989) Analysis of models for extracellular fiber stimulation. IEEE Trans Biomed Eng 36:676–692.
74. Buzsaki G (2004) Large-scale recording of neuronal ensembles. Nat Neurosci 7(5):446–451, May 2004.
75. Thurbon D, Field A, Redman S (1994) Electrotonic profiles of interneurons in stratum pVramidale of the CA1 region of rat hippocampus. J Neurophysiol 71(5): 1948–1958.
76. Rattay R, Richardson NL, Felix H (2001) A model of the electrically excited cochlear neuron. I. Contribution of neural substructures to the generation and propagation of spikes. Hear Res 153:43–63.
77. Hausser M, Spruston N, Stuart G (2000) Diversity and dynamics of dendritic signaling. Science 290(5492):739–744, Oct 2000.
78. Gilja V, Linderman MD, Santhanam G, Afshar A, Ryu SI, Meng TH, et al. (2006) Multiday electrophysiological recordings from freely behaving primates. Proc 28th Ann Int Conf IEEE EMBS, 2006, pp. 4387–4391.

79. Kim KH, Kim JK (2000) Neural spike sorting under nearly 0-dB signal-to-noise ratio using nonlinear energy operator and artificial neural-network classifier. IEEE Trans Biomed Eng 47(10):1406–1411, Oct 2000.
80. Kaiser JF (1990) On a simple algorithm to calculate the energy of a signal. In Proc IEEE Int Conf Acoustic Speech and Signal Processing pp. 381–384.
81. MaragosP, Kaiser JF, Quatieri TF. On amplitude and frequency demodulation using energy operators. IEEE Trans Signal Proc 41, No:4, April 1893.
82. Mukhopadhyay S, Ray GC (1998) A new interpretation of nonlinear energy operator and its efficacy in spike detection. IEEE Trans Biomed Eng 45, 180–187, Feb. 1998.
83. Gibson S, Judy JW, Markovic D (2008) Comparison of Spike-sorting algorithms for future hardware implementation. To Appear in Proc 30th Ann Int Conf IEEE EMBS. 2008 Aug.
84. Simon W (1965) The real-time sorting of neuro-electric action potentials in multiple unit studies. Electroenceph Clin Neurophysiol. 18:192–195.
85. Fukunaga K, Hostetler LD (1975) The estimation of the gradient of a density function, with application in pattern recognition. IEEE Trans Inf Theory 21:180–187.
86. Cheng YZ (1995) Mean shift, mode seeking, and clustering. IEEE Trans Pattern Anal Mach Intell 17(8):790–799, Aug.
87. Comaniciu D, Meer P (2002) Mean shift: a robust approach toward feature space analysis. IEEE Trans Pattern Anal Mach Intell 24(5):603–619, May 2002.
88. Comaniciu D, Ramesh V, Meer P (2001) The variable bandwidth mean shift and data-driven scale selection. In: IEEE International Conference on Computer Vision, I: 438–445.
89. Hall P, Hui TC, Marron JC (1995) Improved variable window kernel estimates of probability densities. Ann Stat.23:1–10.
90. Stuart MacKey R (xxxx) Bio-Medical Telemetry: Sensing and Transmitting Biological Information from Animals and Man. John Wiley & Sons, Inc., New York, August 1998.
91. Deutsch S (1976) A 15-electrode totally implanted time-multiplex telemetry unit. IEEE Trans. Commun 24, 1073–1078, Oct. 1976.
92. Summers G (1970) Transducers for bioimplantable telemetry systems. IEEE IECI-17, 17: 144–150, Apr. 1970.
93. Meindl J (1977) Integrated electron devices in medicine. Int Electron Dev Meet 23: 1–1.
94. Ryan TG, Busacker JW, Hochban RA (1994) Pacemaker telemetry system, U.S Patent 5 350 411.
95. Troyk P, DeMichele G (2000) Inductively-coupled power and data link for neural prostheses using a class-E oscillator and FSK modulation. Proc. IEEE 25th EMBS Conference, vol.4, pp. 3376–3379, Sept. 2003.
96. Loftin SM, McClure KH (2006) Low-power, high-modulation-index amplifier for use in battery-powered device, U.S Patent 7 092 762.
97. Zierhofer CM (2003) Multichannel cochlear implant with neural response telemetry, U.S. Patent 6 600 955.
98. Zierhofer C, Hochmair-Desoyer I, Hochmair E (1995) Electronic design of a cochlear implant for multichannel high-rate pulsatile stimulation strategies. IEEE Trans. Neural Syst Rehabil 3:112–116, Mar. 1995.
99. Dong M, Zhang C, Wang Z, Li D (2004) A neuro-stimulus chip with telemetry unit for cochlear implant. IEEE Int Workshop Biomed Circ Syst, pp. S1/3/INV 9–12, Dec. 2004.
100. W. Liu, Vichienchom K, Clements M, DeMarco SC, Hughes C, McGucken E, Hymayun MS, de Juan E, Weiland JD, Greenberg R (2000) A neuro-stimulus chip with telemetry unit for retinal prosthetic device. IEEE J. Solid-State Circ. 35(10):1487–1497, Oct. 2000.
101. Ghovanloo M, Najafi K (2004) A wideband frequency-shift keying wireless link for inductively powered biomedical implants. IEEE Trans Circ Syst. 51:2374–2383, Dec. 2004
102. Theogaraja L Wyatt J, Rizzo J, Drohan B, Markova M, Kelly S, Swider G, Raj M, Shire D, Gingerich M, Lowenstein J, Yomtov B (2006) Minimally invasive retinal prosthesis. IEEE ISSCC Digest of Technical Papers, pp. 54–55, Feb. 2006.
103. Zhou M, Liu W (2007) A non-coherent PSK receiver with interference-canceling for transcutaneous neural implants. ISSCC Digest of Technical Paper, pp. 156–157, Feb. 2007.

104. Humayun MS, Juan E De, Weiland JD, Dagnelie G, Katona S, Greenberg R, Suzuki S (1999) Pattern electrical stimulation of the human retina. Vision Res 39(15): 2569–2576, July, 1999.
105. Majji AB, Humayun MS, Weiland JD, Suzuki S, D'Anna SA, Juan E De (1999) Long term histological and electrophysiological results of an inactive epi-retinal electrode array implantation in dogs. Invest. Ophthalmol Vis Sci 40(9): 2073–2081, Aug. 1999.
106. Ko WH, Liang SP, Fung CDF (1977) Design of radio-frequency powered coils for implant instruments. Med Biol Eng Comput 15(6):634–640, Nov. 1977.
107. Wang G, Liu W, Sivaprakasam M, Humayun MS, Weiland JD (2005) Power supply topologies for biphasic stimulation in inductively powered implants. IEEE ISCAS 3: 2743–2746.
108. Yang Z, Wang G, Liu W (2006) Analytical calculation of the selfresonant frequency of biomedical telemetry coils. Proc. 28th Ann. Int. Conf. IEEE EMBS, pp. 5880–5883, Sep. 2006.
109. Grover F (1962) Inductance Calculations: Working Formulas and Tables. Dover, New York.
110. Yang Z, Liu W, Basham E (2007) Inductor modeling in wireless links for implantable electronics. IEEE Trans Magn 43: 3851–3860, Oct. 2007.
111. Ferreira J (1992) Analytical computation of ac resistance of round and rectangular litz wire windings. Proc Inst Elect Eng Part B 139(1): 21–25, Jan. 1992.
112. Troyk P, DeMichele G (2003) Inductively-coupled power and data link for neural prostheses using a class-E oscillator and FSK modulation. Proc IEEE 25th EMBS Conference, vol. 4, pp. 3376–3379, Sept. 2003.
113. Zierhofer C, Hochmair-Desoyer I, Hochmair E (1995) Electronic design of a cochlear implant for multichannel high-rate pulastile stimulation strategies. IEEE Trans. Neural Syst Rehabil 3:112–116, Mar. 1995.
114. Liu W, et al. (2000) A neuro-stimulus chip with telemetry unit for retinal prosthetic device. IEEE J Solid-State Circ. 35(10): 1487–1497, Oct. 2000.
115. Ghovanloo M, Najafi K (2004) A wideband frequency-shift keying wireless link for inductively powered biomedical implants. IEEE Trans Circ Syst 51:2374–2383, Dec. 2004.
116. Weiland JD, et al. (2005) Progress towards a high-resolution retinal prosthesis. Proc. IEEE 27th EMBC Conference, pp. 7373–7375, Sep. 2005.
117. Zhou M, et al. (2006) A transcutaneous data telemetry system tolerant to power telemetry interference. Proc. IEEE 28th EMBS Conference, pp. 5884–5887, Sep. 2006.
118. Liu W, Humayun MS (2004) Retinal prosthesis. ISSCC Digest of Technical Paper, pp. 218–219, Feb. 2004.
119. Wang G (2006) Wireless power and data telemetry for retinal prosthesis. Ph.D. dissertation, University of California, Santa Cruz, EE Dept., CA, USA, Mar. 2006.
120. Zhou M, Liu W (2007) A non-coherent PSK receiver with interference-canceling for transcutaneous neural implants. ISSCC Digest of Technical Papers, pp. 156–157, Feb. 2007.
121. Vaughan RG, Scott NL, White DR (1991) The theory of bandpass sampling. IEEE Trans. Signal Process 39: 1973–1984, Sep. 1991.
122. Razavi B (1995) Principles of Data Conversion System design. IEEE Press, New York.
123. Gray PR, Hurst PJ, Lewis SH, Meyer RG (2001) Analysis and Design of Analog Integrated Circuit. Wiley, New York.
124. Federal communications commission, http://www.fcc.gov
125. Deparis N, Loyez C, Rolland N, Rolland P-A (2006) Pulse generator for UWB communication and radar applications with PPM and time hopping possibilities. ISCAS 2006, pp. 661–665, 2006.
126. Saha P, Sasaki N, Kikkawa T (2006) A Single-Chip Gaussian Monocycle Pulse Transmitter Using 0.18 /spl m/m CMOS Technology for Intra/Interchip UWB Communication. IEEE VLSI Symposium on Circuits, June 15–17, 2006, pp. 204–205.
127. Jung B, Tseng YH, Harvey J, Harjani R (2005) Pulse generator design for UWB IR communication systems. IEEE Int Symp Circ Syst (ISCAS '05.), 5:4381–4384, 23–26 May 2005.
128. Bennett WR, Davey JR (1965) Data Transmission. McGraw-Hill, New York.

129. Wentzloff DD, Chandrakasan AP (2006) Gaussian Pulse Generators for Subbanded Ultra-Wideband Trasnmitters. IEEE Trans. Microw Theor Tech 54: 1647–1655, Apr. 2006.
130. Santhanam G, Linderman MD, Gilja V, Afshar A, Ryu SI, Meng T, Shenoy K (2007) HermesB: A continuous neural recording system for freely behaving primates. IEEE Tran Biomed Eng 54(11): 2037–2050, Nov. 2007.
131. Schwartz AB (2004) Cortical neural prosthetics. Annu Rev Neurosci 27: 487–501, July 2004.
132. Borisoff JF, McPhail LT, Saunders JTW, Birch GE, Ramer MS (2006) Detection and classification of sensory information from acute spinal cord recordings. IEEE Trans. Biomed Eng 53(8): 1715–1719, Aug. 2006.
133. Jackson A, Moritz CT, Mavoori J, Lucas TH, Fetz EE (2006) The Neurochip BCI: Towards a neural prosthesis for upper limb function. IEEE Trans. Neural Syst Rehabil Eng 14(2):187–190, June 2006.
134. Strange KD, Hoffer JA (1999) Restoration of use of paralyzed limb muscles using sensory nerve signals for state control of FES-assisted walking. IEEE Trans. Rehabil Eng 7(3), Sep. 1999.
135. Wenzel BJ, Boggs JW, Gustafson KJ, Grill WM (2005) Detecting the onset of hyper-reflexive bladder contractions from the electrical activity of the pudendal nerve. IEEE Trans. Neural Syst Rehabil Eng 13(3), Sep. 2005.
136. Hansen J, Borau A, Rodriguez A, Vidal J, Sinkjaer T, Rijkhoff NJM (2007) Urethral sphincter EMG as event detector for neurogenic detrusor overactivity. IEEE Trans. Biomed Eng 54(7), July 2007.
137. DiMarco AF (2001) Neural prostheses in the respiratory system. J. Rehabil Res Dev 38(6), Nov./Dec. 2001.
138. Mangold S, Keller T, Curt A, Dietz V (2005) Transcutaneous functional electrical stimulation for grasping in subjects with cervical spinal cord injury. Spinal Cord 43:1–13, 2005.
139. Aziz JNY, Genov R, Bardakjian BL, Derchansky M, Carlen PL (2007) Brain-silicon interface for high-resolution in vitro neural recording. IEEE Trans. Biomed Circ Syst 1(1): 56–62, Mar. 2007.
140. Perelman Y, Ginosar R (2007) An integrated system for multichannel neuronal recording with Spike/LFP separation, integrated A/D conversion and threshold detection. IEEE Trans. Biomed Eng 54(1): 130–137, Jan. 2007.
141. Cheney D, Goh A, Xu J, Gugel K, Harris JG, Sanchez JC, Principe JC (2007) Wireless, in vivo neural recording using a custom integrated bioamplifier and the pico system. CNE ′07. 3rd International IEEE/EMBS Conference on Neural Eng, 2007, pp. 19–22, 2–5 May 2007.
142. Harrison RR, Watkins PT, Kier RJ, Lovejoy RO, Black DJ, Greger B, Solzbacher F (2007) A low-power integrated circuit for a wireless 100-electrode neural recording system. IEEE J. Solid-State Circ, 42(1): 123–133, Jan. 2007.
143. Sodagar AM, Wise KD, Najafi K (2007) A fully integrated mixed-signal neural processor for implantable multichannel cortical recording. IEEE Trans. Biomed Eng 54(6): 1075–1088, June 2007.
144. Olsson RH III, Wise KD (2005) A three-dimensional neural recording microsystem with implantable data compression circuitry. IEEE J Solid-State Circ 40: 2796–2804.
145. Gosselin B, Ayoub AE, Sawan M (2007) A mixed-signal multi-chip neural recording interface with bandwidth reduction. In The 2007 IEEE Biomedical Circuits and Systems Conference (BIOCAS), pp. 49–52.
146. Chae M, Liu W, Yang Z, Chen T, Kim J, Sivaprakasam M, Yuce M (2008) A 128-channel 6mW wireless neural recording IC with on-the-fly spike sorting and UWB transmitter. In Dig. Tech. Papers 2008 IEEE Int. Solid-State Circ Conf., San Francisco, CA, Feb. 3–7, 2008, pp. 146–147.
147. Chae MS, Liu W, Sivaprakasam M (2008) Design optimization for integrated neural recording systems. IEEE J Solid-State Circuits 43(9): 1931–1939, Sept. 2008.

148. Chimeno MF, Pallas-Areny R (2000) A comprehensive model for power line interference in biopotential measurements. IEEE Trans Inst Meas 49(3): 535–540, June 2000.
149. Chae M, Kim J, Liu W (2008) Fully-differential self-biased bio-potential amplifier. Electron Lett 44(24): 1390–1391, November 20 2008.
150. Scott MD, Boser BE, Pister SJ (2003) An ultralow-energy ADC for Smart Dust. IEEE Journal of Solid-State Circuits, 38(7): 1123–1129, July 2003.
151. Linderman MD, Gilja V, Santhanam G, Afshar A, Ryu SI, Meng TH, Shenoy KV (2006) Neural recording stability of chronic electrode. Proc. 28th Ann. Int. Conf. IEEE EMBS, pp. 4387–4391, 2006.
152. Hazan L, Zugaro M, Buzsaki G (2006) Klusters, neuroscope, ndmanager: A free software suite for neurophysiological data processing and visualization. J Neurosci Methods, 155(2):207–216, Sep 2006.
153. Maragos P, Kaiser JF, Quatieri T. On amplitude and frequency demodulation using energy Operators. IEEE Trans Signal Process, April 1993.
154. Kim KH, Kim JK (2000) Neural spike sorting under nearly 0-db signal-to-noise ratio using nonlinear energy operator and artificial neural-network classifier. IEEE Trans Biomed Eng, 47(10):1406–1411, Oct. 2000.
155. Yang Z, Chen T, Liu W (2008) Neuron signature based spike feature extraction algorithm for on-chip implementation. Proc. 30th Ann. Int. Conf. IEEE EMBS, pp. 1716–1719, August 2008.
156. Yang Z, Zhao Q, Liu W (2008) Spike feature extraction using informative samples. Advances in Neural Information Processing Systems (NIPS), MIT Press, Cambridge, MA. December 2008.
157. Chen T, Yang Z, Liu W, Chen L (2008) NEUSORT2.0: A multiple-channel neural signal processor with systolic array buffer and channel-interleaving processing schedule. Engineering in Medicine and Biology Society, 2008. EMBS 2008. 30th Annual International Conference of the IEEE, pp. 5029–5032, 20–25 Aug. 2008.
158. Gharpurey R, Kinget P (2008) Ultra Wideband : Circuits, Transceivers and Systems. Springer, New York.
159. Yuce MR, Liu W, Chae MS, Kim JS (2007) A wideband telemetry unit for multi-channel neural recording systems. Ultra-Wideband, 2007. IEEE International Conference on ICUWB 2007, pp. 612–617, 24–26 Sept. 2007.
160. Charles C (2008) An implantable I-UWB transceiver architecture with power carrier synchronization. Circuits and Systems, 2008. IEEE International Symposium on ISCAS, 2008, pp. 1970–1973.
161. Jung B, Tseng YH, Harvey J, Harjani R (2005) Pulse generator design for UWB IR communication systems. Circuits and Systems, 2005. IEEE International Symposium on ISCAS 2005, pp. 4381–4384 Vol. 5, 23–26 May 2005.
162. Webster JG (ed.) (1998) Medical Instrumentation Application and Design. John Wiley & Sons, Inc., New York.
163. Huhta JC, Webster JG (1973) 60-Hz Interference in electrocardiography. IEEE Trans Biomed Eng, Vol. BME-20, pp. 91–101.
164. Thakor NV, Webster JG (1980) Ground-free ECG recording with two electrodes. IEEE Transactions on Biomedical Engineering, Vol. BME-27, No.12, pp. 699–704, Dec. 1980.
165. Webster JG (1984) Reducing motion artifacts and interference in biopotential recording. IEEE Transactions on Biomedical Engineering, Vol. BME-31, No.12, pp. 823–826, Dec. 1984.
166. Piipponen KVT, Sepponen R, Eskelinen P (2007) A biosignal instrumentation system using capacitive coupling for power and signal isolation. IEEE Transactions on Biomedical Engineering, Vol. 54, No. 10, pp. 1822–1828, Oct. 2007.
167. To appear in Ph.D. dissertation of Chae M, Univ. of California, Santa Cruz, EE Department., CA, USA.

168. Weiland JD, Humayun MS (2008) Visual Prosthesis. Proceedings of the IEEE, Vol. 96, No.7, pp. 1076–1084, July 2008.
169. Liu W, Humayun MS (2004) Retinal prosthesis. Solid-State Circuits Conference, 2004. Digest of Technical Papers. IEEE International ISSCC. 2004, Vol. 1, pp. 218–219, 15–19 Feb. 2004.
170. Ortmanns M, Rocke A, Gehrke M, Tiedtke HJ (2007) A 232-channel epiretinal stimulator ASIC. IEEE Journal of Solid-State Circuits, Vol.42, No.12, pp. 2946–2959, Dec. 2007.
171. Rothermel A, Wieczorek V, Liu L, Stett A, Gerhardt M, Harscher A, Kibbel S (2008) A 1600-pixel Subretinal Chip with DC-free Terminals and ±2 V Supply Optimized for Long Lifetime and High Stimulation Efficiency. Solid-State Circuits Conference, 2008. ISSCC 2008. Digest of Technical Papers. IEEE International, pp. 144–602, 3–7 Feb. 2008.
172. Yanai D, Weiland JD, Mahadevappa M, Greenberg RJ, Fine I, Humayun MS (2007) Visual performance using a retinal prosthesis in three subjects with retinitis pigmentosa. Am J Ophthalmol, 143(5):820–827, May 2007. Epub 2007 Mar 23.
173. Zhou M, Liu W (2007) A Non-Coherent PSK Receiver with Interference-Canceling for Transcutaneous Neural Implants. Solid-State Circuits Conference, 2007. ISSCC 2007. Digest of Technical Papers. IEEE International, pp. 156–593, 11–15 Feb. 2007.

Microchip-Embedded Capacitors for Implantable Neural Stimulators

Orlando Auciello

Abstract Miniaturization of microchips for implantation in the human body (e.g., microchip for the artificial retina to restore sight to people blinded by retina photoreceptors degeneration) requires the integration of high-capacitance ($\geq$ 10 μF) energy-storage capacitors into the microchip. These capacitors would be based on high-dielectric constant layers, preferably made of materials that are bioinert (not affected by human body fluids) and are biocompatible (do not elicit adverse reactions in the human body). This chapter focuses on reviewing the work being done at Argonne National Laboratory (Materials Science Division and Center for Nanoscale Materials) to develop high-capacitance microchip-embedded capacitors based on novel high-K dielectric layers (TiAlOx or TiO_2/Al_2O_3 superlattices). The microchip-embedded capacitor provides energy storage and electromagnetic signal coupling needed for neural stimulations. Advances in neural prostheses such as artificial retinas and cochlear implants require miniaturization of device size to minimize tissue damage and improve device/tissue interfaces in the human body. Therefore, development of microchip-embedded capacitors is critical to achieve full-implantable biomedical device miniaturization.

Contents

O. Auciello (✉)
Argonne National Laboratory, Materials Science Division, Argonne, IL 60439, USA;
Center for Nanoscale Materials, Argonne, IL 60439, USA
e-mail: auciello@anl.gov

D.D. Zhou, E. Greenbaum (eds.), *Implantable Neural Prostheses 2*,
Biological and Medical Physics, Biomedical Engineering, DOI 10.1007/978-0-387-98120-8_10,

1 Introduction

Currently, the complementary metal-oxide semiconductor (CMOS) field-effect transistor (FET) made from silicon is one of the most important electronic devices in the integrated-circuits (IC) industry. This has arisen because the performance of CMOS devices has continued to improve over a 40-year time span according to Moore's Law of scaling. The size keeps on decreasing as the number of devices on an integrated circuit increases. However, the scaling law, based on an all-Si technology, is reaching a limit.

The scaling of CMOS transistors has reached a limit imposed by the silicon dioxide (SiO_2) layer, used as a gate dielectric, because in the next generation of CMOS devices with nanoscale gate dimensions, the SiO_2 layer needs to be as thin as 1.4 nm. For this gate thickness, the leakage current through the SiO_2 layer is too large (≤ 1 Acm^{-2} at 1 V) due to direct tunneling of electrons through the SiO_2, so that power dissipation increases to unacceptable values.

It has been determined that SiO_2 needs to be replaced with a physically thicker layer of oxides of a higher dielectric constant (K) material, or "high K" gate oxides [1–8]. The maximum current density in interconnects between transistors resulted in the replacement of aluminum with copper as the conductor. The FET "gate stack," which is the gate electrode and the dielectric layer between the gate and the silicon channel, is now the most serious problem.

Concurrently with the need for high-K dielectric layers for CMOS, there is also a need for high-K dielectric layers for capacitors that can be embedded into implantable microchips to provide energy storage and electromagnetic signal coupling capacitors needed for neural stimulations. Advances in neural prostheses such as artificial retinas and cochlear implants require miniaturization of device package size to minimize tissue damage and improve device/tissue interfaces in the human body.

2 Design and Process Considerations for Oxide Films for Microchip-Embedded Capacitors

2.1 *Materials for High-Dielectric Constant (K) Layers*

Many materials systems are currently under consideration as potential replacements for SiO_2 as the gate dielectric material for sub-0.1-μm CMOS technology. A systematic consideration of the required properties of gate dielectrics indicates that the key guidelines for selecting an alternative gate dielectric are (a) permittivity, band gap, and band alignment to silicon, (b) thermodynamic stability, (c) film morphology, (d) crystalline or amorphous, (e) interface quality, (f) compatibility with the current or expected materials to be used in processing for CMOS devices, (g) process compatibility, (h) defects, and (i) reliability. Many dielectrics appear favorable in some of these areas, but very few materials are promising with respect to all of these guidelines [3].

The candidate high-K materials are Si_3N_4 [9, 10], Al_2O_3 [11–13], Ta_2O_5 [14–16], TiO_2 [17, 18], $SrTiO_3$ [19, 20], ZrO_2 [21, 22], HfO_2 [23–25], $HfSiO_4$, La_2O_3 [26], and α-$LaAlO_3$, Pr_2O_3 [27, 28]. The relationship of static dielectric constant (K) versus band gap of some of the candidate gate oxides is shown in Fig. 1. The materials bonding, electrical behavior, the microstructure, physical properties, and chemistry of these oxides have been investigated as the gate oxide in CMOSFET. However, the oxides materials need to be further optimized or new materials need to be developed.

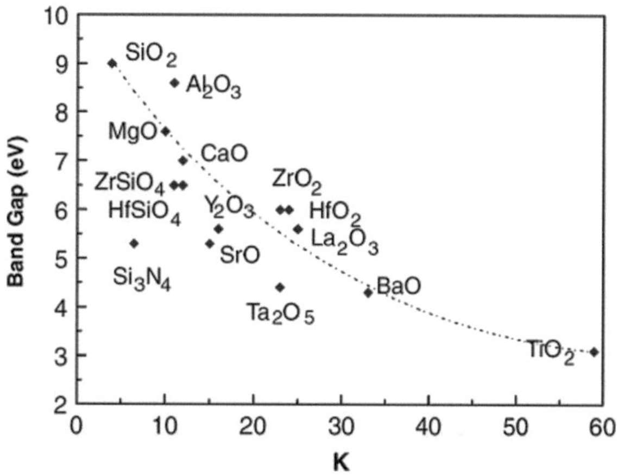

Fig. 1 Static dielectric constant vs. band gap for candidate gate oxides, (Reproduced with permission from [29] Copyright 2000, AVS: Science & Technology of Materials, Interfaces, and Processing)

Among high-dielectric oxide materials, TiO_2 attracted much initial attention, because of its relatively high dielectric constant ($K > 50$). However, its application as gate dielectric was hampered primarily due to its high leakage current as a result of a close zero offset barrier to Si [29]. On the other hand, extensive studies of Al_2O_3 as gate dielectric have recently been performed, since Al_2O_3 exhibits the largest band gap ($Eg = 8.8$ eV) next to SiO_2 with a conduction band offset of 2.8 eV with respect to Si, and thermodynamic stability at the interface with Si substrate. The drawback of Al_2O_3 is its relatively low dielectric constant compared with other leading alternative gate oxide materials.

It has been observed that the conduction-band offset generally increases with decreasing dielectric constant. As a result, a trade-off between the dielectric constant and the offset barrier height is most apparent at intermediate dielectric constants ($12<K<20$). [30]. Materials in that range such as HfO_2 and ZrO_2 and their silicates are the forefront candidates currently investigated for the next generation of gate oxides. However, to meet the long-term development requirement for the next generation of nanoscale CMOS gates and high-K dielectric capacitors, higher dielectric constant is always favored. In addition, it would be ideal if there is a material that can fulfill the double functionality of high-K dielectric for the next generation of nanoscale CMOS devices as explained below, microchip-embedded capacitor layer and hermetic bioinert encapsulating coating for microchip protection for implantable biomedical devices.

We determined that a hybrid structure including TiO_2 and Al_2O_3 might create a novel material with the highest dielectric constant (K > 30) among the amorphous oxides envisioned for the next generation of alternative gate oxides, low tunneling leakage, relatively large barrier offset, and low density of interfacial state to enable its application in the next generation of CMOS nanoelectronics.

For the next generation of integrated circuits, the thickness of the SiO_2 gate (K = 3.9) that has been the gate material since the invention of the integrated circuit will need to be ≤ 1 nm (i.e., 10 Å) thick. Even if a pinhole-free SiO_2 layer could be produced at such thickness, electrons would tunnel from the conduction layer in the Si substrate into the electrode layer (Fig. 2) resulting in a high direct tunneling current that will render the device non-operational. The only solution to this problem is to replace the SiO_2 layer with another dielectric material with high dielectric constant (K). This would allow us to keep the physical thickness of the oxide layer at a suitable value to provide a good insulator layer with high capacitance, low leakage, high voltage break-down field, low interface density of states, and good overall electrical properties.

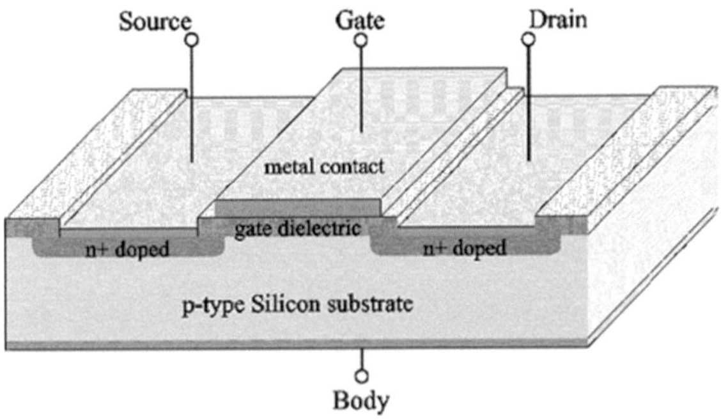

Fig. 2 Schematic of field-effect transistor. As the gate length shortens, the area (*A*) is reduced and the gate thickness (*t*) needs to be reduced to keep the capacitance constant

2.2 TiAlOx or TiO_2/Al_2O_3Superlattice Oxide Layers

Various groups are currently working on alternative high-K layers such as HfO_2, ZrO_2, Al_2O_3 and many others or combinations of those mentioned above. However, these gate oxide candidates exhibit only a short-term development potential and do not have all the properties required for the next generation of high-K layers. Particularly, all of those dielectric oxides mentioned above have lower dielectric constant (K < 22) than the new alloy we are proposing here. Our TiAlOx alloy oxide or TiO_2/Al_2O_3 superlattice layers have the potential for providing a higher dielectric constant (K ≥ 30) than all other alternative layers currently under investigation plus all other critical parameters mentioned above that will put our proposed high-K layers at the forefront of all amorphous gate oxides considered today (Fig. 3.) for the next generation of nanoscale CMOS devices. In fact, recent work performed by our group at Argonne demonstrated a TiAlOx layer with the highest dielectric constant among all amorphous oxides and the thinnest equivalent oxide thickness demonstrated today for amorphous high-K dielectric layers [31].

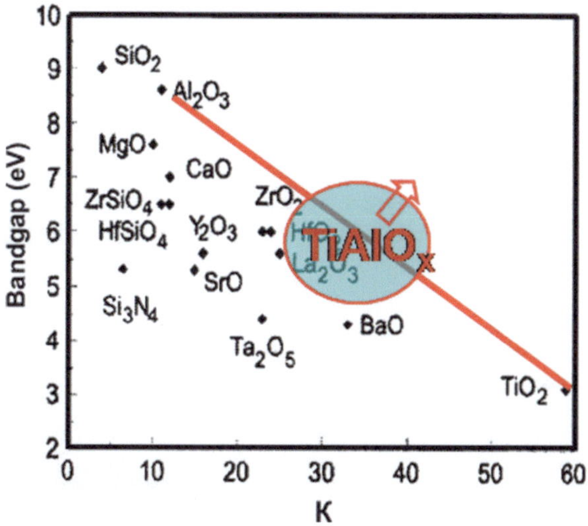

Fig. 3 Next-generation CMOS devices need a breakthrough in high-K films. TiAlOx or TiO_2/Al_2O_3 superlattice layers push the research frontier to enable the next-generation CMOS devices and embedded energy-storage capacitors for implantable microchips

TiAlOx, Al_2O_3, and TiO_2/ Al_2O_3 multilayers were first synthesized by ion beam and magnetron sputter-deposition techniques [31]. TiALOx layers were produced on Si without an SiO_2 interfacial layer by first depositing a TiAl layer (~3-nm thick) by sputter-deposition and then exposing the layer to atomic oxygen at room temperature. The use of atomic oxygen is critical to achieve a fully oxidized TiAl layer without inducing the formation of an SiO_2 layer at the interface, as it occurs when using molecular oxygen. Work done by other groups [1–7], using molecular oxygen, results in the formation of an SiO_2 interfacial layer due to the much lower chemical activity of molecular oxygen, which requires heating the layered structure to $\geq$ 400°C to induce the cracking of the oxygen molecule to induce the chemical reaction for oxidation. A detailed explanation of the difference between molecular and atomic oxygen for the synthesis of high-K dielectrics is presented next.

More recently, we synthesized Al_2O_3 and TiAlOx/Al_2O_3 nanolaminated layers via ALD, using the precursors of trimmethyl aluminum, tetraisopropyl titanium, and water at 300°C. The laminated thin films have the layer-by-layer conformal growth of amorphous Al_2O_3 with a growth rate of 1 Å/cycle and polycrystalline anatase TiO_2 with 0.3 Å /cycle. The leakage current density of applied electric fields was measured for several nanolaminates with different TiO_2 contents and different bilayer, single Al_2O_3 and TiO_2 layer, thicknesses. Leakage current density and permittivity were successfully tailored by adjusting the layer thicknesses in the nanolaminates.

2.3 Synthesis and Deposition Techniques of Oxide Films for Embedded Capacitors

Oxide films can be synthesized by magnetron sputter-deposition or more recently using atomic layer deposition (ALD), which provides a unique growth process capable of growing films one atomic layer at a time.

2.3.1 Sputter-Deposition

Sputter-deposition produces oxide films by creating plasma in front of an oxide target, using RF magnetron sources (see Fig. 4. for example) or ion beams. In both cases, ions from the plasma or from the ion source are accelerated toward the target with kinetic energies up to several hundred eV, or 1–2 keV range. Target atoms are ejected from the target surface and are transferred to the substrate surface to grow the films. To produce the oxide films, oxygen is injected in the chamber during the sputtering process.

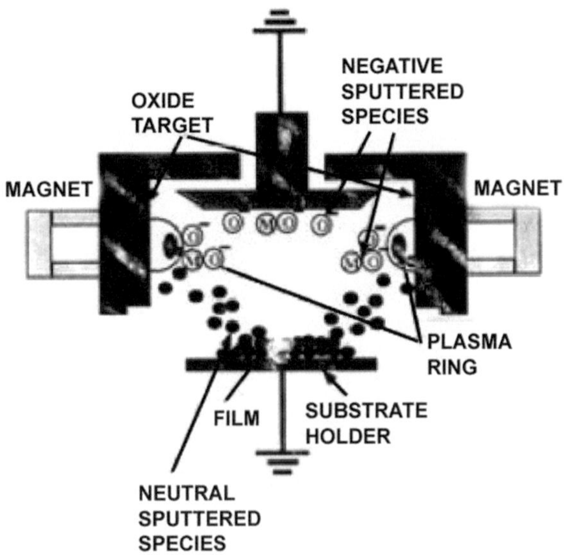

Fig. 4 Schematic of magnetron sputter-deposition concept used to produce oxide films such as the high-K dielectric described in this chapter

Sputter-deposition involves relatively low-cost hardware, and over flat surfaces produces uniform films with small thermal capacity, strong film adhesion, and corrosion resistance. However, sputter-deposition cannot produce films with uniform thickness on high aspect ratio features, such as trenches fabricated on Si substrates, which will be necessary to create large areas for producing large-capacitance capacitors.

2.3.2 Atomic Layer Deposition (ALD)

ALD, also known as atomic layer epitaxy (ALE), has gained more and more attention for ultrathin film deposition for applications in many fields such as optics and microelectronics processing [32]. ALD produces films with excellent conformality and precisely controlled thicknesses. ALD films are deposited layer by layer, so the control will be at the atomic layer level. The process is used to create ultrathin films of unique high-quality, pinhole-free, low stress, and flatness at low temperature ($\geq 300°C$).

The ALD process involves first exposing the substrate surface to a reactant A, which reacts with all of the initial surface sites. Then, after purging away the by-products from reaction A, the surface is exposed to reactant B. This reaction regenerates the initial functional groups and prepares the surface for the next exposure to reactant A. The film is grown to the desired thickness by repeating this ABABAB sequence, resulting in layer-by-layer controlled growth (see Fig. 5. for a schematic of the ALD process). A wide variety of thin films, specifically oxides, can be synthesized using the ALD technique. The metal precursors are the chemicals of ligand-bonded metals MLx (M= Al, W, Ta, Si, etc.; L = CH_3, Cl, F, C_4H_{11}, etc.). Nonmetal precursors usually are oxides, nitrides, sulfides such as H_2O, NH_3, or H_2S.

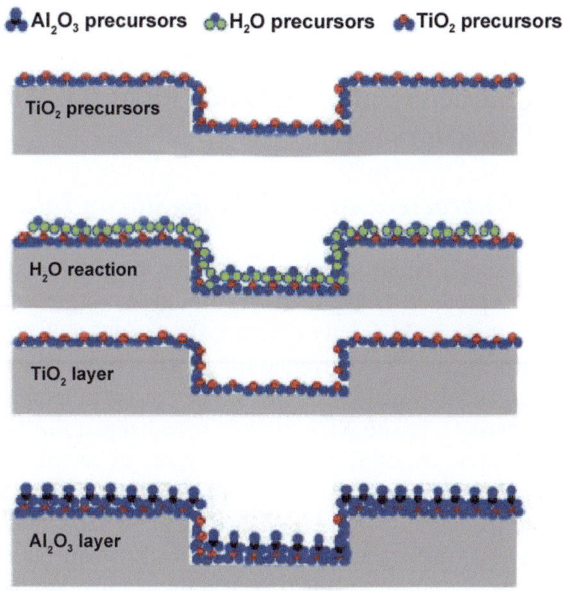

Fig. 5 Schematic illustration of one ALD reaction cycle to grow a layered TiO_2/Al_2O_3 heterostructure

Al_2O_3 layers were synthesized using trimethylaluminum [$Al(CH_3)_3$] (TMA) precursors and H_2O vapor alternative exposure of the substrate surface. A binary reaction for Al_2O_3 was separated into two half-reactions [33]:

$$\text{(A) AlOH} + \text{Al}\,(\text{CH}_3)_3 \rightarrow \text{Al-O-Al}\,(\text{CH}_3)_3 + \text{CH}_4$$

$$\text{(B) Al-O-Al}\,(\text{CH}_3)_2 + 2\text{H}_2\text{O} \rightarrow \text{Al-O-Al}\,(\text{OH})_2 + 2\text{CH}_4$$

The substrate surface is initially covered with hydroxyl (OH) groups through the first half reaction. The hydroxyl groups react with TMA to deposit a monolayer of aluminum-methyl groups and give off methane as a byproduct. Because TMA is inert to the methylterminated surface, further exposure to TMA yields no additional growth beyond one monolayer. In the second half reaction, this new surface is exposed to water regenerating the initial hydroxyl-terminated surface and again releasing methane. The net effect of one AB cycle is to deposit one monolayer of Al_2O_3 on the surface.

To synthesize TiAlOx, Al_2O_3 and TiO_2 layers are deposited alternatively. The precursors of titanium tetrachloride and H_2O were employed alternatively for the synthesis of the TiO_2. A binary reaction for TiO_2 was separated into two half reactions [34]:

$$\text{(A)}\ \text{TiOH}^* + \text{TiCl}_4 \rightarrow \text{TiOTiCl}_3^* + \text{HCl}$$

$$\text{(B)}\ \text{TiCl}^* + \text{H}_2\text{O} \rightarrow \text{TiOH}^* + \text{HCl}$$

Al_2O_3 and TiO_2 layers were synthesized alternatively to obtain TiAlOx thin films.

3 Characterization of Oxide Films for Microchip Embedded Capacitors

3.1 Characterization of Oxide/Silicon Interface and Structure

Studies of TiAl layer oxidation at room temperature with atomic oxygen are important to produce TAO stoichiometric layers without developing an SiO_2 interfacial layer that would greatly decrease the overall dielectric constant of a high-K/SiO_2 heterostructure. TiAl layers were grown at room temperature using sputtering deposition and exposed to atomic oxygen at room temperature, 100°C and 200°C for comparison. In situ XPS analysis revealed that the TiAl metallic layer is fully oxidized upon exposure to atomic oxygen, as revealed by the transformation of the metallic Ti^0 and Al^0 2p peaks into the oxide Ti^{4+} and Al^{3+} peaks (Fig. 6.), which is confirmed by high-resolution cross-section TEM studies in conjunction with EELS analysis, which confirmed an atomically sharp interface between the TAO layer and the Si substrate (Fig. 7.). Highly reactive atomic oxygen can bind to Ti and Al atoms in the metallic TiAl alloy layer with sufficient activation energy, even at room temperature. According to the Ellingham diagram, the TiAl alloy would oxidize preferentially at room temperature (RT), via oxygen binding to Ti and Al atoms, thus inhibiting the formation of an SiO_2 interface.

3.2 Electrical Performance of High-K Dielectric Oxide

The electrical properties of TAO are characterized by measuring the capacitance vs. voltage (C vs V) (Fig. 8.). The C vs V curve corresponds to the thinnest (0.4 nm

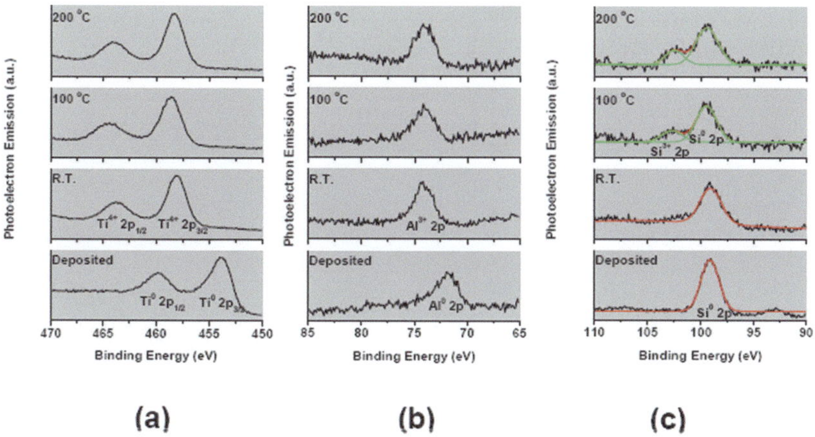

Fig. 6 XPS spectra of TiAl alloy layer, showing the transformation of an as-deposited metallic $Ti_{0.75}Al_{0.25}$ layer into fully oxidized $Ti_{0.75}Al_{0.25}O_3$ film at RT, 100°C, and 200°C for comparison. (**a**), (**b**), and (**c**) show the inhibition of an SiO_2 interface formation when the oxidation is produced via exposure of the TiAl films to atomic oxygen at room temperature

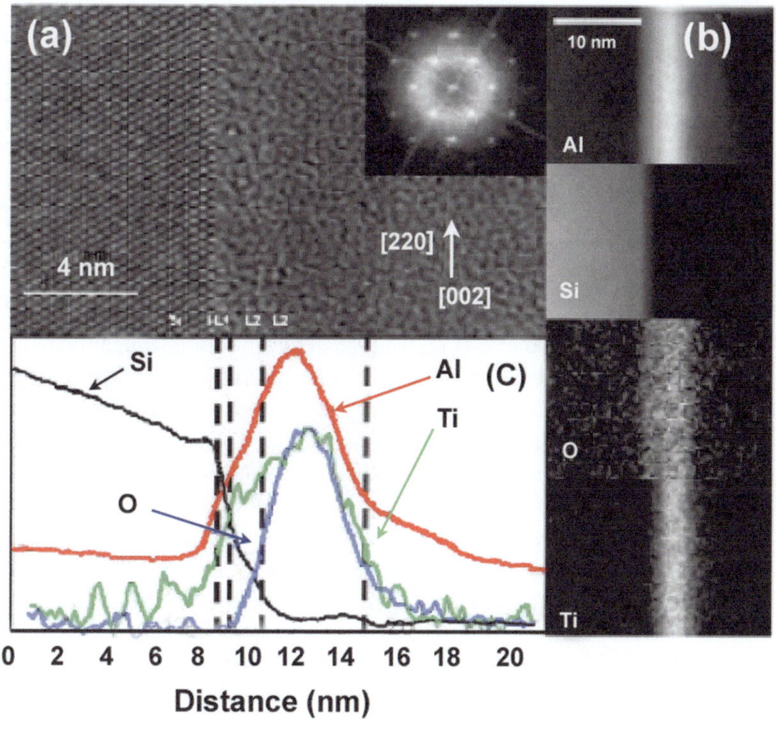

Fig. 7 High-resolution TEM microstructural plus EELS analysis of the same TAO layer on Si as in the XPS analysis shown in Fig. 6.

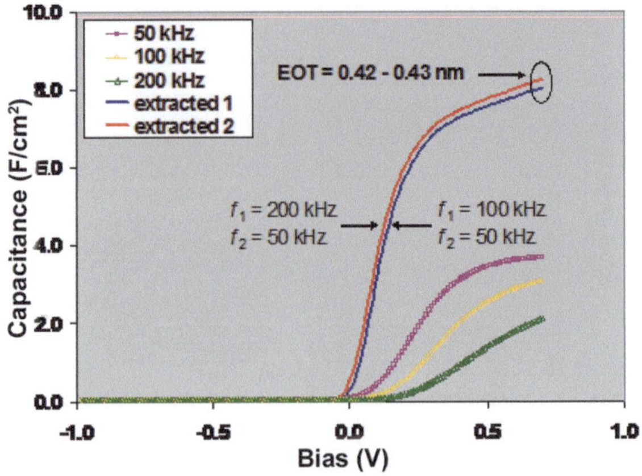

Fig. 8 C vs. V curve for a 3-nm $Ti_{0.75}Al_{0.25}O_3$ layer on Si

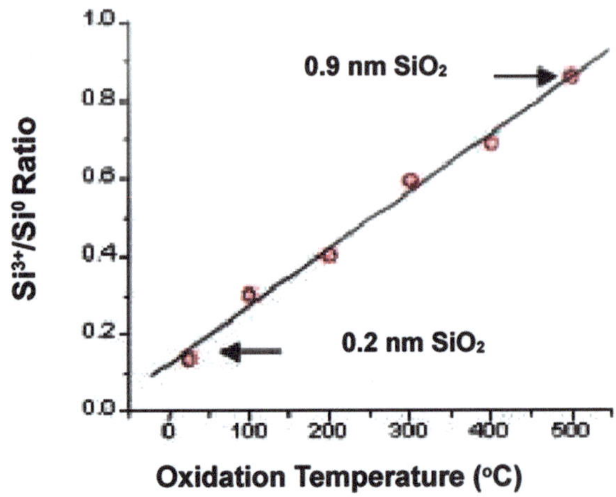

Fig. 9 Equivalent oxide thickness as a function of physical thickness of $Ti_{0.75}Al_{0.25}O_3$

equivalent SiO_2 oxide thickness) (Fig. 9.) high-K layer demonstrated today among all amorphous high-K oxides. The ratio of the K-line SiO_2 to K-line TAO can be precisely measured from the slope of the EOT-tox plot not shown here. The permittivity of the $Ti_{0.75}Al_{0.25}O_3$ layers estimated by this method is as high as 30, which surpasses the values of most popular high-K dielectrics, such as Al_2O_3, HfO_2, ZrO_2, and their alloyed oxides and is among the highest values that have been reported for the binary and ternary amorphous oxides. In addition to the highest permittivity demonstrated for amorphous high-K oxides, the 3-nm $Ti_{0.75}Al_{0.25}O_3$ layer exhibits a high capacitance density of 7.7–8.3 $\mu F/cm^2$, achieved mainly due to the suppression of SiO_2 interfacial layer formation. This high capacitance density will enable the fabrication of energy-storage capacitors with up to $\geq 10\ \mu F$.

The metallic Al at the Al/TiO_2 interface preferentially bonds with oxygen atoms, which explains the reduction of TiO_2 into Ti_2O_3 and TiO. With the second oxidation process, the top Al layer was converted into Al_2O_3 completely, showing only a single Al^{3+}peak. Since the TiO_2 reduction happened only at the interface where

Al metal is present, the thin Ti_2O_3 and TiO layer reoxidized into TiO_2 during the second oxidation process. Thanks to the good barrier properties of TiO_2 and Al_2O_3 and the low temperature process, the formation of an SiOx interlayer was practically inhibited, compared with one made by the prior method at 500°C. As a result, EOT $\sim$ 1.1 nm was achieved on the $Al_2O_3/TiO_2/Al_2O_3$ nanolaminate-based MOS structure, approximately 3 times lower than the one produced with a single oxidation process at high temperature. The measurements of C vs. V and leakage current vs. V for nanolaminated capacitors are shown in Fig. 10, showing promising electrical properties of high-K nanolaminates.

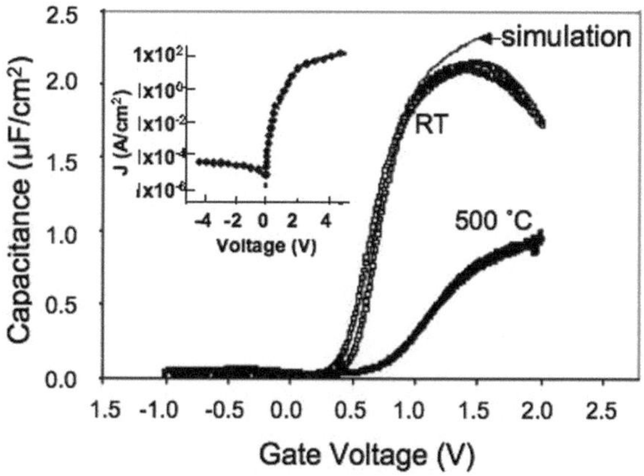

Fig. 10 Comparisons of C-V and I-V curves of the $Al_2O_3/TiO_2/Al_2O_3$ stack fabricated by the 2-step method at room temperature with the stack made by direct oxidation at 500°C

4 Challenges for Oxide Films as High-K Dielectric Films for Microchip-Embedded Capacitors

The applications of oxide coatings to the next generation of nanoscale high-K dielectric gate oxide for CMOS and for energy storage microchip-embedded capacitors require that a high-K material has high breakdown voltage, low leakage current, and is suitable for integration with Si-based CMOS devices. The TiAlOx or Al_2O_3/TiAlOx-layered heterostructure being investigated and developed by our group shows some promising results. The TiAlOx high-K-based capacitor does not require the presence of a hermetic coating for the capacitor to function properly, since it would work for a microchip encapsulated in a hermetic package implanted in the body. However, since TiAlOx may be used alone or combined with another coating such as Al_2O_3 and UNCD (ultrananocrystalline diamond) to provide an alternative hermetic encapsulating coating for implantable microchips (see Chapter 3 for details), the TiAlOx material must be biocompatible, mechanically stable, and provide electrical insulation and corrosion protection. Although promising results have been demonstrated, some challenges remain.

For application as a high-K dielectric layer for nansocale CMOS usable for future high-performance retina microchips, it is necessary to understand the TiAlOx/Si interface to reduce the density of charge-trapping states at that interface, which is currently the main limitation for insertion into a commercial CMOS device.

Research should focus on investigating a novel TiAlOx alloy oxide, TiO_2/Al_2O_3, or $TiAlOx/Al_2O_3$ superlattice layers on Si substrates with minimal or no SiO_2 interfacial layer formation for high-K dielectric nanoscale CMOS gates, high-K layer for embedded energy-storage capacitors for the retina microchip, and as bioinert coatings.

For application as a high-K dielectric layer for microchip-embedded energy-storage capacitors, the TiAlOx layer has to sustain a high breakdown voltage $\geq$ 20 V) and provide low leakage for high operating voltages.

For application as a hermetic encapsulating coating, the challenges are to determine what composition yields the lowest leakage current in a saline environment, demonstrate a reliable coating performance, and eventually determine if it can be integrated with other coatings in a heterostructure hermetic coating layer.

The ongoing work on the development of high-K oxide TiAlOx has been focused on those issues, and it is hoped that they will not hinder the application of this multifunctional material.

5 Conclusions

Research on the new multifunctional TiAlOx and TiO_2/Al_2O_3 multilayer has demonstrated that these materials can provide high-K dielectric layers with record low leakage current and the thinnest equivalent oxide thickness for application as a nanoscale CMOS gate. In addition, these oxide layers are showing high capacitance density that, combined with microfabrication of large arrays of high aspect ratio ridges, will provide the basis for producing capacitors with $\geq$ 10 μF capacitance required for the retinal microchip.

The multifunctional thin films discussed in the chapter may find application in a broad range of implantable biomedical devices, including a new generation of neural prostheses.

Acknowledgments The author wishes to acknowledge support from the U. S. Department of Energy, BES-Materials Science for work in the Materials Science Division, under contract W-31-109-ENG-38. The work at the Center for Nanoscale Materials and at the Electron Microscopy Center for Materials Research at Argonne National Laboratory was supported by the U.S. Department of Energy-Office of Science under Contract No. DE-AC02-06CH11357 by U Chicago Argonne, LLC. The author also acknowledges the many colleagues and postdoctorals who have made substantial contributions to the work discussed in this chapter over the years, namely: J.A. Carlisle, W. Fan, B. Kabius, R. Baragiola, and E.A. Irene.

References

1. Robertson J (2004) High dielectric constant oxides. Eur Phys J Appl Phys 28:265.
2. Robertson J (2006) High dielectric constant gate oxides for metal oxide Si transistors. Rep Prog Phys 69: 327.

3. Wilk G, Wallace RM, Anthony JM (2001) High-gate dielectrics: Current status and materials properties considerations. J Appl Phys 89:5243.
4. Wallace RM, Wilk GD (2003) High-κ dielectric materials for microelectronics. Crit Rev Solid State Mater Sci 28:231.
5. Wallace RM, High dielectric constant gate oxides (private communication).
6. Huff H, Gilmer D (eds) (2004) High K Gate Dielectrics. Berlin: Springer.
7. Houssa M (ed) (2003) High Dielectric Constant Materials: VLSI MOSFET Applications. London: IOP.
8. Demkov AA, Navrotsky A (eds) (2005) Materials Fundamentals of Gate Oxides. Dordrecht: Springer.
9. Sugizaki T, Kohayashi M, Ishidao M et al. (2003) Novel Multi-bit SONOS Type Flash Memory Using a High-k Charge Trapping Layer. Symposium on VLSl Technology Digest of Technical Papers, p. 27.
10. Sim H, Samantaray CB, Lee T et al. (2004) Electrical and structural characteristics of high-k gate dielectrics with epitaxial Si_3N_4 interfacial layer on Si(111). Jpn J Appl Phys Part 1 43(12):7926.
11. Klein TM, Niu D, Epling WS et al. (1999) Evidence of aluminum silicate formation during chemical vapor deposition of amorphous Al_2O_3 thin films on Si(100). Appl Phys Lett 75:4001.
12. Gusev EP, Copel M, Cartier E et al. (2000) High-resolution depth profiling in ultrathin Al_2O_3 films on Si. Appl Phys Lett. 76(2):176.
13. Chin A, Wu YH, Chen SB et al. (2000) High quality La_2O_3 and Al_2O_3 gate dielectrics with equivalent oxide thickness 5–10 Å. Tech Dig VLSI Symp, p. 16.
14. Roy PK, Kizilyalli IC (1998) Stacked high-gate dielectric for gigascale integration of metal–oxide–semiconductor technologies. Appl Phys Lett 72:2835.
15. Lu Q, Park D, Kalnitsky A et al. (1998) Leakage current comparison between ultrathin Ta_2O_5 films and conventional gate dielectrics. IEEE Electron Device Lett 19:341.
16. Fleming RM, Lang DV, Jones CDW et al. (2000) Defect dominated charge transport in amorphous Ta_2O_5 thin films. J Appl Phys 88:850.
17. Kadoshima M, Hratani M, Shimamoto Y et al. (2003) Rutile-type TiO_2 thin film for high-k gate insulator. Thin Solid Films 424(2):224.
18. Bera MK, Maiti CK (2007) Charge trapping properties of ultra-thin TiO_2 films on strained-Si. Semicond Sci Technol 22:774.
19. Yu Z, Ramdani J, Curless JA et al. (2000) Epitaxial oxide thin films on Si (001). J Vac Sci Technol B 18:2139.
20. McKee RA, Walker FJ, Chisholm MF (1998) Crystalline oxides on silicon: The first five monolayers. Phys Rev Lett 81:3014.
21. Robertson J, Xiong K, Falabretti B (2005) Point defects in ZrO_2 high K gate oxide. IEEE Trans Device Mater Reliab 5(1):84.
22. Huang AP, Fu RKY, Chu PK et al. (2005) Plasma nitridation and microstructure of high-k ZrO_2 thin films fabricated by cathodic arc deposition. J Crys Growth 277:422.
23. Nishikawa T, Otsuka T, Morita K (2002) Reduction of leakage current by HfO_2 high K dielectric film stacked on the ferroelectric layer of a MFIS structure. Integr Ferroelectr 48(1):41.
24. Chatterjee S, Kuo Y, Lu J et al. (2006) Electrical reliability aspects of HfO_2 high-k gate dielectrics with TaN metal gate electrodes under constant voltage stress. Microelectron Reliab 46(1):69.
25. Aoyama T, Sugita Y, Morisaki Y et al. (2002) CMOSFETs using HfO_2 High-k gate dielectrics. Proc Symp Semicond Integr Circ Technol 63:6.
26. Iwai H et al. (2002) Advanced gate dielectric materials for sub-100 nm CMOS. Tech Digest Int Electron Devices Meeting (IEEE).
27. Busani T, Devine RA (2005) The importance of network structure in high-K dielectrics: $LaAlO_3$, Pr_2O_3, and Ta_2O_5. J Appl Phys 98:044102.

28. Shao QY, Li AD, Cheng JB et al. (2005) Growth behavior of high k $LaAlO_3$ films on Si by metalorganic chemical vapor deposition for alternative gate dielectric application. Appl Surf Sci 250(1)4:14.
29. Robertson J (2000) Band offsets of wide-band-gap oxides and implications for future electronic devices. J Vac Sci Technol B 18:1785–1791.
30. Osburn CM, Kim I, Han SK et al. (2002) Vertically scaled MOSFET gate stacks and junctions: How far are we likely to go? IBM J Res Dev 46:299.
31. Auciello O, Fan W, Kabius B et al. (2005) New TiAl alloy high-K dielectric layer for next generation integrated circuit gates. Appl Phys Lett 86:1.
32. Tripp MK, Fabreguette F, Herrmann CF et al. (2005) Multilayer coating method for x-ray reflectivity enhancement of polysilicon micro-mirrors at 1.54 Å wavelength, Micromachining Technology for Micro-Optics and Nano-Optics III. Johnson EG, Nordin GP, Suleski TJ (eds) Proceedings of SPIE Vol. 5720 (SPIE), Bellingham, WA, 241.
33. Dillon AC, Ott AW, George SM et al. (1995) Surface chemistry of Al_2O_3 deposition using $Al(CH_3)_3$ and H_2O in a binary reaction sequence. Surf Sci 322:230.
34. Lakomaa EL, Haukka S, Suntola S (1992) Atomic layer growth of TiO_2 on silica. Appl Surf Sci 60/61:742.

An Effective Design Process for the Successful Development of Medical Devices

Mike Colvin

Abstract The most important point in the successful development of a medical device is the proper overall design. The quality, safety, and effectiveness of a device are established during the design phase. The design process is the foundation of the medical device and will be the basis for the device from its inception till the end of its lifetime. There are domestic and international guidelines on the proper steps to develop a medical device. However, these are guides; they do not specify when and how to implement each phase of design control. The guides also do not specify to what depth an organization must go as it progresses in the overall developmental process. The challenge that faces development organizations is to create a design process plan that is simple, straightforward, and not overburdening.

This chapter gives an overview of the design and development of a medical device. It also gives some guidance and advice about the developmental process. Finally, it gives some case examples of pitfalls that real companies have experienced in their medical device development.

Contents

M. Colvin (✉)
Boston Scientific Neuromodulation Corporation, Valencia, CA 91355, USA
e-mail: michael.colvin@bsci.com

D.D. Zhou, E. Greenbaum (eds.), *Implantable Neural Prostheses 2*,
Biological and Medical Physics, Biomedical Engineering, DOI 10.1007/978-0-387-98120-8_11,

1 Introduction

The first and most important point in the development of a medical device is the proper overall design. It is believed that the intrinsic quality, safety, and effectiveness of a device are established during the design phase. FDA statistics show that 30% of all medical device recalls are a result of inadequate design control [1]. The design process is the foundation and skeleton that is the basis of a medical device from its inception to the end of its lifetime. The design process is a living and an iterative process; it evolves and improves with time.

There is no universal design process that will work for every medical device. However, the requirements set forth for design controls in the FDA Quality System Regulation (Chapter 21 Code of Federal Regulations Part 820) and ISO Standard 13485:2003 do construct an effective medical device design process. However, it is a guide; it does not cover how and when to implement each phase of design control. It also does not specify to what depth an organization must go to as it progresses in the overall development process. The challenge that faces development organizations is to create a design process plan that is simple, straightforward, and not overburdening.

2 The Design Control Process for the Development of Medical Devices

2.1 Overview of the Design Control Process

Design control is part of a larger system of checks and balances that the medical device manufacturer should already have in place as part of their overall development plan. Definition and timing of the process are important because formal design control guidelines are not intended to apply to the research and feasibility stages of product development. The difficult question for any organization (especially the less mature) is determining when research and feasibility ends, and development begins. A comprehensive product development process (PDP) with specific milestones and development phases that include clearly defined metrics can minimize any confusion that arises.

The overview of design control is shown in the FDA's waterfall diagram [2] in Fig. 1.

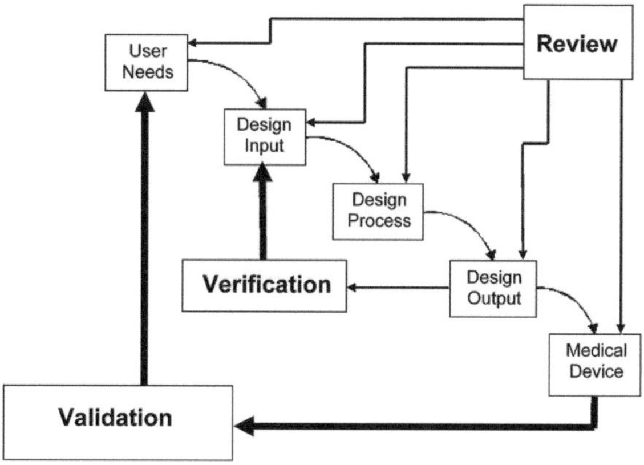

Fig. 1 The FDA's waterfall design control process

The waterfall design process basically shows that a device begins with user needs. This leads to a design input, then to a design process, followed by the design output, and finally a finished medical device. During each of these phases, there is a review that takes place, and there is verification and a validation of the device during the design output and finished medical device phases, respectively.

A more detailed product design and development phase's timeline is shown in Fig. 2. It is broken down into the concept phase, followed by the requirements and definition phase, followed by the design and verification phase, followed by the validation phase, and finally the design transfer phase. The various reviews that should occur are also shown (covered in more detail later in text).

An example of a medical device development tree detailing the processes, decisions, and documents that should occur is shown in Fig. 3. As can be seen, there are various levels of processes, decisions, and documentation that should occur in the development of a medical device. The process may involve several iterations in order to develop a successful medical device. The approval sections are of particular importance and should not be skipped.

2.2 Research and Development Phase

The distinction between research and development is of utmost importance and often hard to delineate. Research activities may be undertaken in an effort to determine new business opportunities or basic characteristics for a new product. It may be reasonable to develop a rapid prototype to explore the feasibility of an idea or design, for example, prior to developing design input requirements (covered below).

Formal design control is not required for feasibility/research studies. However, once it is decided that a design concept is feasible and can be developed, i.e., initial results indicate that the device works as intended, an assessment must be made to determine the adequacy of the existing design. The biggest challenge for

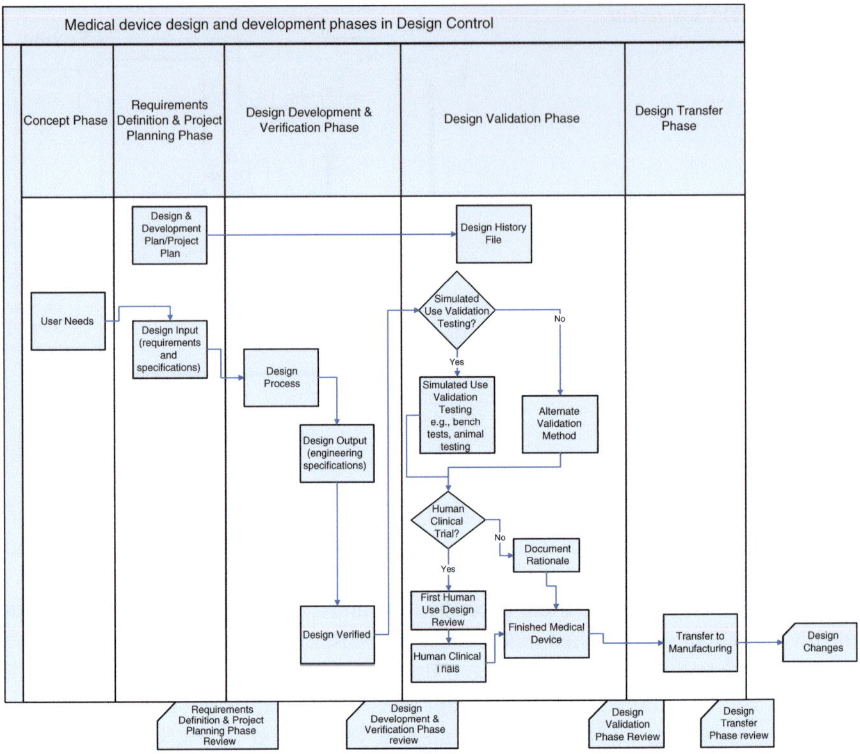

Fig. 2 Medical device design and development phases in design control

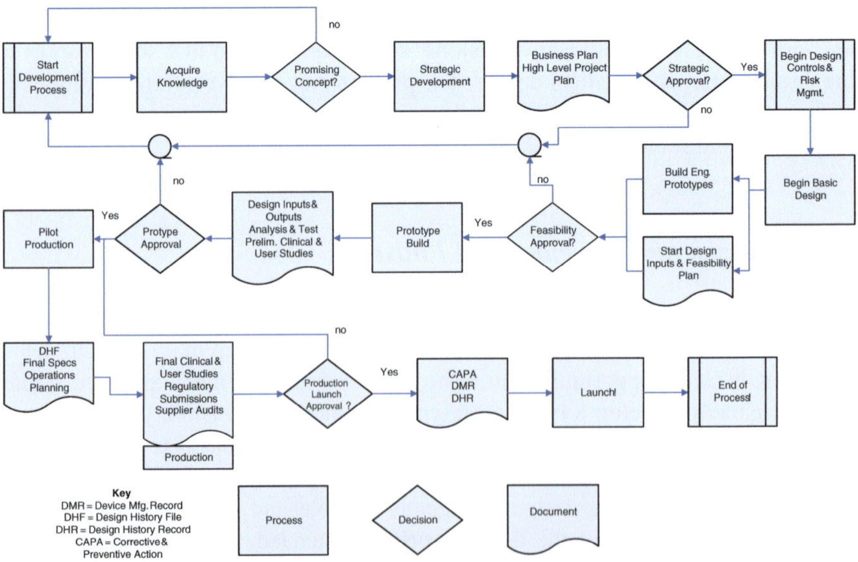

Fig. 3 Medical device development tree

the manufacturers of medical devices is to understand how to implement and integrate federal, legislative, and regulatory requirements properly within their existing quality systems, especially into design control.

Manufacturers should avoid the danger of equating the prototype design with a finished product design. The objective at this point is to demonstrate "proof of concept." Prototypes manufactured during the feasibility/research phase lack safety features and additional functions necessary for a finished device and are developed under conditions which prevent adequate design controls, e.g., in a prototype clean-room. On the other hand, many design organizations can never get out of the research mode; they constantly are tweaking the feasibility design in a genuine attempt to "improve" the device performance or add features, resulting in unnecessary development delays and increased costs. It is very easy to keep adding "enhanced features" to the design of a medical device. It is critical and timely in the development of a medical device that the organization is aware and in agreement when the process shifts from research to development.

In the development process, a device is designed to meet design input requirements (through design verification) and user (or) clinical needs (through design validation). Testing begins in the laboratory using animal and/or bench tests. Once the design and operating parameters have been subject to adequate preclinical tests, the developer may wish to conduct an initial limited study (called feasibility clinical studies/trials) in humans to confirm the design and operating specifications before beginning an extensive clinical study (called pivotal clinical studies/trials). The initial limited study may indicate that minor or major changes in the device are necessary before proceeding. It may also indicate that the device does not meet expectations and the project may be terminated. The performance of the device in the initial limited study may also be used to establish the parameters for the larger clinical study, such as sample size and the metrics of measurement.

2.3 General Design Control Philosophy

Federal and Government regulatory bodies are judicious by developing and enforcing regulations to guard against inadequately developed (i.e., unsafe) or poorly designed (i.e., poor performance) medical products. Organizations often overreact to these regulations and create overly elegant and overly detailed processes that they cannot follow practically. They often implement an overly strict design control and quality policy too early in the development process. Over zealous internal compliance audits that focus on the finer points and not on the big picture often exacerbates the problem. This typically results in more, not fewer, less important requirements. The overall requirements become overwhelming and stifle progress. Ironically, the net result is that the organization can get into trouble with regulatory agencies because of they cannot follow their own self-imposed requirements.

Companies should not implement detailed processes/procedures during the product research phase. During the research phase, you are making unique product designs that require highly skilled and creative people at this phase. Product development requires a unique blend of visionary and detail-oriented individuals. It is more about craftsmanship; the goal at this point in time is not to attempt a high-volume production.

Detailed processes/procedures are best implemented well within the development phase and continue on with the transfer from R&D to the manufacturing line. Relegating product feasibility to overly detailed processes/procedures is more likely to achieve mediocrity than excellence, by oppressing creativity. On the other hand, poor planning and too little detail lead to vague and undefined targets within the R&D development team. Start-up organizations with breakthrough technologies led by a visionary focus are particularly vulnerable to having too little details.

The ideal process should provide a framework that enables the organization to document what they do in a format that follows the applicable and necessary regulations and will ensure that the quality, safety, effectiveness, regulatory, and business needs are met.

It is important to realize that the design processes may apply to various facets of the organization having different styles and timelines. Such facets are related to products, services, and software, as well as to manufacturing processes. Organizations must find the balance between looking at the small and big picture and include enough, but not over, detail to ensure consistency and efficiency in designing a successful product.

2.4 Design and Development Planning

The objective of this phase is to come up with a design and development plan that spans across all design control phases (i.e., from design input to design validation and design transfer). The design and development plan is also called the project plan in some organizations.

In this phase, the design and development plan must be defined and should cover all phases of design control [3–4]. The following are examples of what should be covered. It should include a description of the goals and objectives of the overall design and development project, i.e., what is to be developed. Included should be a definition and delineation of organizational responsibilities with respect to each phase during the design and development; this should also include interface with any contractors. It should identify the major tasks to be undertaken, deliverables for each task, and individual or organizational responsibilities (staff and resources) for completing each task. In this phase, scheduling of major tasks and milestones to meet overall project time constraints should be delineated. In addition, major reviews and decision points should be spelled out.

The design and development phase should also include the selection of reviewers, the composition of review teams, and procedures to be followed by reviewers. Typically, the review team should at least include members from the following disciplines: R&D, regulatory affairs, quality assurance, production/manufacturing, and marketing. In addition it should include controls for design documentation.

Design and development planning enables the management to exercise greater control over the design and development process by clearly communicating policies, procedures, and goals to members of the design and development team and providing a basis for measuring conformance to quality system objectives.

Design and development activities should be specified at the level of detail necessary for carrying out the design and development process. The extent of design and development planning is dependent on the size of the organization and the size and complexity of the product to be developed. Some organizations may have documented policies and procedures which apply to all design and development activities. For each specific design and development program, such organizations may also prepare a plan which spells out the project-specific elements in detail and incorporates the general policies and procedures by reference. Other organizations may develop a comprehensive design and development plan, which is specifically tailored to each individual project.

2.5 Design Input

Design input should occur formally late in the research phase or early in the development phase [2, 5]. Design input requirements (DIR) are derived from user needs and are expressed in engineering terms. Design input establishes the requirements that will ensure the device will meet the needs of the intended users. Design input is undoubtedly the most important element in the development of a successful medical device. Design input forms the basis for all design and development activities; it is also the standard document on which all design–verification and design-validation activities are based.

Companies usually require a business-marketing/technical review and approval of the design input requirement. This ensures that they are properly investing their precious resources in developing the proposed medical device. Unfortunately, organizations often cut short this part of the project because they rush to get into development in order to meet a perceived window of opportunity in the marketplace. However, the result is often an insufficient DIR, ultimately resulting in unsatisfactory outcomes such as major development delays due to redesign/respecification, unacceptable device performance, overspending or project cancellations.

Hence, spending enough time developing a solid DIR will minimize significant and unnecessary changes during design and development and help in speeding the time to introduce the product to market. The DIR also helps to delineate the distinction between the research/feasibility phase from the design and development phase.

Design input requirements must be comprehensive. This may be quite difficult for manufacturers who are implementing design control for the first time. It may be helpful to realize that design input requirements fall into three broad types. Virtually every product will have requirements of all three types as illustrated in Table 1 below:

In order to develop an adequate DIR, a number of sources of information should be considered. Some of these are DIR of similar previous medical device(s), overall market analysis/growth potential, analysis of competitors' devices, expected product

Table 1 Types of design-input requirements

	Functional requirements	Performance requirements	Interface requirements
Definition	Specifies what the device does	Specifies how much or how well the device must perform	Specifies device characteristics that are critical for compatibility with external systems
Contents	Operational capabilities of the device, the inputs for the device, and the resultant outputs of the device	– Speed – Strength – Response times – Accuracy – Range and Limits of Operation – Use environment (temperature, humidity, shock, vibration, electromagnetic compatibility)	– User and/or patient interface – Characteristics outside the control of developers – Characteristics mandated by external systems

lifetime, and current medical practices. System considerations ensuring compatible operation with other medical devices to achieve the medical need(s)/purpose(s) also should be included. Review of complaints, field performance and manufacturing issues of previous similar medical device(s) should be added.

Other factors that should be considered for a proper DIR are human factors input for the new medical device, the environment in which the medical device must function (temperature, electromagnetic compatibility, ambient light and acoustic noise, hermeticity, etc.), new technologies, therapies and results of research activities, customer input, marketing/business needs and considerations, domestic and foreign regulatory strategy, domestic and foreign reimbursement strategies, and is the product/procedure too invasive?

Decisions will be needed to resolve conflicting information sources and arrive at a concise description of the medical factors, user needs, and purpose(s) to be addressed by the medical device. Following these decisions, it is necessary to define, from a black box perspective, the requirements of the medical device in measurable engineering terms. The actual requirements in the DIR will fall into two broad categories: the standards that the product shall meet such as sterility, biocompatibility, International electrical standards and American National Standards Institute (ANSI), etc., and those specific to the medical device (i.e., functional, performance, and interface requirements).

In general, the DIR is adequate if the design verification and validation test plans can be written and risk-management activities can be completed. Hence, if the product was developed exactly as described in the DIR, the overall residual risk would be acceptable.

Device development from the earliest stages requires active involvement of practicing clinicians. Clinician/inventors are frequently involved in creation of the device concept and are often integral members of the design team performing the majority of the early animal studies. Through this involvement, the clinician/inventor obtains intimate knowledge of device performance and failure modes. Safety concerns during first clinical use and pilot phase mandate participation by these clinician/inventors. The clinician/inventors frequently take leadership roles and have equity positions in the company developing the device. These interests present important conflicts of interest which must be addressed to ensure patient safety, data integrity, and public trust in the process. Many institutions have set up formal processes to address these conflicts of interest.

2.5.1 Case Studies of Design Input Requirements

The following are two case examples where the DIR was insufficient:

Case 1

A small class III medical device manufacturer was developing a novel in vivo diagnostic combination device (combination biological and a device). At the completion of the initial design input process, it was determined that the device could not be sterilized by traditional sterilization methods. The design team decided to continue with the device development, testing it in vitro for the interim until a sterilization method could be developed. Several hundreds of thousand dollars and after a four-year delay, a novel technique was finally developed to sterilize the device. It was a close call; the company had spent many years and millions of dollars in development on a device that potentially could not be sterilized. The risk was double-fold, they might have never developed the novel process, and or it may have never received regulatory approval.

Case 2

A medium class III medical device company wanted to improve on a class of implants with a proven track record of long-term use. The improvements were to be a less invasive surgical technique, a more reliable device, and a change in the implantation site. The DIR was carefully assembled. Part of the process was to have a couple of key opinion leaders weigh in on the design; this occurred with positive physician feedback and the project was funded and a team was assembled. After developing this device for over two years and spending over 10 million dollars, the project was cancelled. The reason, a survey was conducted with a group of physicians within this specialty. The consensus among the group was that they would never implant a design of this type in patients. All of this time and money was lost because insufficient market analysis was conducted.

2.6 Design Output

Design output is the result of each stage of the design and development process and consists of the device, its packaging, and labeling [2, 5]. The finished design output is the design output at the end of the design-validation phase prior to design transfer. The finished design output is usually represented by the following: engineering drawings used to make the device that also contain the performance specifications, labeling (including the instructions for use for the patient/clinician), packaging specifications, and manufacturing specifications. The above contents make up the device master record (DMR). The DMR is the basis for manufacturing the device.

2.7 Design Review

These ongoing meetings (a minimum of 2 formal design reviews are required, however many more are recommended) occur throughout the design process. They verify that the development of the design continues to meet the requirements outlined in the design input stage [2, 5].

Some of the design reviews that may be conducted are:

- Project development kick-off review
- Design input requirements review
- Design verification phase review
- First human use review
- Design-validation phase review and
- Design-transfer phase review

It is important to have a cross-functional team (engineering, regulatory, QA) perform the review. At least one member of the design review should be independent (i.e., a person who is not involved in the project). Members of the scientific advisory board and experienced, technically astute engineers independent of the development organization are priceless to weigh in and should be included.

This is one area where the momentum of the design process/project can be pushed-on, redirected, slow-downed, or even stopped. These meetings are critical for the project team to do a careful "sanity check." This could have prevented the result in Case 2 in the design input section.

2.8 Design Verification

The design verification identifies whether or not the design output requirements continue to meet the design input requirements [2, 5]. This stage should ideally involve persons who are not responsible for the design under review. Results are documented in the design history file (explained in section below). It verifies that the product was made correctly according to the design input requirements. The DIR is, of course, the basis for all verification activities. Four guiding principles will enable success at this point:

(1) Test samples should be virtually identical to the first human use devices (including sterilization and packaging). If the test samples are not identical, an equivalency analysis should be made.

(2) The verification/validation activities must attempt assertively to uncover issues rather than seeking to merely demonstrate that the medical device meets the DIR.
(3) All design issues uncovered must be addressed. If design changes are made, an appropriate subset (regression testing) of the original verification work must be repeated. If design changes are not needed for some issues, or issues are not completely corrected, risk management must be applied to ensure that any residual risk is acceptable.
(4) Verification plans have a large element of art. Experienced, technically astute test engineers independent of the development organization are invaluable to weigh in.

Three additional principles apply only to the verification activities:

(1) Verification is carried out against every DIR requirement.
(2) Verification activities are not limited to testing only. Design analysis can be an appropriate alternative to testing in some applications. However, avoid, at all costs, the lure to reduce the scope of these activities because of some logical explanation as to why testing is unnecessary.
(3) Every requirement in the DIR must be verified even if the requirement is carried over from a previous device. Whenever possible, perform step stress testing to failure, this is used to ensure there is adequate design margin in the device. This discussion addresses primarily the final medical device. However, the expectation is that similar verification activities will also be carried out for all critical subsystem and custom component specifications.

2.9 Design Validation

Design validation follows successful design verification, and may be part of the overall risk analysis [2]. This includes testing of production units under actual or simulated use conditions. Validation demonstrates that all verifications have been completed and that only safe and effective devices are produced. It confirms that the right product was made.

Validation activities are carried out against the medical need(s)/purpose(s) of the user. If the medical device is an enhancement of a previous medical device (and the same functional and environmental requirements are contained in the DIR as the previous device), it may be necessary only to validate the new medical need(s)/purpose(s) of the user.

Simulated use of the product should precede actual human use. The simulations should include stressing the device by using it at nominal as well as at the device limit conditions. Reasonable misuse conditions should be included in the analysis.

All labeling, packaging, and training materials should be included in the validation to ensure that everything stated reflects accurate operation of the device.

The individuals performing the simulated use, including the labeling and training, should be very knowledgeable of the operation of the device, of similar devices, and current medical practice. An attempt to uncover device issues, including practical misuse, is necessary at this point.

If human use is included in the validation, it should be representative of the range of use the product will encounter, for example large and small hospitals, clinics, key opinion leaders, garden variety physicians, and typical environments (home, emergency vehicle and hospital, etc.), and all specified therapies, etc.

2.9.1 Case Studies of Design Verification and Validation

The following are two case examples where the design verification and validation was insufficient:

Case 1

A large medical pacemaker supplier had approved devices (pacemakers) on the market that had a seal with a propensity to leak. Moisture would leak into the devices causing device failure. This could cause loss of consciousness, heart failure, or death of the patient. The product was recalled. This failure mode should have been caught in real-time or accelerated-aging soak testing. It is obvious that this testing was "cut-short" in order to meet deadlines.

Case 2

A large medical implantable cardioverter defibrillator (ICD) supplier had approved devices (ICD) on the market. A project was put in place to narrow the diameter of the pacing leads; this was an attractive option for this medical specialty. The leads were redesigned with thinner insulation and conductor wires. However, after regulatory approval and market release it was found that the leads would fatigue and fracture in the patient causing inappropriate shocks or loss in therapy all-together. This could cause loss of consciousness, heart failure, or death. The product was recalled. This failure mode should have been detected in fatigue testing. Again, it is obvious that the product was placed into the market without the proper animal or bench testing.

2.10 Design Transfer

The design transfer addresses the transfer of a design to production and review of those specifications [2, 5]. It ensures that the finished production devices have the same qualities established during the design and development phase. Design transfer should be thought of as the ongoing interaction between manufacturing and the R&D design team throughout the development of a medical device (i.e., including design changes), as opposed to a point in time at the end of the development phase. Design transfer should accomplish the following objectives:

- Identify and find solutions to designs that are difficult to manufacture or result in a low product/process yield.

- Perform a process failure modes and effects analysis (FMEA) and use it to drive process improvements as well as process verification/validation activities. Review and update the FMEA as process changes occur.
- Identify manufacturing processes/tooling/inspections, etc., that introduce or fail to detect defects.
- Build the validation test samples using the final design revision, the same people, the same processes/tooling and the same components/materials, same facility, as will be used to manufacture the first human use devices.
- End up with a manufacturing process that replicates successfully the design an unlimited number of times without introducing unacceptable risk.

2.11 Design Changes

Design change [2, 5] is the process and documentation which shows that all modifications to the design after the design review have been identified, reviewed, and approved. If changes are made, they must be verified and, if necessary, validated again; this is a critical step that often is not given enough importance and has resulted in many product recalls.

2.11.1 Case Studies of the Design Changes Process

The following are two case examples where the design changes process was insufficient:

Case 1

A large implantable cardioverter defibrillator company made two "noninvasive corrections" to the device firmware. This resulted in the device to fail the "Skipped-Charge Shock" check and resulted in a sensor noise anomaly. The product was recalled. An example of what was considered a "noninvasive correction" to the firmware resulted in a product recall. This likely could have been caught in proper design reviews and design-verification activities.

Case 2

A large medical implantable cardioverter defibrillator (ICD) supplier had approved devices (ICD) on the market. A design change had been made to narrow the diameter of the leads. The product was approved and released to market. It was found that the leads would fatigue and fracture in the patient. In some cases the fractures were linked to perforating the patient's heart. This could cause loss of consciousness, heart failure, or death. The product was recalled. Another example of insufficient animal or bench fatigue testing.

2.12 Risk and Hazard Analysis

These activities are required during most phases of development. The extent of this testing is proportional to the risks associated with the product. If risks are unacceptable, the device may need to be redesigned, or warning labels should be attached. It is important that any changes do not introduce new risks or hazards. Though not usually thought considered of as part of design controls, risk management is included in ISO Standard 13485:2003 and the basic elements are addressed in the preamble to the FDA Quality System Regulations. Risk and hazard analysis ensures four things:

1. If the device operated precisely as defined by the instructions for use, it is safe.
2. The implementation of the DIR in hardware/software does not adversely affect safety.
3. Manufacturing of the device (including critical subsystems and custom components) does not adversely affect safety.
4. Sufficient labeling and training information is included with the medical device to enable a medical professional or patient to use the device safely.

2.13 Design History File

The design history file is a compilation of all the records that were generated as part of the design and development activities undertaken to develop the device. It is unusual for all design history documents to be filed in a single location. For example, the file may include e-mail, memoranda, or may reside in design engineers laboratory notebooks. The design history file can also be a list index pointing to all the records. The intent is simple; it is so that the manufacturer have access to the information when needed.

3 Conclusion

The first and most important point in the development of a medical device is the overall design. The process is described well in the FDA Quality System Regulation and ISO Standard 13485:2003. However, the difficulty most companies will encounter is implementation and execution of a proper design-control system. The goal should be to implement and execute a design process that is simple, straightforward, and not overburdening.

Many companies will struggle with where research ends and development begins. In the research phase, testing is carried out on unique product designs that require highly skilled people. During this phase, detailed processes/procedures should not be implemented. Too much process detail leads to an exaggerated number of tasks and deliverables that require the expenditure of unnecessary resources to track progress; this adds additional unnecessary time to the project schedule. Detailed processes/procedures are best implemented well within the development phase.

Design input is the most important element in the development of a successful medical device. Design input forms the basis of all development activities, it is also the standard for which all verification and validation activities are based. Unfortunately, organizations often short-cut this part of the project because they rush to get into the development phase. However, the result is often an insufficient design input requirements, ultimately resulting in unsatisfactory outcomes such as major development delays for redesign/respecification, unacceptable device performance, overspending, or project cancellations. In the long run, spending enough time (including feasibility studies) developing a solid design input requirements should actually speed the time to introduce the product to market and minimize future problems in the field and even possible product recalls.

References

1. Federal Register, 61–195, 1996
2. FDA Design Control Guidance for Medical Device Manufacturers, U.S. Food and Drug Administration, Center for Radiological Health, March 11, 1997.
3. Six Sigma for Medical Device Design, Jose Justiniaro, Venky Gopalaswamy, 13–34, 2004, CRC Press, Boca Raton, FL.
4. Practical Design Control Implementation for Medical Devices, Jose Justiniaro, Venky Gopalaswamy, 2005, Taylor & Francis, London.
5. Design and Development of Medical Electronic Instrumentation: A Practical Perspective of the Design, Construction, and Test of Medical Devices, David Prutchi, Michael Norris, 97–145, 2005, John Wiley E Sons, Inc., New York.

Subject Index

D.D. Zhou, E. Greenbaum (eds.), *Implantable Neural Prostheses 2*,
Biological and Medical Physics, Biomedical Engineering, DOI 10.1007/978-0-387-98120-8,

O

P

MIX
Papier aus verantwortungsvollen Quellen
Paper from responsible sources
FSC® C105338

If you have any concerns about our products,
you can contact us on
ProductSafety@springernature.com

In case Publisher is established outside the EU,
the EU authorized representative is:
Springer Nature Customer Service Center GmbH
Europaplatz 3, 69115 Heidelberg, Germany

Printed by Libri Plureos GmbH
in Hamburg, Germany